INTERNATIONAL
REVIEW OF CYTOLOGY

VOLUME 76

INTERNATIONAL

Review of Cytology

EDITED BY

G. H. BOURNE
St. George's University School of Medicine
St. George's, Granada
West Indies

J. F. DANIELLI
Worcester Polytechnic Institute
Worcester, Massachusetts

ASSISTANT EDITOR
K. W. JEON
Department of Zoology
University of Tennessee
Knoxville, Tennessee

VOLUME 76

1982

ACADEMIC PRESS   *A Subsidiary of Harcourt Brace Jovanovich, Publishers*
New York   London
Paris   San Diego   San Francisco   São Paulo   Sydney   Tokyo   Toronto

ACADEMIC PRESS, INC.
111 Fifth Avenue, New York, New York 10003

United Kingdom Edition published by
ACADEMIC PRESS, INC. (LONDON) LTD.
24/28 Oval Road, London NW1 7DX

LIBRARY OF CONGRESS CATALOG CARD NUMBER: 52–5203

ISBN 0–12–364476–3

PRINTED IN THE UNITED STATES OF AMERICA

82 83 84 85   9 8 7 6 5 4 3 2 1

Contents

Moderately Repetitive DNA in Evolution

ROBERT A. BOUCHARD

Structural Attributes of Membranous Organelles in Bacteria

CHARLES C. REMSEN

Separated Anterior Pituitary Cells and Their Response to Hypophysiotropic Hormones

CARL DENEF, LUC SWENNEN, AND MARIA ANDRIES

What Is the Role of Naturally Produced Electric Current in Vertebrate Regeneration and Healing?

RICHARD B. BORGENS

Metabolism of Ethylene by Plants

JOHN H. DODDS AND MICHAEL A. HALL

List of Contributors

Numbers in parentheses indicate the pages on which the authors' contributions begin.

MARIA ANDRIES (225), *Laboratory of Cell Pharmacology, Department of Pharmacology, School of Medicine, Campus Gasthuisberg, Katholieke Universiteit Leuven, B-3000 Leuven, Belgium*

RICHARD B. BORGENS (245), *The Institute for Medical Research, San Jose, California 95128*

ROBERT A. BOUCHARD (113), *Department of Biology, B-022, University of California, San Diego, La Jolla, California 92093*

CARL DENEF (225), *Laboratory of Cell Pharmacology, Department of Pharmacology, School of Medicine, Campus Gasthuisberg, Katholieke Universiteit Leuven, B-3000 Leuven, Belgium*

JOHN H. DODDS (299), *Department of Plant Biology, University of Birmingham, Birmingham B15 2TT, England*

S. J. FLINT (47), *Department of Biochemical Sciences, Princeton University, Princeton, New Jersey 08540*

MICHAEL A. HALL (299), *Department of Botany and Microbiology, University College of Wales, Aberystwyth, Dyfed, Wales*

ANN S. HENDERSON (1), *Department of Human Genetics and Development, Columbia University College of Physicians and Surgeons, New York, New York 10032*

CHARLES C. REMSEN (195), *Center for Great Lakes Studies and Department of Zoology, University of Wisconsin–Milwaukee, Milwaukee, Wisconsin 53201*

MAXINE F. SINGER (67), *Laboratory of Biochemistry, National Cancer Institute, National Institutes of Health, Bethesda, Maryland 20205*

LUC SWENNEN (225), *Laboratory of Cell Pharmacology, Department of Pharmacology, School of Medicine, Campus Gasthuisberg, Katholieke Universiteit Leuven, B-3000 Leuven, Belgium*

Cytological Hybridization to Mammalian Chromosomes

ANN S. HENDERSON

Department of Human Genetics and Development, Columbia University College of Physicians and Surgeons, New York, New York

I. Introduction

The purpose of this article is to compare and bring up to date procedures related to the use of molecular hybridization in mammalian chromosomes. The technique of cytological hybridization was developed in several laboratories (Gall and Pardue, 1969; John *et al.*, 1969; Buongiorno-Nadelli and Amaldi, 1970). The opportunity to visualize nucleic acid:chromosomal DNA hybridization has resulted in a large number of studies which have assigned specific genes or chromosomal sequences. Further, the localization of satellite DNA sequences or sequences involving the rDNA complex in primates and other mammals has allowed comparative studies on evolution of the chromosome complement to be made at the molecular level. Other studies have used cytological hybridization to demonstrate the chromosomal integration of DNA from exogenous sources, such as viral DNA or DNA inserted as the result of DNA-mediated transformation. These and some other investigations that have used cytological

1

TABLE I

Topic	References
I. Review articles: general discussion of technique	Pardue and Gall (1972, 1975); Jones (1972); Steffensen and Wimber (1972); Hennig (1973); Steffensen (1977); Atwood (1979)
II. Chromosomal localization of repetitive nontranscribed sequences	
A. Review articles	Jones et al. (1972); Jones (1973, 1977); Macaya et al. (1977); Miklos and John (1979)
B. Localization of satellite DNA sequences in human chromosomes to centromeres of many chromosomes; heterochromatic regions on the Y-chromosome	
1. Satellite I	Evans et al. (1974a); Jones et al. (1974); Gosden et al. (1975)
2. Satellite II	Jones and Corneo (1971); Evans et al. (1974a); Gosden et al. (1975)
3. Satellite III	Jones et al. (1973); Evans et al. (1974a); Saunders (1974); Gosden et al. (1975); Moar et al. (1975)
4. Satellite IV	Evans et al. (1974a); Gosden et al. (1975)
C. Localization of other satellite DNA sequences in the human chromosome complement	
1. Satellites A, B, C, and D: A has no chromosomal specificity; B is localized to nucleolar DNA; C is homologous to centromeric regions of chromosome 9 and D and G group; D is homologous to centromeric region of 9	Chuang and Saunders (1974); Saunders (1974)
2. "H" (Hoechst); specific to centromeric regions of chromosomes 1, 3, 9, D and G group, Y "Eco RI fragment"; specific to centromeric regions of chromosomes 1, 3, 7, 10, 19	Manuelidis (1978)

(*Continued*)

2

TABLE I—*Continued*

Topic	References
3. "1.687" = satellite I?; homologous to portions of Y chromosome; centromeric regions of chromosomes 1, 8, 9, and some secondary constrictions, telomeric regions; "1.696" = satellite II?; specific to centromeric regions of chromosomes 1, 16, some secondary constriction regions; "1.700" = satellite III?; specific to chromosomes 1, 9, Y, and others; "1.703 and 1.714" contains middle repetitive and repetitive DNA; specific for centromeric and telomeric regions	Marx *et al.* (1976)
D. Localization of Y-specific DNA in human complement	Evans *et al.* (1974a); Bostock *et al.* (1978)
E. Localization of "repetitious DNA" in the human complement	
Homologous to heterochromatic regions	Saunders *et al.* (1972);
Homologous to Q-positive bands	Sanchez and Yunis (1974)
F. Comparative studies in primates	
1. Review—comparative study of centromeric DNA distribution in primates	Jones (1973)
2. Localization of human satellite III in higher primates	Mitchell *et al.* (1977)
3. Repetitive DNA in the orangutan; comparison of distribution in people and other primates	Gosden *et al.* (1978)
G. Localization of repetitive DNA in other mammals	
1. Primary satellite fraction to the centromeric regions of mouse chromosomes	John *et al.* (1969); Jones (1970); Jones and Robertson (1970); Pardue and Gall (1970); Kuo and Hsu (1978)
2. Primary satellite to large marker chromosomes in methotrexate-resistant mouse cells	Bostock and Clarke (1980)
3. Repetitive DNA to heterochromatic regions in mouse chromosomes	Ahnstrom and Najarajan (1974)
4. Centromere-specific DNA in the rat	Prescott *et al.* (1973)
5. Repetitive DNA in the rat	Sealy *et al.* (1981)
6. Repetitive DNA in the field vole to heterochromatic regions, particularly on sex chromosomes	Arrighi *et al.* (1970); Yasmineh and Yunis (1975)

(Continued)

TABLE I—*Continued*

Topic	References
7. Repetitive DNA of four satellite fractions in the hedgehog. Fraction I is specific for centromeres, telomeres, and intercalary heterochromatin; Fraction II is as fraction I, but shows specificity with respect to Q-banding pattern; Fractions III and IV are homologous to Q-positive bands	Willey and Yunis (1975)
III. Localization of sequences that transcribe ribosomal RNA	
A. rDNA sites to the short arms of the human D and G-group chromosomes	Henderson *et al.* (1972); Evans *et al.* (1974b)
B. Comparative studies of rDNA location among primates	Henderson *et al.* (1974a,b, 1976a, 1977, 1979a); Warburton *et al.* (1975); Gosden *et al.* (1978)
C. In the mouse	Henderson *et al.* (1974c, 1976c); Elsevier and Ruddle (1975)
D. In other mammals, including the rat, kangaroo, bat, field vole, Chinese hamster	Hsu *et al.* (1974)
IV. Localization of other components of the rDNA complex	
A. 5.8 S	Henderson *et al.* (1979b, 1980a)
V. Localization of genes that transcribe 5 S RNA	
A. On human chromosomes at the terminal region of lq	Steffensen *et al.* (1974, 1975); Atwood *et al.* (1975)
B. In the chromosome complement of other primates; localization to a region homologous by banding pattern to human chromosome lq	Henderson *et al.* (1976b); Warburton *et al.* (1976a)
VI. Localization of genes that transcribe mRNAs	
A. Histone mRNA in the human chromosome complement	Yu *et al.* (1977); Szabo *et al.* (1978)
B. "Heterogeneous RNA"; localized to light G-bands	Yunis *et al.* (1977); Yunis and Tsai (1977)
C. Confirmation of genetic assignments for mouse α and β globin in the mouse	Henderson *et al.* (1978)
D. Confirmation of genetic assignment for human α globin	Gerhard *et al.* (1981)
E. Localization of human insulin gene	Harper *et al.* (1981)
VII. Studies of association of viruses with mammalian cells	
A. Adenoviruses. The cellular distribution of various adenoviruses has been compared between	

(Continued)

4

TABLE I—*Continued*

Topic	References
1. Transformed vs nontransformed cells	McDougall *et al.* (1972); Loni and Green (1973)
2. Permissive vs nonpermissive cells	Moar and Jones (1975)
3. Transformed vs tumor cells	Dunn *et al.* (1973)
B. SV40	
1. Association with the nucleolus	Geuskens and May (1974)
2. Sequence in defective SV40 is homologous to repetitive regions in host cell chromosomal DNA	Segal *et al.* (1976)
C. EBV	
1. Intracellular localization	zur Hausen *et al.* (1972); Moar and Klein (1978)
2. Chromosomal localization in transformed cell lines	Henderson (1982)
D. Localization of murine sarcoma virus to centromeric heterochromatin and other regions of chromosomes	Loni and Green (1974)
E. Intracellular localization of *Visna* virus in sheep	Brahic and Haase (1978); Brahic *et al.* (1981)
F. Localization of SSLV in virus-infected cells	Kaufman *et al.* (1979)
G. Localization of Polyoma virus DNA	
1. In PML brain tissue	Dörries *et al.* (1979);
2. In inducible lines of polyoma-transformed cells	Neer *et al.* (1979)
H. Localization of AKR-MuLV-specific RNA in AKR mouse cells	Godard and Jones (1979)
I. Localization of *Herpes simplex* I DNA in mouse L cells	Henderson *et al.* (1981a)
VIII. Integration of DNA introduced via DNA-mediated transformation; localization to specific chromosomes	Robins *et al.* (1981a,b); Henderson *et al.* (1982)
IX. Intracellular localizations	
A. 4 S, 5 S, and "pulse-labeled" RNA in Chinese hamster preparations	Buongiorno-Nadelli and Amaldi (1970); Amaldi and Buongiorno-Nadelli (1971)
B. Localization of globin mRNA sequences	Harrison *et al.* (1974); Conkie *et al.* (1974)
X. Other types of cytological hybridization studies	
A. Determination of rDNA connections during acrocentric association	Henderson *et al.* (1973)
B. Stable dicentric chromosome in human complement identified by presence of intercalary rDNA	Warburton *et al.* (1973)

(*Continued*)

5

TABLE I—*Continued*

Topic	References
C. Metacentric fragment in Cat-eye syndrome identified as containing rDNA	Johnson *et al.* (1974)
D. Quantitation of rDNA	Henderson and Atwood (1976); Warburton *et al.* (1976b); Wolgemuth-Jarashow *et al.* (1976)
E. Relationship of rDNA to acrocentric satellite association	Henderson and Atwood (1976); Warburton *et al.* (1976b); Warburton and Henderson (1979)
F. Relationship of rDNA to silver stain	Miller *et al.* (1978); Warburton and Henderson (1979)
G. Low level amplification demonstrated in primates	Wolgemuth-Jarashow *et al.* (1977; 1979); Wolgemuth *et al.* (1980)
H. Distribution of human satellite III in leukemic patients	Prosser *et al.* (1978)

hybridization to mammalian chromosomes as a viable experimental tool are summarized in Table I.

In principle, physical mapping of genes or any DNA sequence to chromosomes can be accomplished most simply and directly by hybridization to chromosomal DNA. On a practical level, the use of cytological hybridization has been primarily limited to reiterated sequences. The localization of genes of low multiplicity has seldom succeeded, whereas reiterated sequences in the genome can be located with poorly defined probes of low specific activity, small numbers of chromosome spreads, and suboptimal hybridization conditions. There is, however, no known theoretical restriction to the formation of any hybrid with chromosomal DNA. Mapping of single or low-copy genes is feasible if large numbers of chromosome spreads are analyzed and the hybridization is carried out under optimal conditions. Hybridization using cloned DNA probes has not been fully tested at this time, but should simplify procedures for hybridization and analysis. Chromosomal regions that code for mRNA have been localized using cloned probes in reasonable autoradiographic exposure times (Gerhard *et al.*, 1981; Harper *et al.*, 1981).

Optimal conditions for cytological hybridization will be considered within the framework of this article. These include general methodology, as well as kinetic considerations and methods for analysis. Of corollary interest is the use of grain count analysis for quantitation of repetitive sequences. Most of the techniques have been described previously, but are reiterated here for the purpose of comparison. A comparison of techniques from various laboratories is useful in determining maximum conditions in untried experiments, particularly those where the chromosomal DNA is not highly reiterated, amplified, or endoreplicated. The appropriate conditions for cytological hybridization are of particular importance at the present time since the continuing availability of homogeneous cloned probes will provide an avenue for the study of mammalian cytogenetics on a more sophisticated level.

The general approach is designed for investigators who are familiar with techniques of molecular hybridization in general, but may be unfamiliar with hybridization to fixed cellular preparations. The conditions discussed are subject to change with the rapid advancement of techniques applicable for use with cytological hybridization, but are those known at the present time to maximize the sensitivity of the reaction while minimizing the effects of nonspecific binding.

II. General Discussion

The use of cytological hybridization for determination of low-multiplicity sequence organization in mammalian chromosomes is complicated by the tremendous amount and diversity of DNA packaged into the chromosome, as well as the fact that the hybridization reaction per se involves annealing to a partially denatured nucleoprotein complex of extreme diversity. False localizations occur. The presence of minor contaminants in the probe can result in hybridization to unforeseen reiterated regions (Szabo *et al.*, 1977; Henderson *et al.*, 1978). This should be alleviated by the use of homogeneous cloned probes, but the complexity of cloned DNA may be such that a portion of the probe is homologous to reiterated chromosomal regions; this results in a similar problem. A probe binding to a reiterated region will produce more grains than other portions of the probe hybridized to less reiterated regions, thus confounding the assignment. The correct interpretation would be apparent from kinetic studies since the concentration in the probe of any fraction or partial fraction that hybridizes with a chromosomal region can be calculated from the time course of the reaction to that region. To accomplish this type of analysis requires large numbers of slides to be scored and analyzed in order to obtain the points required to establish the time course of the reaction. Such procedures are extremely time consuming and seldom employed. Problems in analysis can be further compounded by the tenacious binding of probe to chromosomal protein or other cellular components.

The feasibility of mapping chromosomal DNA is established, however, and under appropriate experimental conditions, preferential labeling is expected. The criteria must agree in other respects with the hypothesis that preferential labeling is caused by the intended probe to chromosomal DNA. Criteria which must be met to avoid false assignments are as follows: (1) the number of specific sequences preferentially labeled should not exceed the number inferred from reassociation kinetics, nor from saturation of filter-bound DNA; (2) the efficiency of grain formation calculated from the specific activity of the probe, exposure time, and site-associated grains must not exceed the sensitivity of the emulsion; and (3) the time course of the hybridization reaction must be consistent with the probe concentration close to that known to be present in the hybridizing solution.

A. Efficiency of Hybridization

Theoretically, the overall efficiency for cytological hybridization (E_t) is the product of the proportion of sites hybridized, a geometric factor, and an emulsion factor. In practice E_t is defined as the ratio of grains to expected disintegrations assuming that the chromosomal DNA is hybridized to saturation, i.e., E_t is the product of the emulsion efficiency (E_e) and the percent hybridization (E_h). E_e for isotopes commonly used in hybridization experiments has been determined from independent experiments. The efficiency of any isotope with respect to the emulsion is dependent on the nature and thickness of the cytological preparation (Cleaver, 1967; Rogers, 1973). Assuming that chromosomal preparations have a thickness of 1 μm, E_e for ^{3}H is about 10%. For ^{125}I, the emulsion efficiency has been determined directly as 23% (Ada et $al.$, 1966), but experiments using cytological hybridization have found lower values. Szabo et $al.$ (1977) calculated the E_e for ^{125}I as 13%, although the authors believe this to be an underestimate. Our estimates result in a value of about 20%.

The hybridization efficiency (E_h) can be determined by inference from E_t in a given experiment where $E_t = E_e E_h$. The relationship between measurable parameters of the hybridization reaction can be used to make an empirical determination of the feasibility of assignment since E_t has defined limits (Atwood et $al.$, 1976a). The parameters are expressed as $E_t = C(SWD)^{-1}$, where C is a constant defining the units of the equation (in this case, C is 4.18×10^{14} or Avagadro's number divided by the number of minutes per day times 10^6), S is the specific activity of the probe in dpm/μg, W is the molecular weight of the site(s) hybridized, and D is the reciprocal grains per day. A given experiment is considered feasible if $E_t \leq E_e$. A nomogram illustrating the relationship of grain formation to site-size and specific activity of the probe is given in Fig. 1. As seen, localization of comparatively small regions of chromosomal DNA is theoretically successful even with low levels of efficiency.

The efficiency of hybridization is probably variable depending on the type of

Fig. 1. Nomogram illustrating the rate of grain formation as a function of experimental parameters. The number of days required to produce one grain intersects a line between the specific activity and molecular weight. The overall efficiency was assumed to be 3%.

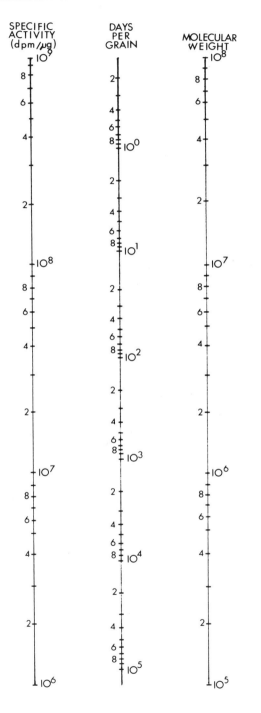

TABLE II

Comparison of Hybridization Efficiency

Tissue/ organism	Probe	E_e (%)	E_h (%)	Reference
Xenopus nuclei	³H-labeled 18/28 S rRNA	Assumes 10	7	Jones (1972, recalculated from Gall and Pardue, 1969)
Sheep choroid plexus cells	[³H]cDNA from Visna virus RNA	10	100	Brahic and Haase (1978)
Mouse fetal liver cells	[³H]cDNA from globin mRNA	Assumes 10	5	Harrison et al. (1974)
HeLa cell nuclei	Short-pulse [³H]RNA	Assumes 10	1–2	John et al. (1969)
Drosophila polytene chromosomes	³H-labeled 5 S RNA	Assumes 10	3–5	Steffenson and Wimber (1972)
	¹²⁵I-labeled rRNA	13	12	Szabo et al. (1977)
	¹²⁵I-labeled 5 S rRNA	13	9	Szabo et al. (1977)
	[³H]poly(U)	—	19	Barbera et al. (1979)
Mouse chromosomes	[³H]cDNA from mouse satellite DNA	Assumes 10	20	Jones et al. (1973)
Human chromosomes	¹²⁵I-labeled histone mRNA	—	1.5–2	Yu et al. (1977)

Human chromosomes	[³H]cDNA from 28 S rRNA	—	3-4	Malcolm et al. (1977)
	³H-labeled nick-translated plasmid derived rDNA	—	70	Malcolm et al. (1977)
	[³H]cRNA from transcription of plasmid derived rDNA	—	70	Malcolm et al. (1977)
	^{125}I-labeled 28 S rRNA	20	5-15	Cote et al. (1980)
	^{125}I-labeled 5 S rRNA	20	5-15	Cote et al. (1980)
	^{125}I-labeled 18/28 S rRNA	20	15 ± 5 if $1 N = 220$	Henderson et al. (unpublished data from 10 experiments)
		20	95 ± 28 if $1 N = 35$	
	^{125}I-labeled 5 S RNA	20	14 ± 4 if $1 N = 220$	Henderson et al. (unpublished data from 7 experiments)
		20	41 ± 11 if $1 N = 75$	
Rat chromosomes	^{125}I-labeled growth hormone plasmid DNA	20	18 ± 4	Henderson [unpublished data from 7 experiments of Robins et al. (1981a,b)][a]

[a] The efficiency may be higher in these experiments. The slides were developed at the point where the site was identifiable, rather than a time where the majority of potential grains could be realized. In addition, high grain density over some of the sites probably resulted in an underestimate of the total grains.

chromosome studied as well as the complexity of the site. Estimates have varied from 5% (John *et al.*, 1969) to 100% (Brahic and Haase, 1978). E_h, as estimated in various laboratories, is compared in Table II. At the present time, most estimates made for hybridization efficiency, particularly in mammalian chromosomes, must be considered relative. There are several reasons why this is the case. First, grains are usually counted from autoradiographic preparations exposed to a point where the site is identifiable. The proper estimate would be one based on exposure continued to a time where the number of grains realized approaches saturation. Second, the estimation of E_h requires a prior knowledge of the percentage of the genome involved, obtained from filter or solution hybridizations. The hybridization levels for ribosomal RNA (rRNA) or 5 S RNA are normally used as standards since the data are simplest to analyze. These genes may be polymorphic with respect to gene number among mammalian chromosomes that carry these genes (Evans *et al.*, 1974b; Warburton *et al.*, 1976a; Henderson *et al.*, 1976c, 1979a, 1980a,b). Individual differences have also been reported from filter hybridization data (Bross and Krone, 1973; Atwood *et al.*, 1976b). This can result in large deviations in site-specific grains over chromosomes prepared from different individuals. The range of gene numbers reported, however, is far in excess of individual differences. For example, in estimates of human rDNA, the saturation values vary between 0.0035% of the genome or less than 50 genes per haploid complement (Young *et al.*, 1976; Cote *et al.*, 1980), and 1% of the genome or about 1000 copies (Johnson *et al.*, 1975). We, and others (Bross and Krone, 1972, 1973; Gaubatz *et al.*, 1976), have found values of 0.03 to 0.04% giving an estimate of 220–300 genes per haploid complement. This may be an overestimate since ribosomal RNA preparations are usually contaminated with other RNA species. The calculation of < 50 copies must be too low, however, since the E_h for rRNA:chromosomal DNA in many of our experiments would result in E_h > 100% (see Table II). The use of human 5 S DNA as a standard is subject to the same criticism since estimates of 50 copies (Cote *et al.*, 1980) to 2000 copies (Hatlen and Attardi, 1971) have been reported. Finally, variation in efficiency exists between slides, depending on the method of preparation and age of the slide. Thus, efficiency estimates reported from various laboratories are variable depending on the means of slide preparation, age of the slide, time of autoradiographic exposure, and the standard used for determining hybridization efficiency.

The fact that all available sites may not be occupied at saturation is a basic difference between cytological hybridization and other types of hybridization measurements. The failure to fill these sites can be attributed to several causes (see Barbera *et al.*, 1979, for discussion). Denaturation of chromosomal DNA is probably never complete and it is not known whether there is differential denaturation depending on the complexity of the site. Further, loss of chromosomal DNA occurs during denaturation and there is steric hindrance from the presence of chromosomal proteins. There is also the possibility of undetectable reassociation of localized regions of chromosomal DNA.

There is some compensation for low hybridization efficiency. Autoradiography is a very sensitive technique and an extremely low number of localized disintegrations can be detected with time. For example, if a molecule of molecular weight 10^6 and specific activity 8×10^8 dpm/μg is used for hybridization, successful localization would detect about 10^{-4} dpm. Efficiency estimates are usually based on the formation of RNA:chromosomal DNA hybrids or hybrids formed between short lengths of single-stranded DNA to chromosomal DNA. Under appropriate conditions, the use of double-stranded DNA should allow the formation of hybrids on both strands of the chromosomal DNA and may amplify the signal by the formation of concatenates.

B. Quantitation Using Cytological Hybridization

Quantitation of gene sites using cytological hybridization is dependent on proportionality between the density of autoradiographic grains and the

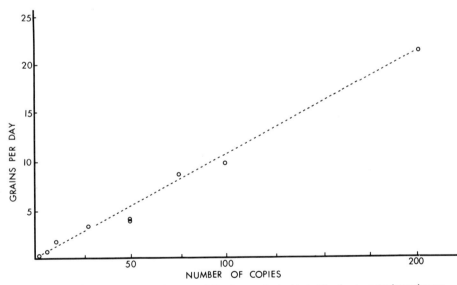

Fig. 2.   Quantitation using grain counts following cytological hybridization to metaphase chromosomes with various gene copy numbers. The chromosomal integration of various amounts of human growth hormone (HGH) DNA into rat chromosomes was achieved by DNA-mediated transformation (Robins *et al.*, 1981a,b). The number of integrated copies for each transformation event was monitored independently (see Robins *et al.*, 1981a) and ranged from 2 to 200 copies, or approximately 12 to 1200 kb DNA. Grain counts over HGH DNA-bearing chromosomes were made following cytological hybridization of ^{125}I plasmid HGH DNA with a specific activity of 8×10^8 dpm/μg. Grains per day is calculated per chromatid.

number of genes or defined DNA sequences present. The relationship between grain counts and expected sites is linear. In preparations of mouse testicular cells, grain counts have been made over spermatids (1C), fibroblast and Sertoli cells (2C), and over spermatocytes in first meiotic prophase (4C) following hybridization with ^{125}I-labeled rRNA. The rRNA bound was proportional to the expected DNA content (Wolgemuth-Jarashow et al., 1976). These results were confirmed by Szabo et al. (1977). Direct proportionality was demonstrated between grain counts and expected number of genes following hybridization of 5 S RNA to Drosophila nuclei ranging from diploid through several degrees of polyteny. Grain counts and the expected number of viral or proviral genomes per cell is also proportional (Brahic and Haase, 1978; Brahic et al., 1981). This proportionality holds true for hybridization to chromosomes. Figure 2 shows grain count data obtained from hybridization of human growth hormone DNA to rat cell chromosomes that had incorporated various amounts of human growth hormone DNA into chromosomes as the result of DNA-mediated transformation (Robins et al., 1981a,b). Linearity is observed between approximately 12 to 1200 kb of integrated DNA.

  If the average efficiency (E_t) for a given set of experiments is known, then quantitation can be made on the basis of grain counts. The calculation of relative copy number (N) uses a modification of the formula described above: $N = C(WE_tDS)^{-1}$. Here W is the molecular weight of a single gene or discrete sequence.

III. Methodology

  Conditions for the hybridization in situ reaction using radioactive probes have been discussed in several reviews (Pardue and Gall, 1972, 1975; Hennig, 1973). Many of the methods currently in use have been derived empirically from other types of hybridization systems. For cytological hybridization, fixed tissue is usually carried through the following procedures: (1) removal of endogenous RNA through the use of high concentrations of RNase, (2) denaturation of chromosomal DNA, (3) application of the probe under annealing conditions, (4) removal of unbound or nonhomologously bound probe, and (5) autoradiography. It is usually necessary to band chromosomes for identification at some point in the procedures. Differences will be noted in methodology to be used for cytological hybridization and that used for other types of hybridization studies. These result from the fact that some compromise is made between optimal hybridization conditions and maintaining some semblance of chromosome morphology. A suggested protocol for cytological hybridization is given in Fig. 3.

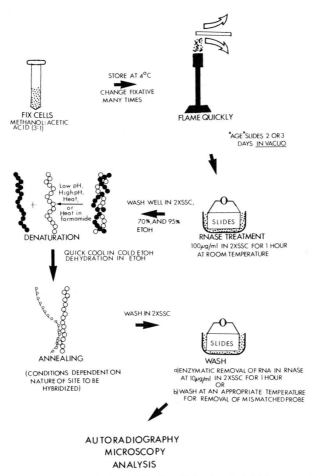

FIG. 3.   One protocol for cytological hybridization. Following the initial treatment with RNase to remove endogenous RNA, the slides are washed many times in 2 × SSC, followed by 3 changes each of 70% ethanol and 95% ethanol and air drying. Denaturation methods have been discussed in the text. One means of terminating denaturation is to plunge the slides into ice-cold 70% ethanol. This is followed by washes in 70 and 95% ethanol, and air drying. Annealing conditions are also discussed in the text. After the final RNase treatment, or other procedures to remove mismatched probe, the slides are again washed extensively in 2 × SSC. One method is to suspend the slides in a large container of 2 × SSC with stirring and changes of 2 × SSC. After washing, the slides are again dehydrated in 70 and 95% ethanol before air drying.

A. SLIDE PREPARATIONS

The preparation of slides is the most ambiguously defined of any of the procedures used, but it is perhaps the most important step. It is the stability and cleanliness of the slide preparations that result in the best hybridizations. It is easiest to begin with actively growing cells in tissue culture. Lymphocytic cultures, particularly PHA-stimulated lymphocytes, produce the best preparations. Methods for obtaining metaphase plates have been described (Hamerton, 1971). Special techniques, such as procedures for obtaining elongated chromosomes in late prophase or early metaphase, have been reviewed by Yunis *et al.* (1977). Exposure to colcemid should be as brief as possible. One to 2 hours is the usual time for typical lymphocyte cultures. This may require some prior synchronization for long-term cultured cells. Time in hypotonic solutions should also be brief for lymphocytic cultures; 5 minutes is probably more than adequate. A longer time in hypotonic solution (20 to 30 minutes) may be necessary for fibroblast cultures. The longer exposure results in fewer cells that fix with intact cellular membranes or cytoplasm. The presence of cytoplasm surrounding the chromosomes results in higher background levels. Proteinase K (Brahic and Haase, 1978) and H_2O_2 (Seabright, 1971) have been used to eliminate some of the cytoplasm surrounding chromosomes. The cells are then fixed, preferably in acid:alcohol. The usual fix for mammalian chromosome preparations is methanol:acetic acid (3:1) although methanol:acetone has also been found to be effective (Moar and Klein, 1978). Definite problems can arise when other types of fixation are used. Acids of high molarity can result in damage or uncontrolled denaturation of DNA, as well as destruction of chromosomal proteins that maintain the general morphology of the chromosome (Steffensen, 1977). Since many of the fixatives used for cellular sectioning fall into this category, fixatives for embedded cells should be chosen with care. The method of fixation can also affect the autoradiographic emulsion causing chemography or desensitization. Two commonly used fixatives, formalin and glutaraldehyde, can cause desensitization. A further disadvantage of formalin is that the fixative interferes with the controlled denaturation of the DNA. Godard and Jones (1979), in comparing fixation methods for studying complex RNA *in situ,* have shown that low concentrations of glutaraldehyde are acceptable for hybridization experiments. When acid–alcohols are used for fixation, the cells should remain in fixative at 4°C for several days with many changes of the fixative. One effect of acid fixation is removal of protein; multiple changes not only ensures that a maximum effort has been made to extract chromosomal proteins, but it also improves the morphology.

Cells in fixative are dropped onto slides and dried very rapidly in a flame. Marx *et al.* (1976) have suggested that the temperature of the slide not exceed skin temperature. The use of a heated plate for drying does not result in good preparations for hybridization. After preparation, the slides should be kept dry in a vacuum desiccator. No matter how much care is taken, the efficiency of hybridization will be reduced in older slides. No real quantitation has been done with respect to what constitutes an older slide, but some attempt should be made to use the preparations within 1 month. The probable result of time in storage is degradation of DNA. Steffensen (1977) has recommended that slides be stored at −20°C or lower. If this is done, the slides should be brought to room temperature in the presence of desiccant. We have found, at least in the New York City area, that the presence of atmospheric moisture will seriously effect the hybridization efficiency.

Microscope slides or coverslips must be extremely clean. Slides provided by some vendors may contain residual chemicals from processing. Various methods for cleaning slides have been suggested, but washing well in detergent or acid, followed by rinse in running water, is probably adequate. We have found that Gold Seal precleaned slides as shipped are adequate and further washing is unnecessary. The use of subbing solution such as gelatin:chrome alum has been used for slide preparations in some cases (Pardue and Gall, 1975), but this step is superfluous in the preparation of mammalian chromosomes, and may prevent adequate spreading of metaphase plates.

B. Chromosome Identification

The most commonly used methods for identification of chromosomes are banding patterns produced as the result of binding of quinacrine or quinacrine mustard, and G-banding produced by protelytic enzymes. The use of quinacrine to produce fluorescent Q-bands has been described (Caspersson *et al.*, 1970). The simplest G-banding method is to expose the chromosomes to commercial trypsin for times between 10 seconds and several minutes (Seabright, 1971). The time in trypsin is dependent on the age of the slide and numerous other unknown variables making this a trial and error step. An alternative, devised in several laboratories, that alleviates the time spent in determining unknown variables and produces consistent results is as follows: the usual procedures are modified to include heating the slides at 56°C for 10 minutes in neutral phosphate buffer (0.02 M) prior to banding. Banding is unexpectedly achieved in a solution containing 20% methanol, 70% neutral phosphate buffer, 8% commercial trypsin, and 2% Giemsa in about 5 to 7 minutes at room temperature. Q- or G-banding patterns can be produced before or after hybridization procedures. There are disadvantages to the type of banding used as well as the time of banding. Banding prior to hybridization requires that photographs of the same metaphase plate be made before hybridization and after the autoradiogram is developed. This is a large expenditure in time and a tedious procedure. If photographs of G-banded preparations are made prior to hybridization, the use of oils or xylenes should be avoided, since this results in higher background levels and variable hybridization efficiencies (Warburton *et al.*, 1976b); photography must be done with a high-dry lens. The alternative, as used in Q-banding, is to place a coverslip over the preparation with a buffered solution between the coverslip and slide. For G-banded preparations, this can result in swollen chromosomes during photography and morphological distortion after the hybridization procedures are completed. In Q-banded preparations, the banding is revealed following fluorescent microscopy; this may effect denaturation of the chromosomal DNA. Q-banding prior to hybridization may also result in higher background levels. The reasons for this are not apparent. One problem is that it is difficult to remove quinacrine from chromosomes even with stringent washing procedures. Following G-banding, the stain is easily removed in 70% ethanol. The advantage to any banding procedure prior to hybridization is that a record is obtained of each plate, so that morphological distortion occurring as the result of stringent denaturation procedures can be tolerated. This is of importance where maximum conditions are required.

There are also problems with banding after denaturation and hybridization procedures. Denaturants not only denature chromosomal DNA, but also chromosomal protein. Thus, stringent denaturation cannot be followed by G-banding. Chandler and Yunis (1978), however, have described a G-banding method for prometaphase chromosomes following hybridization. Methods for Q-banding after hybridization have also been devised (Sawin *et al.*, 1978; Lawrie and Gosden, 1980). Banding after hybridization may be dependent on the use of mild denaturation in order that the chromosomes maintain the capacity to band. The preferred approach should be selected on the basis of the degree of sensitivity required for a given experiment.

C. PREPARATION OF RADIOACTIVE PROBES

The most effective means for obtaining probes of high specific activity is by radioactive labeling *in vitro*. Some of the methods that have been used successfully for cytological hybridization are given in Table III.

D. STAINING OF CHROMOSOMES

The choice of stain is dependent on whether it is compatible with autoradiographic procedures and whether the stain allows the grain density to be visable over the stained background. Specific staining procedures for cytological hybridization have been given in Hennig (1973). Stains compatible with dipping autoradiography have been discussed by Thurston and Joftes (1963).

TABLE III
METHODS FOR PREPARATION OF RADIOACTIVE PROBES

Method	Isotope	Specific activity (dpm/μg)[a]	Yield expressed as percentage input of template[a] (%)
Iodination of single-stranded RNA or DNA as described by Commerford (1971), and modified in several laboratories (Prensky et al., 1973; Yu et al., 1974; Prensky, 1976)	^{125}I	5×10^8	60
Iodination of double-stranded DNA by allowing iodination to occur under partial reannealing conditions for DNA	^{125}I	Up to 2.5×10^8	60
Transcription of double-stranded DNA to RNA using E. coli RNA polymerase (Geuskens and May, 1974)	^{3}H	1.5×10^8	4 to 10
Transcription of complementary DNA from RNA using reverse transcriptase (Kacian and Myers, 1976)	^{3}H	2×10^8	25
Nick-translation of double-stranded DNA (Schachat and Hogness, 1973)	^{125}I	Up to 2×10^9	60 to 80

[a] Figures are from our laboratory data.

E. REMOVAL OF ENDOGENOUS RNA AND OTHER CELLULAR COMPONENTS

The removal of endogenous RNA is usually accomplished by incubation of the slides in 50 to 100 μg/ml RNase in $2 \times$ SSC (SSC = 0.15 M NaCl, 0.015 M Na-citrate) for a period of 1 to 2 hours at room temperature. The failure to remove endogenous RNA when studying transcriptively active regions results in a reduction in the degree of hybridization. Szabo et al. (1975) and Steffensen (1977) have reported that little or no hybridization of 5 S RNA to polytene chromosomes occurs without the removal of RNA. We have found that grain counts over rDNA-containing chromosomes in the human complement are reduced to 38% in untreated slides as compared with slides treated with RNase prior to hybridization. Obviously, slides should be washed well following RNase digestion, particularly if RNA probes are to be used in later steps.

Amylase has also been used in prehybridization steps as a means of reducing background by removing ''sticky'' components from the slide. This step is probably unnecessary in the usual case (Hennig, 1973). Theoretically, removal of as many chromosomal proteins as possible should result in a higher hybridization efficiency. Some portion of the histone proteins are removed during fixation; other proteins are removed during G-banding procedures. We have attempted to further extract chromosomal proteins from fixed cellular preparations by the use of proteolytic enzymes and by extraction of fixed chromosomes in 4 M urea:2% SDS for 1 to 2 minutes. (It should be noted at this point that stringent use of these procedures will remove everything from the slide.) Further extraction of protein did not increase the hybridization efficiency. One possible reason for this, as yet untested, is that chromosomal DNA is also removed during these procedures. Deproteination methods have only been investigated for efficiency in the formation of rRNA:human chromosomal DNA hybrids and might be effective when used prior to the formation of other types of hybrids, or used under more stringent conditions.

F. Denaturation of Chromosomal DNA

Methods for denaturation of chromosomal DNA have been devised in several laboratories (Gall and Pardue, 1969; Jones, 1972; Hennig, 1973; Steffensen, 1977; Singh *et al.*, 1977; Szabo *et al.*, 1977). No distinction is usually made between denaturation methods for endoduplicated DNA of polytene chromosomes and mammalian DNA. Most procedures were devised based on *Drosophila* salivary gland chromosomes or highly repetitive sequences within the mammalian genome. Four denaturation methods have been used successfully. These are incubation in the presence of high pH, low pH, high temperature or formamide at various concentrations and temperatures. Some conditions and methods for chromosomal DNA denaturation from different laboratories are compared in Table IV. The use of any means of denaturation must take into account at least two factors. These are the effectiveness of the denaturation procedure with respect to the sequence to be localized and the effect of denaturation procedures on the chromosomal morphology. Obviously, the most stringent forms of denaturation will destroy recognizable morphology of the chromosomes. In theory, if a prior photograph of the banded preparation is obtained, the chromosomal morphology may be distorted, yet useful following hybridization procedures. In practice, one must be able to relocate a given plate after developing the autoradiogram; this is adequately tedious for chromosomes that have maintained some morphological characteristics, but virtually impossible where there are no recognizable outlines of the chromosome for identification. For example, the use of high concentrations of NaOH will assuredly result in effective denaturation, but rather horrible chromosome preparations. In this case, some compromise is made and a lower concentration of alkali is used [(0.07 N NaOH for short periods of time has been used successfully (Pardue and Gall,

1970)]. The best chromosomal morphology is obtained using acidic conditions (0.2 N HCl for 20 minutes) but the hybridization is less effective (Gall *et al.*, 1971; MacGregor and Kezer, 1971; Jones, 1972). Singh *et al.* (1977) found that heat denaturation at 100°C in 0.1 × SSC gives excellent hybridization irrespective of the base composition of the probe. Some expertise and manipulation of time at this heat may be necessary for different preparations. Further, the use of 100°C is probably not adequate for denaturation of more complex nucleoproteins, so that this method might be restricted to identification of repetitive sites in the genome. Formamide in SSC at various concentrations has also been used as an effective denaturant (Steffensen and Wimber, 1972; Henderson *et al.*, 1972; Singh *et al.*, 1977). Formamide is useful in preventing interstrand scissions since denaturation can occur at a lower temperature. As seen in Table IV, there are discrepancies with respect to the most efficient method, depending on the laboratory. Such discrepancies may be procedural, or simply reflect localized differences in chromosomal DNA composition at different target sites along the chromosome, or between chromosome complements, e.g., endoreplicated chromosomal DNA as compared with diploid chromosome sequences. A comparative study by Singh *et al.* (1977) of denaturation methods may be useful as reference, particularly in choosing conditions for localizing repetitive sequences of various base compositions.

We have tested standard denaturation procedures for hybridization of ^{125}I-labeled rRNA to human chromosomes (Table V). The method that gave the highest efficiency and most consistent results was denaturation in 95% formamide:1 × SSC for periods of at least 1 hour (Fig. 4). This was true whether formamide was used alone or in combination with other means of denaturation. The morphology of the chromosomes is not unduly disturbed by this treatment, but stringent treatment with formamide precludes banding the chromosomes after the denaturation procedure and a photographic record of the chromosome plates prior to denaturation is necessary. One of the problems in working with formamide solutions, of particular importance in the study of chromosomes, is that the product as supplied by the dealer often contains ammoniacal compounds that severely effect morphology of the chromosomes and standardization of the hybridization, as well as lead to DNA loss. The deleterious effects are reduced by lowering the pH of the formamide solution to below 7, and maintaining the pH at approximately 7 during the denaturation process.

As discussed by Singh *et al.* (1977), there are many factors which effect a change in grain density as the result of denaturation. For example, treatment in strong alkali results in significant loss of DNA (Comings *et al.*, 1973). Barbera *et al.* (1979) have compared DNA loss in polytene chromosomes with UV spectrophotometry following various denaturation procedures and the effect of DNA loss on the hybridization efficiency. The greatest loss of chromosomal DNA occurred during denaturation in 0.1 N NaOH, but high DNA loss also

TABLE IV

COMPARISON OF DNA DENATURATION PROCEDURES

Organism/ tissue	Probe	Method of denaturation	Temperature (°C)	Time	Results	Reference
D. hydei chromosomes	Total transcript— [³H]cRNA	90% formamide in 0.1 × SSC	65	2.5 hours	This method gave a higher representation of radioactivity than denaturation by boiling in 0.1 × SSC at 100°C for 30 seconds	Alonso *et al.* (1974)
Mouse, quail metaphase chromosomes	Comparison of AT-rich and GC-rich transcripts (cRNA)	Heat in 0.1 × SSC	100	5 seconds to 5 minutes	Heat is the most effective of the methods used for denaturation	Singh *et al.* (1977)
		90% formamide in 1 × SSC	65	2.5 hours	This method is better than denaturation in acid or base, but is not as effective as boiling	
		0.07 N NaOH	rt[a]	1 to 3 minutes	May give higher retention of binding than use of formamide, but is variable; there is a possible loss of DNA and discrimination against less repetitive DNA	
		0.2 N HCl	rt	20 minutes	The hybridization is low and there may be discrimination against GC-rich DNA; may damage bases	

D. hydei polytene chromosomes	Total transcript—[³H]cRNA	0.07 *N* NaOH	25	2 minutes	DNA loss = 36% DNA denaturation = 82%	Barbera *et al.* (1979)
		90% formamide in 0.1 × SSC	65	2.5 hours	DNA loss = 30% DNA denaturation = 79%	
		2 × SSC at pH 12	25	5 minutes	DNA loss = 25% DNA denaturation = 66%	
		0.24 *N* HCl	37	30 minutes	DNA loss = 50% DNA denaturation = 18% Lowest hybridization efficiency	
		Heat, 0.1 × SSC	100	30 seconds	DNA loss = 10% DNA denaturation = 51% Good hybridization efficiency	
		60% formamide in 2×10^{-4} *M* EDTA	55	4 minutes	DNA loss = 10% DNA denaturation = 79% Highest hybridization efficiency	
		0.1 × SSC in 4% formaldehyde	100	30 seconds	DNA loss = 0 DNA denaturation = 100% No hybridization	
D. melanogaster polytene chromosomes	5 S RNA	Methods listed are those where the annealing conditions were 50% formamide: 2 × SSC; other information is in Szabo *et al.* (1977)			Hybridization (%)	Szabo *et al.* (1977)
		0.2 *N* HCl	rt	20 minutes	100	
		0.2 *N* H$_2$SO$_4$	rt	20 minutes	120	

(Continued)

TABLE IV—Continued

Organism/ tissue	Probe	Method of denaturation	Temperature (°C)	Time	Results	Reference
		0.2 N HCl/80% formamide	rt	20 minutes	150	
		95% formamide in 0.1 × SSC	60	2.5 hours	110	
		0.01 N NaOH	rt	2 minutes	290	
Primate cell cultures	[³H]EBV DNA	0.2 N HCl	rt	20 minutes	Either concentration of NaOH was superior to denaturation in HCl, but the lower concentration resulted in better morphology	Moar and Klein (1978)
		0.07 N NaOH		3 minutes		
		0.05 N NaOH		3 minutes		
Sheep chorid plexus cells	[³H]cDNA from Visna virus	0.2 M HCl followed by heat at 70°C in 2 × SSC followed by treatment with proteinase K	rt	20 minutes	Treatment increased the efficiency at least 2-fold	Brahic and Haase (1978)
			70	30 minutes		
			37	15 minutes		
Human chromosomes	¹²⁵I	0.2 M HCl	rt	20 minutes	Formamide resulted in more uniform distribution, but HCl resulted in a lower background. The values for mean grains were the same	Cote et al. (1981)
		70% formamide: 1.8 × SSC	70	5 to 10 minutes		

[a] rt, Room temperature.

TABLE V
COMPARISON OF DENATURATION METHODS FOR THE FORMATION OF rRNA : CHROMOSOMAL
DNA HYBRIDS[a]

Denaturation method	Grain counts as percentage of control	
	Individual A	Individual B
A. 95% formamide : 1 × SSC at 70°C for 60 minutes	100	100
B. 0.07 N NaOH for 2 minutes at 38°C	74	—
C. 0.2 N HCl for 30 minutes at 38°C	38	42
D. 0.01 × SSC for 2 minutes at 100°C	25	18
A + B	97	100
A + C	95	100
A + D	63	75
B + C	50	50
B + D	49	42
C + D	52	28
A + B + C + D	63	80

[a] The number of metaphase plates counted for each pont was 25 to 30. Denaturation was terminated in each case by plunging the slides into cold 70% ethanol. The specific activity of the rRNA was 3×10^7 dpm/μg; slides were exposed for 4 days.

occurred following incubation at 65°C in 90% formamide:1 × SSC. As expected, less stringent means of denaturation results in the least loss of DNA. What type of DNA is lost, or whether the loss is random throughout the genome is not known. The perfect denaturant for chromosomal DNA is obviously yet to be found, i.e., one that is effective, but results in low DNA loss.

Chromosomal DNA retains a constant number of sites available for hybridization even under conditions that would favor renaturation. Chromosomal preparations placed under annealing conditions for periods of 5 hours and longer show the same efficiency of hybridization as nontreated slides (Alonso, et al. 1974; Szabo et al., 1975, 1977; Barbera et al., 1979; Henderson, unpublished results). This is a strong argument that the chromosomal DNA is irreversibly denatured and probably remains so during long periods of hybridization. The possibility of undetected "snapback" exists, however. It is probable that the major denatured state is retained as the result of denaturation of surrounding protein binding to the glass slide so that the chromosomal strands are no longer in close juxtaposition. Fixation alone in acid solutions results in some denaturation and probably influences further denaturation of the chromosomal DNA. Acetic acid has been used as an effective denaturant for the localization of

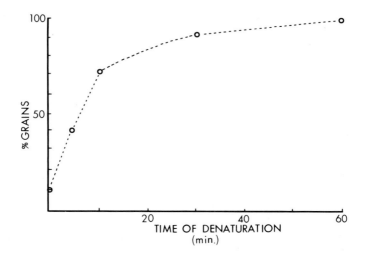

FIG. 4. Denaturation of chromosomal DNA. The percentage grains resulting from hybridization of ^{125}I-labeled rRNA to human chromosomes following increasing times of denaturation of chromosomal DNA is given. Although the major proportion of hybridization occurs with 60 minutes denaturation time, there is a small increment in the number of grains as the time of denaturation is continued up to about 2 hours. Correction has been made for background grains. The concentration of ^{125}I-labeled rRNA used in these experiments was 2 μg/ml; the specific activity was 1 $\times$ 10^8 dpm/μg and autoradiographic exposure was for 3 days. All hybridizations were in 50% formamide:3 $\times$ SSC at 42°C. At least 20 metaphase plates were analyzed in determination of each point.

repetitive AT-rich sequences in mammalian chromosomes (Shapiro *et al.*, 1978).

None of the denaturation methods described here has been tested on a quantitative basis for small nonrepetitive DNA sequences simply because the work involved could be enormous. Until this is accomplished, the best approach is trial and error based on the type of experiment.

G. ANNEALING CONDITIONS

The conditions for hybridization per se that have been subjected to analysis are primarily those where RNA is used as a probe. These are usually the same conditions as used for filter hybridizations. Commonly used procedures are annealing in 2 $\times$ SSC at 68°C or incubation in 50 to 70% formamide in various salt concentrations at 38 to 55°C. The actual conditions chosen should be dependent on experimental variables including salt concentration, time of hybridization and the concentration, length, and complexity of the probe.

On a practical level, an appropriate volume of hybridizing solution adequate to fill the coverslip area is applied to slides, the area is covered with a clean coverslip, and the slides are placed in a constant temperature apparatus in a moisture chamber to prevent evaporation. The simplest moisture chamber is a plastic petri dish containing filter paper saturated with the hybridization buffer. The slides are supported above the surface of the filter paper. More complicated procedures have been devised. For example, the coverslip can be sealed with rubber cement if evaporation is a problem and, assuredly, more elaborate moisture chambers can be devised. A covered petri dish is probably adequate to prevent evaporation, however, even if the coverslip is not sealed.

For reference, the conditions we use for hybridization to human chromosomes is included here (see also Henderson *et al.*, 1981). The procedure has only been tested in part for efficiency and should be used as a guideline, rather than taken as is. Following denaturation in 95% formamide: 1 × SSC for 2.5 hours at 70°C, the slides are plunged into ice cold 70% ethanol. After several washes in cold 95% ethanol, the slides are allowed to dry under moisture-free conditions. The hybridization probe is applied ice-cold to the slides, and covered with a coverslip. The slides are transferred immediately to an incubator at temperatures between 40 and 45°C, depending on the estimated optimal temperature for the formation of a given hybrid. Double-stranded DNA probes are denatured in 0.1 N NaOH at room temperature for 15 minutes, neutralized, and mixed with cold formamide and SSC to a final concentration of 3 × SSC:50% formamide. Less complex probes are brought to 50% formamide:3 × SSC in the cold. The concentration of probe used in the hybridization solution is dependent on the reiteration of the site to be localized. Ribosomal rRNA sites or 5 S RNA sites can be localized effectively at a RNA input of 1 to 2 μg/ml, with other conditions as discussed below. For the localization of chromosomal DNA sequences expected to be of low reiteration, the concentration of the probe used is between 50 and 100 ng/ml; the hybridization temperature is 45°C and the time of incubation is 16 hours.

Various aspects of the hybridization reaction have been investigated for the formation of RNA:chromosomal DNA hybrids (Alonso *et al.*, 1974; Szabo *et al.*, 1975, 1977; Singh *et al.*, 1977; Barbara *et al.*, 1979). The time course of the reaction, concentration of the probe, and resulting saturation levels have been determined under a variety of conditions. It is assumed, but not proven, that studies of reiterated sequences can be directly related to the localization of any chromosomal DNA sequence. Although some of the complementary sequences are inaccessible to the probe, the reaction rate should not be affected. Reaction rates, however, are slower *in situ* presumably due to diffusion hindrance in locally high DNA concentrations (Szabo *et al.*, 1977). As with all steps in cytological hybridization, optimal conditions for annealing must be determined on the basis of incomplete theory and experimental manipulation.

1. Salt Concentration

The usual range of salt concentrations has been tested for cytological hybridization, but only for hybridization to repetitive sequences. Salt concentrations from 1 to 5 × SSC in formamide gave essentially the same results in our experiments for the formation of ^{125}I-labeled rRNA:chromosomal DNA hybrids. Other studies have used salt concentrations between 2 and 6 × SSC (Steffensen, 1977). The effect of salt concentration should be more carefully defined for localization of smaller chromosome regions. Hybridizations proceed more rapidly and to a greater extent in higher salt concentrations (Gillespie and Spiegelman, 1965), but specificity can be diminished at high ionic strength (Church and McCarthy, 1968).

2. Optimal Temperature

Estimates for the optimal temperature of renaturation (T_{opt}) in cytological hybridizations should be based on known qualities of the hybrid complex. T_{opt} is known to be broad for hybridization to mammalian DNA. For most types of hybridization reactions, T_{opt} can be estimated as the midpoint of transition of the melting temperature (T_m) less 15 to 30°C (Marmur and Doty, 1961; McConaughy et al., 1969). Within this range, the most suitable reaction conditions for hybridization may not necessarily be those that give the highest rates of reaction or extent of hybridization. For example, higher temperatures will result in greater specificity, but long incubation at high temperatures can result in chain breakage or depurination. At lower temperatures, duplexes may form with greater mismatching. Moar et al. (1975) have noted that most cytological hybridization experiments have used annealing conditions where the temperatures are below the T_{opt} required for the formation of all possible RNA:chromosomal DNA hybrids. Values of T_m and T_{opt} for hybrids formed in situ are compared in Table VI. The data are far from complete, but the T_{opt} is in the expected range. The values for T_m are consistent with the formation of true hybrids in situ, since the thermal stability of any nucleic acid duplex is sensitive to the presence of mismatched base pairs. The T_m and T_{opt} are known to be influenced by a variety of factors, however, and optimal conditions should be established for individual cases.

If, as expected, the formation of cytological hybrids shares features in common with other types of hybridization systems, particularly filter hybridization, then an estimate of T_{opt} can be made from published values for T_{opt} or T_m of a given hybrid. The T_{opt} for filter hybrids has been shown to be the same as that for hybrids in situ, at least for hybridizations of repetitive sequences to human and mouse chromosomes (Moar et al., 1975), and measured T_m values in situ are within the range expected for a given hybrid (Szabo et al., 1977; Cote et al., 1980; Henderson, unpublished data). The T_{opt} for hybridization of [3H]cDNA from Visna virus RNA to whole cells, however, was estimated to be 15 to 20°C lower for hybrids in situ (Brahic and Haase, 1978). Since this figure was obtained from hybridization to whole cells where the degree of specificity is less, some precaution should be taken in interpreting these data. It does suggest that deviations from the expected values might occur.

TABLE VI
Comparison of Optimal Temperature and Melting Temperature

Organism/tissue	Probe	T_{opt} Known T_{opt}	T_{opt} Cytological hybridization	T_m Known T_m	T_m Cytological hybridization	References
Cultured primate cells	[³H]EBV DNA		50°C in 6 × SSC: 50% formamide	66.5°C from filter hybridization in 50% formamide: 6 × SSC		Moar and Klein (1978)
Sheep cells	[³H]cDNA from *Visna* virus	45°C from filter hybrids in 50% formamide: 600 mM NaCl	20 to 30°C in 50% formamide: 600 mM NaCl	55°C, liquid hybridization in 50% formamide: 3 × SSC	50°C in 50% formamide: 3 × SSC	Brahic and Haase (1978)
Mouse, human chromosomes	[³H]cRNAs from satellite DNAs of mouse and human DNA	58°C in 3 × SSC for mouse; 55°C in 3 × SSC for human satellite III	50 to 60°C in 3 × SSC	Mouse: 74°C Human satellite III: 70°C in 1 × SSC		Moar et al. (1975)
Drosophila polytene chromosomes	¹²⁵I-labeled 5 S RNA				55°C in 50% formamide: 2 × SSC	Szabo et al. (1977)
Human chromosomes	¹²⁵I-labeled 5 S RNA				53°C in 50% formamide: 2 × SSC	Cote et al. (1981)
	¹²⁵I-labeled rRNA				73°C in 50% formamide: 2 × SSC	Cote et al. (1981)
	¹²⁵I-labeled rRNA		50°C in 50% formamide: 3 × SSC			Henderson (unpublished data)

RNA will preferentially anneal to double-stranded DNA to form R-loops in the presence of formamide at temperatures near the denaturation temperature of DNA (Thomas *et al.*, 1976). For this reason, a wide range of temperatures have been tested for the cytological hybridization reaction. ^{125}I-labeled rRNA has been hybridized to human chromosomes at temperatures between 40 and 90°C. The grain density over rDNA-containing chromosomes is maximum at 50°C, but this is also the temperature where maximum background levels are found (Fig. 5). The results suggest that some compromise regarding the T_{opt} may be necessary in predicted long autoradiographic exposures to avoid high background levels. The choice of temperature should take into consideration the nature of the site to be hybridized, the possibility of adventitious binding, and maximum saturation conditions.

3. *Time and Concentration of Probe*

The most critical aspects of the hybridization reactions are the appropriate time and concentration necessary to achieve saturation at a specific site. Cytological hybridization to highly repetitive sites can be achieved quickly (Jones, 1972). In ribosomal RNA:chromosomal DNA hybridizations, incubations should be continued at least 1 to 2 hours under typical experimental conditions for maximal binding (Fig. 6; and Szabo *et al.*, 1977). For comparison, EBV DNA annealed to saturation in DNA of cultured cells in 6 hours (Moar and Klein, 1978). Most cytological hybridizations are continued for some hours even though the hybridization is expected to be rapid. Extending the period for hybridization apparently

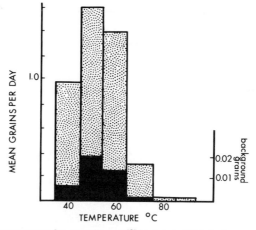

FIG. 5.   Optimal temperature for annealing of ^{125}I-labeled rRNA to human chromosomal DNA. Hybridization was for 12 hours at the temperatures indicated. Other conditions are given in Fig. 4. Stippled bars are mean grains per day per chromosome over the satellited regions of human acrocentric chromosomes; solid bars are the calculated background over this region.

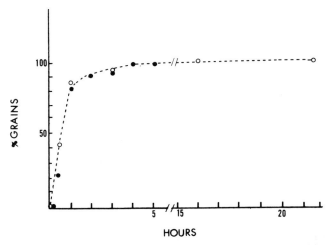

HOURS

FIG. 6.   The proportion of rRNA bound to human chromosomes as a function of time of annealing. Two experiments are given here: ○, the specific activity of the ^{125}I-labeled rRNA was 1.5×10^8 dpm/μg with an exposure time of approximately 2 days; ●, the specific activity of the probe was 1.3×10^8 dpm/μg, exposure was for 3 days. Other conditions are given in Figs. 4 and 5.

has no effect on hybrid formation under conditions where the probe is of appropriate concentration. Moar and Klein (1978) have suggested doubling the time at plateau. Periods of incubation in excess of 18 hours, however, can lead to further loss of chromosomal DNA (Barbera *et al.*, 1979).

A concentration curve for binding of rRNA to human chromosomal DNA is given in Fig. 7. The DNA of the diploid nucleus is 6×10^{-6} μg per cell or 12×10^{-6} μg per metaphase plate. If 10^5 cells are hybridized, then the total amount of DNA per slide is between 0.6 and 1.2 μg. The probe is in vast excess; the concentration of input RNA at saturation is about 1 μg/ml (see also Szabo *et al.*, 1977). The saturation of chromosomal rDNA occurs at a somewhat lower concentration of probe than that we, and others, have determined for the saturation of filter-bound DNA (Bross and Krone, 1972, 1973; Gaubatz *et al.*, 1976). This can be due to many factors, including the fact that fewer sites are available for hybridization.

The time required for saturation of a given site is directly proportional to the complexity of the target DNA. Under conditions of saturating probe concentration, one can predict appropriate reaction times that will increase the probability of hybridization to the intended target. The amount of probe hybridized approaches saturation according to the equation $G_0/G_s = 1 - e^{-kC_r t}$ (Szabo *et al.*, 1977), where G_0 is the background grains, G_s is the number of grains at saturation, C_r the concentration of RNA, t is time, and k is the rate constant. The mean number of grains from a series of experiments using rRNA:human chromosomal

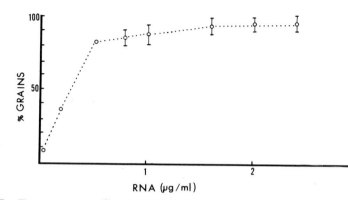

FIG. 7. The proportion of ^{125}I-labeled rRNA bound to human chromosomes as a function of concentration of the probe. Each point represents data from at least 2 separate experiments and in some cases, 4 experiments. The data are combined here for presentation, but each experiment produced an adequate saturation curve. Experimental conditions are given in Figs. 4 and 5.

DNA hybrids is given as a function of $C_r t$ in Fig. 8. Here $C_r t_{\frac{1}{2}}$ is estimated as 48×10^{-3} mole-sec/liter, a value the same as that reported by Szabo *et al.* (1977) for hybridization of rRNA to *Drosophila* polytene chromosomes and close to that reported by Cote *et al.* (1980) for hybridization of 28 S rRNA to human leukocyte nuclei. These values reflect a reaction rate about 10 times slower than that using DNA immobilized to nitrocellulose filters (Szabo *et al.*, 1977; Cote *et al.*, 1981). $C_r t_{\frac{1}{2}}$ *in situ* values have also been reported for 5 S RNA: leukocyte nuclear DNA (2×10^{-3} mole-sec/liter) (Cote *et al.*, 1980), histone mRNA:human chromosomal DNA (5×10^{-3} mole-sec/liter) (Yu *et al.*, 1977), and a $C_0 t_{\frac{1}{2}}$ *in situ* value for [³H]cDNA of *Visna* virus RNA hybridized to cellular preparations (1.5×10^{-1} mole-sec/liter) (Brahic and Haase, 1978). These figures also reflect a slower rate *in situ*. Such values cannot be compared until correction for differences in renaturation rate resulting from different reaction conditions is made. $C_r t_{\frac{1}{2}}$ is expected to be proportional to the complexity of the probe (Birnstiel *et al.*, 1972), and is proportional for corrected $C_r t_{\frac{1}{2}}$ values *in situ* based on the few determinations that have been made.

Where one of the components of the hybridization reaction is immobilized, pseudo-first-order kinetics are expected, i.e., first order kinetics are approximated as the reaction approaches equilibrium at rates proportional to the concentration of the excess component. As expected, the cytological hybridization reaction approximates a first-order reaction. Using a special case of the reciprocal plot, data from $C_r t$ analysis can be used to test the order of the reaction based on reaction kinetics (Britten, 1968). The plot, (ln fraction single-stranded)$^{-1}$ vs $C_r t$ will be linear if the reaction follows first-order kinetics.

Figure 9 shows that this is the case; the relationship is linear in the normal $C_r t$ range used for cytological hybridization.

The length of the probe inversely affects the diffusion rate and the hybridization efficiency. Brahic and Haase (1978) showed that reducing the probe from 500 to 50 nucleotides for cytological hybridization resulted in approximately a 3-fold increase in hybridization efficiency. The probe length should be measured particularly following radioactive labeling *in vitro,* although stringent procedures for labeling, such as iodination or nick-translation, will result in reasonably short probes. The length of the probe is also important since one means of augmenting the strength of the hybridization signal is to enlarge the probe with specific or nonspecific sequences. If this approach is considered, some concession may have to be made to the loss of hybridization efficiency in increasing the hybridization signal.

We have tested the effect of probe length by comparing hybridization efficiency *in situ* of rRNA of normal length (about 18 S after iodination) to rRNA cross-linked covalently using nitrogen mustard (Malka, 1978). The covalently linked molecules were 35 S and 62 S after iodination. Hybridization *in situ* and filter hybridization experiments showed that the number of counts bound was higher (2 to 3 times) using the larger probes, but saturation was not achieved

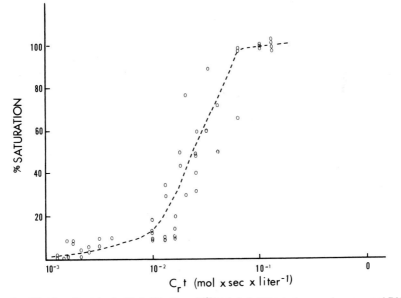

FIG. 8.   Kinetics of cytological hybridization of ^{125}I-labeled rRNA to human chromosomal DNA. The data are obtained from five experiments, using slides from 14 individuals.

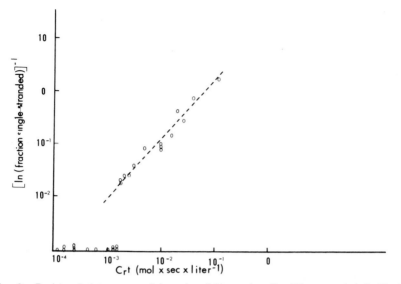

FIG. 9.   Reciprocal plot as a test of the order of the reaction. Conditions are given in Fig. 8.

under any conditions. Increasing the length of the probe resulted in disproportionately reduced hybridization. This is due to diffusion rates, and possibly steric hindrance from trailing of unhybridized portions of the probe. Tereba *et al.* (1979) have suggested another approach to increase the amount of probe bound. Their probe consisted of several RAV-O viral RNA molecules attached by poly(A) · poly(BrdUrd) hybrids to ^{125}I-labeled DNA from sea urchin sperm. Between 5 and 20% of the DNA forms networks upon renaturation increasing the strength of the signal. Incubation of poly(BrdUrd)-containing DNA with the viral RNA resulted in the attachment of 1 to 10 RNA molecules.

The analyses discussed here should be evaluated with the reservation that the data are generated from grain counts of relatively small samples. Further, recording of grains by laboratory personnel may be subjective. A computer analysis of a very large chromosome sample would be necessary to obtain data of sufficient precision for a formal comparison between cytological hybridization and other methods of hybridization.

H. Background; Nonspecific Binding by the Probe

One of the major problems in the use of autoradiographic detection of chromosomal hybrids is the potential for nonspecific grains. Nonspecific grains can result from several causes: (1) the binding of reiterated portions of the probe to unintended sites; (2) failure to remove probe that is nonspecifically bound or mismatched to chromosomal DNA; (3) sticking of the probe to tissues or portions of the glass slide; and (4) poor autoradiographic procedures. Localization of small chromosomal regions may be dependent on the relative absence of background from all causes. On a practical level, this is one type of hybridization where the background levels are open to public inspection. The presence of background grains detracts from the argument, in spite of the significance of a given assignment.

As mentioned, nonrandom annealing of the probe to other than the intended site can be avoided by appropriate hybridization conditions. The presence of mispaired or partially bound probe is usually eliminated following hybridization procedures by either selective enzymatic digestion, or heating to temperatures sufficient to remove mismatched probe, while retaining matched hybrids. Following hybridization with RNA probes, partially bound or nonspecifically bound probe is removed by mild RNase digestion (usually 5 to 10 μg/ml RNase in 2 $\times$ SSC at room temperature for 1 hour). In principle, this will remove loosely paired structures, but some mismatched probe may still be stable. In DNA:chromosomal DNA hybridizations, it is more difficult to dissociate nonspecific binding. Enzymatic removal with S_1 nuclease or micrococcal nuclease has been used but is not always successful. Enzymatic digestion of plasmid derived DNA probes is probably not successful due to reassociation of the plasmid "tails." This may also remove part of the augmented hybridization signal. One procedure for removing nonspecific binding by DNA probes is to include a wash at a temperature adequate to remove adventitiously bound probe. Cote et al. (1980) have suggested incubating the slides for 5 to 10 minutes in 50% formamide:2 $\times$ SSC at a temperature 5°C above the hybridization temperature or 20 to 30 minutes at the hybridization temperature. Other investigators have used incubation in 2 $\times$ SSC for 55°C for 1 hour (Malcolm et al., 1977) and 2 $\times$ SSC for 15 minutes at 65–68°C (Gerhard et al., 1981) to dissociate nonspecific hybrids. Some attention should be given, however, to the nature of the hybrid in determination of appropriate temperatures for dissociation. We have found that incubation following hybridization of plasmid-derived probes is effective at 45 to 50°C in 2 $\times$ SSC for 10 minutes, but higher temperatures or time of incubation can result in some loss of specific label. Extensive washes in 2 $\times$ SSC and dehydration in ethanols usually follow these procedures.

Totally random grains can also affect the validity of site localization. These result from sticking of the probe to dirt on the slides or partially denatured proteins, extraneous sources of radiation, and improper darkroom technique. Nonspecific sticking of the probe can be reduced by including a large excess of nonradioactive carrier such as bacterial nucleic acids in the probe mixture. Although this is effective to some extent, no quantitative measurement of the degree of effectiveness has been made. Another method is to coat the slides with Denhardt's (1966) solution or a modification of this solution. This involves coating the slide with serum albumin, Ficoll, and polyvinylpyrridone and has been found to be effective in reducing background levels over cellular preparations (Brahic and Haase, 1978). We have tested various combinations of coatings for slides and found them to be ineffective in reducing background levels over metaphase plate preparations, although coating with Denhardt's solution has been recommended by other investigators as an effective procedure (see Gerhard *et al.*, 1981). The best way to avoid background is meticulous attention to detail during fixation, hybridization, and autoradiographic procedures.

I. Autoradiography

Several excellent reviews on autoradiography are available (Cleaver, 1967; Miller, 1970; Rogers, 1973). Essentially a photographic emulsion a few microns thick is applied to the surface of the slide. Grains are formed during development as the result of activation of silver bromide to metallic silver. The sensitivity of an emulsion is dependent on the type of radioisotope, the thickness of the cytological preparation, and the type of emulsion. The highest efficiency is produced by low energy β particles. This includes the isotopes ^{3}H, ^{14}C, ^{32}P, and ^{125}I (where the ejected particle behaves as a β particle; the main difference being the mode of origin). In practice, ^{14}C is not used for cytological hybridization because the specific activity of precursors is not adequate. ^{32}P has a distinct disadvantage in that γ-rays are also recorded which result in a tracking of grains from the source and a distinct loss in specificity. The use of ^{3}H and ^{125}I are preferred, based on track length in the emulsion and the high specific activities that can be obtained. With any isotope, grain density following autoradiography is highest over the source and decreases symmetrically on each side of the source with increasing distance. The rate of decrease is isotope dependent, and determines the resolution. Most disintegrations from ^{3}H will be expressed as grains at a distance of 0.5 μm from the source. There is greater scatter from the source when ^{125}I is used. 55% of the grains resulting from ^{125}I decay will be the same distance from the source as ^{3}H. The majority of the remaining grains will fall within a 3.5 μm radius, although 13% could extend to a radius of 16 μm (Ertl *et al.*, 1970; Prensky *et al.*, 1973). On a practical level, the majority of grains produced from ^{125}I disintegration are found to be reasonably site-specific. All specific grains should be considered for quantitative purposes, however. Figure 10 illustrates a computer analysis of grain-to-chromatid distance following hybridization *in situ* of 5 S RNA to human chromosomes. This analysis shows that using a distance of 2.5 μm, or 25% the length of chromosome 1, results in obtaining all grains associated with the site. Increasing the grain-to-chromatid distance does not result in an increase in site-specific grains.

Scintillators in emulsion has been used to intensify the autoradiographic

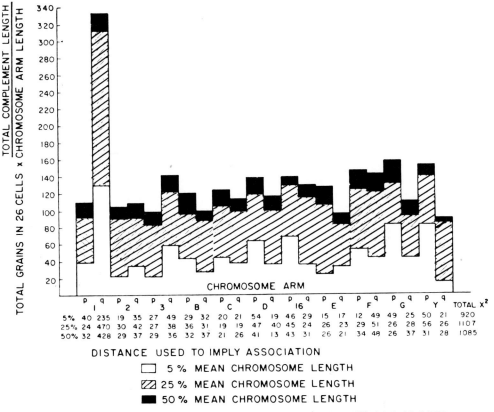

FIG. 10. Computer analysis of grain density following hybridization of ^{125}I-labeled 5 S RNA to human chromosomes. Grains at a distance greater than 25% of the chromosome length do not contribute to the specific labelling. (Reprinted from Warburton *et al.*, 1978.)

signal. There are often dramatic increases in efficiency reported (Durie and Salmon, 1975; Kopriwa, 1979; Stanulis *et al.*, 1979; Fischer, 1979). We have been unable to obtain any increase in efficiency using a series of scintillators under various conditions. This may simply reflect the fact that the emulsion efficiency is approaching maximum. Woodcock *et al.* (1979) have discussed several reasons why present methods for using scintillators with autoradiographic emulsions may not be successful under the same conditions that work well with gel fluorography. The fluor cannot be successfully mixed with the emulsion, nor does an "undercoat or overcoat" containing the fluor penetrate adequately for the emulsion to be responsive to the fluor. This problem could be solved if an appropriate solvent for use with autoradiographic emulsions and fluors is found. Most organic solvents used for this purpose are ineffective or greatly increase background levels.

J. OTHER METHODS FOR DETECTING CYTOLOGICAL HYBRIDS

1. *Hybridization to DNA or RNA in Whole Tissue or Tissue Sections*

One of the earliest uses of cytological hybridization was to DNA of fixed whole cells (Buongiorno-Nadalli and Amaldi, 1970; Gall and Pardue, 1969; John *et al.*, 1969). Localization of DNA or RNA *in situ* in cellular preparations requires special attention to methods of fixation. Sections embedded in paraffin and other embedding media do not allow diffusion of the probe. Specific procedures to obtain the highest efficiency for hybridization to RNA *in situ* have been discussed by Moar and Klein (1978) and Godard and Jones (1979). Appels *et al.* (1979) have devised a procedure for hybridization prior to embedding the tissue in hydroxymethacrylate which allows the distribution of label to be followed in serial sections.

2. *Other Methods for Detection of Hybrids in Situ*

Two primary methods for the detection of hybrids by methods other than autoradiography have been devised. The first, fluorescence, has been successfully used for gene localization in polytene chromosomes and to follow the distribution of various DNA sequences in cellular preparations (Rudkin and Stollar, 1977; Stuart and Potter, 1978; Bauman *et al.*, 1980). Indirect fluorescence is obtained by reacting RNA:DNA hybrids against an antibody to poly(rA)·poly(dT) (antihybrid serum), and then to anti-IgG tagged with rhodamine (Rudkin *et al.*, 1977). A modification uses direct fluorescence where the RNA or DNA is tagged directly with rhodamine; this method has been used successfully for intracellular localization of adenovirus DNA (Bauman *et al.*, 1980, 1981). The sensitivity of the latter method has been calculated as somewhat less than autoradiography when probes of reasonably high specific activity are available. A distinct advantage of fluorescent detection is that the preparations can be viewed immediately.

A second method of detection has been used successfully in detecting hybrids with scanning electron microscopy. This is of importance because the autoradiographic emulsion efficiency for electron microscopy is lower due to the magnification and the limitation on the quantity of silver that can be effectively distributed in the autoradiographic emulsion (Rogers, 1973). Manning *et al.* (1975) have described a method for attaching biotin to RNA using a cytochrome c bridge. Polymethacrylate spheres are then covalently attached to avidin. The biotin RNA complex is hybridized to denatured DNA, and a portion become labeled by attachment of the spheres due to the strong interaction between biotin and avidin. The efficiency of this method using mammalian chromosomes has not been determined.

IV. Problems and Prospectives

One of the primary factors restricting the use of cytological hybridization to other than repetitive sequences has been "bad press." This has resulted from two causes: the imposition of possible limitations from saturation filter hybridizations using eukaryotic DNA, and the failure of investigators using cytological hybridization to establish adequate criteria for assignment of DNA sequences *in situ*.

Filter hybridization has had its share of problems. The large size and complexity of the eukaryotic genome led to the question of whether DNA sequences present as unique sequences or with low reiteration could be detected by this method. The primary criticism was that reaction rates for other than repetitive sequences would be too low for detection. It is true that reaction rates determined from filter hybridizations cannot be directly related to rates for solution hybridizations. The means by which DNA strands are immobilized on filters are not well-defined, but the DNA may entangle by coiling, resulting in restriction of free diffusion of strands (Spiegelman *et al.*, 1973) and leading to reduced nucleation rates. Of relevance to cytological hybridization was whether unique sequences could be detected at all under usual reaction conditions. It was subsequently demonstrated that filter hybridizations did include annealing by small DNA sequences in the presence of a vast excess of noncomplementary DNA (Shoyab *et al.*, 1974). The advantages, and limitations, of using filter hybridizations in detection of small DNA sequences have been discussed (Sambrook *et al.*, 1975). While many kinetic factors are expected to be similar between filter hybrids and cytological hybrids, the means of detection and analysis are dissimilar, and a comparison between two less than perfectly defined hybridization systems was unfortunate. It should be noted that reaction rates for cytological hybridizations are lower than those observed using filter hybridizations. If this is found to be a consistent feature of cytological hybridizations in further studies, it may represent the only real restriction in detection of small sequences within the chromosomal complement.

A more serious problem has arisen as the result of cytological hybridization experiments. Inadequate means of obtaining sufficiently radioactive probes, lack of methods for analysis, and the unforeseen problem of minor contaminants in probe preparations have led to reports of assignments that were contrary to established theoretical and kinetic limitations (Price *et al.*, 1972, 1973). The primary objection to such studies was that the specific activity of the probe was not adequate to produce the desired assignment (Bishop and Jones, 1972; Pren-

sky and Holmquist, 1973; Atwood *et al.*, 1975a). The confusion that resulted,
however, was instructive and led to the establishment of guidelines to be used for
hybridization to chromosomes (Atwood *et al.*, 1976a; Szabo *et al.*, 1977).
These guidelines are of particular importance where the probe is either uninten-
tionally or intentionally heterogeneous. As mentioned, the presence of unexpected reiterated sequences in a probe preparation is not a trivial matter when
low-multiplicity sequences are sought. We have tested the validity of assignment
of low-multiplicity genes using known sites for α and β globin in the mouse
(Henderson *et al.*, 1978). Using reasonably stringent conditions, it was possible
to assign the appropriate regions of the chromosomes, i.e., β globin to chromosome 7; α globin to chromosome 11. Other regions were significantly labeled,
however. The primary source of contamination in the probe was assumed to be
ribosomal RNA and appropriate measures were taken to compete out binding of
rRNA. This should have been satisfactory, but the rDNA sites were still significantly labeled. In retrospect, the grain density over the rDNA-containing
chromosomes was probably due to aggregated 5.8 S RNA present in the probe.
A prior knowledge of these sites simplified the analysis, but unknown sites that
label in addition to the intended site will severely complicate analysis. Most
problems can theoretically be solved by the use of appropriate kinetic conditions, but these experiments are time-consuming and the results obtained with
respect to appropriate time and concentration for specific assignments are based
on the less than accurate, and perhaps subjective, use of grain counts. $C_r t$ comparisons between hybrids *in situ* and those formed on filters are instructive, but
the use of $C_r t$ analysis from filter hybridizations, as is usually the case, may
not result in the most accurate comparison. Many of these problems have been
or will be solved by the availability of homogeneous probes, appropriate methods
for obtaining probes of high specific activity, more precise kinetic guidelines
for cytological hybridization (particularly where cloned probes are used) and the
use of computer assistance for analysis.

Other problems have been considered within this article. The most exasperat-
ing problem, that of slow detection by autoradiography, is most simply solved at
the present time by patience on the part of the investigator. Other problems are
not so simply solved. Reasonable estimates of hybridization efficiency cannot be
established in the absence of appropriate data from other types of hybridization
experiments. Further, there is a lack of consistency between results from various
laboratories in the quality and efficiency of cytological hybridizations. More
practical are problems related to maximum conditions for denaturation and annealing. Methods for better denaturation, while maintaining adequate fixation for
chromosome morphology, are needed to increase the number of sites available
for hybridization. The option for more controlled conditions during fixation
should be considered, since fixation indubitably influences later denaturation of
the DNA. Finally, the use of the light microscope for detection hinders the

degree of resolution that can be obtained. It would be useful to devise methods for using autoradiography in conjunction with low-power electron microscopy.

The possibilities for the use of cytological hybridization at the present time are enormous. As probes become available, the mammalian chromosome can be dissected, at least at the gross molecular level. Homogeneous probes are available or will be available for mapping transcriptive and other sequences. The integration of viral-specific DNA or other DNAs from exogenous sources can be followed at the chromosomal level. With more precise kinetic data and means of analysis, other problems can be considered such as the chromosomal distribution of mixtures of mRNA species from cell-cycle stages, different stages of differentiation or between cells of various types.

Nucleic acid hybridization has been a valuable research approach for two decades. Cytological hybridization, as one of the viable offshoots of conventional hybridization, has only been used successfully since 1969. It is still a young science, attended by the trappings of incomplete integrity, but will become more valuable as careful studies are undertaken to determine conditions that will lead to more efficient hybridization and recognition of hybrids. There are still many technical problems in recognition of, particularly, unique sequences. It is anticipated that these problems will be solved by rigorous experimental technique and analysis.

Acknowledgments

I am indebted to my colleagues, Drs. K. C. Atwood, R. Goodman, D. Warburton, M. T. Yu, and Ms. S. Ripley for encouragement and for research data and help in preparation of this manuscript. I would also like to thank Ms. Lois Purcell, Ms. Lena Lofstrom, and Mr. Stephen Quirk for excellent assistance. This work was supported in part with grants from the National Institutes of Health, Elsa U. Pardee Foundation, the March of Dimes-Birth Defects Foundation, the National Science Foundation, and a Scholar Award from the Leukemia Society of America. S. Karger AG, Basel has given permission for the use of Fig. 10.

References

Ada, G. L., Humphrey, J. H., Askonas, B. A., McDevitt, H. O., and Nossal, G. J. V. (1966). *Exp. Cell Res.* **41,** 557–572.

Ahnström, G., and Natarajan, A. T. (1974). *Hereditas* **76,** 316–320.

Alonso, C., Helmsing, P. J., and Berendes, H. D. (1974). *Exp. Cell Res.* **85,** 383–390.

Amaldi, F., and Buongiorno-Nardelli, M. (1971). *Exp. Cell Res.* **65,** 329–334.

Appels, R., Steffensen, D. M., and Craig, S. (1979). *Exp. Cell Res.* **124,** 436–441.

Arrighi, F. E., Hsu, T. C., Saunders, P., and Saunders, G. F. (1970). *Chromosoma* **32,** 224–236.

Atwood, K. C. (1979). *In* "Concepts of the Function of DNA, Chromatin and Chromosomes" (A. S. Dion, ed.), pp. 187–209. Year Book Medical Publ. Chicago, Illinois.

Atwood, K. C., Henderson, A. S., Kacian, D., and Eicher, E. (1975a). *Cytogenet. Cell Genet.* **14,** 279–281.

Atwood, K. C., Yu, M. T., Johnson, L. D., and Henderson, A. S. (1975b). *Cytogenet. Cell Genet.* **15**, 50–54.

Atwood, K. C., Yu, M. T., Eicher, E., and Henderson, A. S. (1976a). *Cytogenet. Cell Genet.* **16**, 372–375.

Atwood, K. C., Gluecksohn-Waelsch, S., Yu, M. T., and Henderson, A. S. (1976b). *Cytogenet. Cell Genet.* **17**, 9–17.

Barberá, E., Caliani, M. J., Pagés, M., and Alonso, C. (1979). *Exp. Cell Res.* **119**, 151–162.

Bauman, J. G. J., Wiegant, J., Borst, P., and van Duijn, P. (1980). *Exp. Cell Res.* **128**, 485–490.

Bauman, J. G. J., Wiegant, J., and van Duijn, P. (1981). *J. Histochem. Cytochem* **29**, 227–246.

Birnstiel, M. L., Sells, B. H., and Purdom, I. F. (1972). *J. Mol. Biol.* **63**, 21–39.

Bishop, J. O., and Jones, K. W. (1972). *Nature (London)* **240**, 149–150.

Bostock, C. J., and Clarke, E. M. (1980). *Cell* **19**, 709–715.

Bostock, C. J., Gosden, J. R., and Mitchell, A. R. (1978). *Nature (London)* **272**, 324–328.

Brahic, M., and Haase, A. T. (1978). *Proc. Natl. Acad. Sci. U.S.A.* **75**, 6125–6129.

Brahic, M., Stowring, L., Ventura, P., and Haase, A. T. (1981). *Nature (London)* **292**, 240–242.

Britten, R. J. (1968). *Carnegie Inst. Year Book* **67**, 332–335.

Bross, K., and Krone, W. (1972). *Hum. Genet.* **14**, 137–141.

Bross, K., and Krone, W. (1973). *Hum. Genet.* **18**, 71–75.

Buongiorno-Nardelli, M., and Amaldi, F. (1970). *Nature (London)* **225**, 946–948.

Caspersson, T., Zech, L., Johansson, C., and Modest, E. J. (1970). *Chromosoma* **30**, 215–227.

Chandler, M. E., and Yunis, J. J. (1978). *Cytogenet. Cell Genet.* **22**, 352–356.

Chuang, C. R., and Saunders, G. F. (1974). *Biochem. Biophys. Res. Commun.* **57**, 1221–1230.

Church, R. B., and McCarthy, B. J. (1968). *Biochem. Genet.* **2**, 55–73.

Cleaver, J. E. (1967). *In* "Frontiers of Biology" (A. Neuberger and E. L. Tatum, eds.), Vol. 6. North-Holland Publ., Amsterdam.

Comings, D. E., Avelino, E., Okado, T. A., and Wyandt, H. E. (1973). *Exp. Cell Res.* **77**, 469–493.

Commerford, S. L. (1971). *Biochemistry* **10**, 1993–1999.

Conkie, D., Affara, N., Harrison, P. R., Paul, J., and Jones, K. (1974). *J. Cell Biol.* **63**, 414–419.

Cote, B. D., Uhlenbeck, O. C., and Steffensen, D. M. (1980). *Chromosoma* **80**, 349–367.

Denhardt, D. T. (1966). *Biochem. Biophys. Res. Commun.* **23**, 641–645.

Dörries, K., Johnson, R., and ter Meulen, U. (1979). *J. Gen. Virol.* **42**, 49–57.

Dunn, A. R., Gallimore, P. H., Jones, K. W., and McDougall, J. K. (1973). *Int. J. Cancer* **11**, 628–636.

Durie, B. G. M., and Salman, S. E. (1975). *Science* **190**, 1093–1095.

Elsevier, S. M., and Ruddle, F. H. (1975). *Chromosoma* **52**, 219–228.

Ertl, H. H., Feinendegen, L. E., and Heiniger, H. J. (1970). *Phys. Med. Biol.* **15**, 447–456.

Evans, H. J., Gosden, J. R., Mitchell, A. R., and Buckland, R. A. (1974a). *Nature (London)* **251**, 346–347.

Evans, H. J., Buckland, R. A., and Pardue, M. L. (1974b). *Chromosoma* **48**, 405–426.

Fischer, H. A. (1979). *J. Histochem. Cytochem.* **27**, 1527–1528.

Gall, J. G., and Pardue, M. L. (1969). *Proc. Natl. Acad. Sci. U.S.A.* **64**, 600–604.

Gall, J. G., Cohen, E. H., and Polan, M. L. (1971). *Chromosoma* **33**, 319–344.

Gaubatz, J., Prashad, N., and Cutler, R. G. (1976). *Biochim. Biophys. Acta* **418**, 358–375.

Gerhard, D. S., Kawasaki, E. S., Bancroft, F. C., and Szabo, P. (1981). *Proc. Natl. Acad. Sci. U.S.A.* **78**, 3755–3759.

Geuskens, M., and May, E. (1974). *Exp. Cell Res.* **87**, 175–185.

Gillespie, D., and Spiegelman, S. (1965). *J. Mol. Biol.* **12**, 829–842.

Godard, C., and Jones, K. W. (1979). *Nucleic Acids. Res.* **6**, 2849–2861.

Gosden, J. R., Mitchell, A. R., Buckland, R. A., Clayton, R. P., and Evans, H. J. (1975). *Exp. Cell Res.* **92**, 148–158.

Gosden, J. R., Lawrie, S., and Seuanez, H. (1978). *Cytogenet. Cell Genet.* **21**, 1–10.

Hamerton, J. L. (1971). "Human Cytogenetics." Academic Press, New York.

Harper, M. E., Ulrich, A., and Saunders, G. F. (1981) *Proc. Natl. Acad. Sci. U.S.A.* **78.** 4458–4460.

Harrison, P. R., Conkie, D., Paul, J., and Jones, K. (1974). *FEBS Lett.* **32**, 109–112.

Hatlen, L., and Attardi, G. (1971). *J. Mol. Biol.* **56**, 535–553.

Henderson, A. S. (1982). In preparation.

Henderson, A. S., and Atwood, K. C. (1976). *Hum. Genet.* **31**, 113–115.

Henderson, A. S., Warburton, D., and Atwood, K. C. (1972). *Proc. Natl. Acad. Sci. U.S.A.* **69**, 3394–3398.

Henderson, A. S., Warburton, D., and Atwood, K. C. (1973). *Nature (London)* **245**, 95–97.

Henderson, A. S., Warburton, D., and Atwood, K. C. (1974a). *Chromosoma* **44**, 367–370.

Henderson, A. S., Warburton, D., and Atwood, K. C. (1974b). *Chromosoma* **46**, 435–441.

Henderson, A. S., Yu, M. T., Eicher, E. M., and Atwood, K. C. (1974c). *Chromosoma* **49**, 155–160.

Henderson, A. S., Atwood, K. C., and Warburton, D. (1976a). *Chromosoma* **59**, 147–155.

Henderson, A. S., Atwood, K. C., Yu, M. T., and Warburton, D. (1976b). *Chromosoma* **56**, 29–32.

Henderson, A. S., Eicher, E., Yu, M. T., and Atwood, K. C. (1976c). *Cytogenet. Cell Genet.* **17**, 307–316.

Henderson, A. S., Warburton, D., Megraw-Ripley, S., and Atwood, K. C. (1977). *Cytogenet. Cell Genet.* **19**, 281–302.

Henderson, A. S., Yu, M. T., and Atwood, K. C. (1978). *Cytogenet. Cell Genet.* **21**, 231–240.

Henderson, A. S., Warburton, D., Megraw-Ripley, S., and Atwood, K. C. (1979a). *Cytogenet. Cell Genet.* **23**, 213–216.

Henderson, A. S., Yu, M. T., and Milcarek, C. (1979b). *Cytogenet. Cell Genet.* **23**, 201–207.

Henderson, A. S., Wolf, L., and Megraw-Ripley, S. (1980a). *Cytogenet. Cell Genet.* **28**, 136–139.

Henderson, A. S., Moskowitz, G., and Warburton, D. (1980b). *Hum. Genet.* **54**, 83–85.

Henderson, A. S., Yu, M. T., and Silverstein, S. (1981). *Cytogenet. Cell Genet.* **29**, 107–115.

Henderson, A. S., Robins, D., and Ripley, S. (1982). Submitted.

Hennig, W. (1973). *Int. Rev. Cytol.* **36**, 1–44.

Hsu, T. C., Spirito, S. E., and Pardue, M. L. (1975). *Chromosoma* **53**, 25–36.

John, H. A., Birnstiel, M., and Jones, K. W. (1969). *Nature (London)* **223**, 582–587.

Johnson, L. D., Harris, R. C., and Henderson, A. S. (1974). *Hum. Genet* **21**, 217–219.

Johnson, L. K., Johnson, R. W., and Strehler, B. L. (1975). *J. Mol. Cell. Cardiol.* **7**, 125–133.

Jones, K. W. (1970). *Nature (London)* **225**, 912–915.

Jones, K. W. (1972). In "New Techniques in Biophysics and Cell Biology" (R. H. Pain and B. J. Smith, eds.), Vol. 1, pp. 29–66. Wiley, New York.

Jones, K. W. (1973). *J. Med. Genet.* **10**, 273–281.

Jones, K. W. (1977). In "Molecular Structure of Human Chromosomes" (J. J. Yunis, ed.), pp. 295–326. Academic Press, New York.

Jones, K. W., and Corneo, G. (1971). *Nature (London) New Biol.* **233**, 268–271.

Jones, K. W., and Robertson, F. W. (1970). *Chromosoma* **31**, 331–345.

Jones, K. W., Prosser, J., Corneo, G., Ginelli, E., and Bobrow, M. (1972). In "Modern Aspects of Cytogenetics: Constitutive Heterochromatin in Man" (R. A. Pfeiffer, ed.), pp. 45–61. Schattauer, Stuttgart.

Jones, K. W., Prosser, J., Corneo, G., and Ginelli, E. (1973). *Chromosoma* **42**, 445–451.

Jones, K. W., Purdom, I. F., Prosser, J., and Corneo, G. (1974). *Chromosoma* **49**, 161–171.
Kacian, D. L., and Myers, J. C. (1976). *Proc. Natl. Acad. Sci. U.S.A.* **73**, 2191–2195.
Kaufman, S., Gallo, R., and Miller, N. (1979). *J. Virol.* **30**, 637–641.
Kopriwa, B. (1979). *J. Histochem. Cytochem.* **27**, 1524–1526.
Kuo, M. T., and Hsu, T. C. (1978). *Chromosoma* **65**, 325–334.
Lawrie, S. S., and Gosden, J. R. (1980). *Hum. Genet.* **53**, 371–373.
Loni, M. C., and Green, M. (1973). *J. Virol.* **12**, 1288–1292.
Loni, M. C., and Green, M. (1974). *Proc. Natl. Acad. Sci. U.S.A.* **71**, 3418–3422.
Macaya, G., Thiery, J. P., and Bernardi, G. (1977). *In* "Molecular Structure of Human Chromosomes" (J. J. Yunis, ed.), pp. 35–58. Academic Press, New York.
McConaughy, B. L., Laird, C. D., and McCarthy, B. J. (1969). *Biochemistry* **8**, 3289–3295.
McDougall, J. K., Dunn, A. R., and Jones, K. W. (1972). *Nature (London)* **236**, 346–348.
MacGregor, H. C., and Kezer, J. (1971). *Chromosome* **33**, 167–182.
Malcolm, S., Williamson, R., Boyd, E., and Ferguson-Smith, M. A. (1977). *Cytogenet. Cell Genet.* **19**, 256–261.
Malka, D. (1978). Ph.D. Dissertation, Columbia University, New York.
Manning, J. E., Hershey, N. D., Broker, T. R., Pelligrini, M., Mitchell, H. K., and Davidson, N. (1975). *Chromosoma* **53**, 107–117.
Manuelidis, L. (1978). *Chromosoma* **66**, 23–32.
Marmur, J., and Doty, P. (1961). *J. Mol. Biol.* **3**, 585–594.
Marx, K. A., Allen, J. R., and Hearst, J. E. (1976). *Chromosoma* **59**, 23–42.
Miklos, G. L. G., and John, B. (1979). *Am. J. Hum. Genet.* **31**, 264–280.
Miller, D. A., Berg, W. R., Warburton, D., Dev, V. G., and Miller, O. J. (1978). *Hum. Genet.* **43**, 289–297.
Miller, O. J. (1970). *Adv. Hum. Genet.* **1**, 35–130.
Mitchell, A. R., Seuanez, H. N., Lawrie, S., Martin, D. E., and Gosden, J. R. (1977). *Chromosoma* **61**, 345–358.
Moar, M. H., and Jones, K. W. (1975). *Int. J. Cancer* **16**, 998–1007.
Moar, M. H., and Klein, G. (1978). *Biochim. Biophys. Acta* **519**, 49–64.
Moar, M. H., Purdom, I. F., and Jones, K. W. (1975). *Chromosoma* **53**, 345–359.
Neer, A., Baran, N., and Manor, H. (1977). *Cell* **11**, 65–71.
Pardue, M. L., and Gall, J. G. (1969). *Proc. Natl. Acad. Sci. U.S.A.* **64**, 600–604.
Pardue, M. L., and Gall, J. G. (1970). *Science* **168**, 1356–1358.
Pardue, M. L., and Gall, J. G. (1972). *In* "Molecular Genetics and Developmental Biology" (M. Sussman, ed.), pp. 341–349. Prentice-Hall, New York.
Pardue, M. L., and Gall, J. G. (1975). *In* "Methods in Cell Biology" (D. M. Prescott, ed.), Vol. 10, pp. 1–16. Academic Press, New York.
Prensky, W. (1976). *In* "Methods in Cell Biology" (D. M. Prescott, ed.), Vol. 13, pp. 121–152. Academic Press, New York.
Prensky, W., and Holmquist, G. (1973). *Nature (London)* **241**, 44–45.
Prensky, W., Steffensen, D., and Hughes, W. L. (1973). *Proc. Natl. Acad. Sci. U.S.A.* **70**, 1860–1864.
Prescott, D. M., Bostock, C. J., Hatch, F. T., and Mazrimas, J. A. (1973). *Chromosoma* **42**, 205–213.
Price, P. M., Conover, J. H., and Hirschhorn, K. (1972). *Nature (London)* **237**, 340–342.
Price, P. M., Hirschhorn, K., Gabelman, N., and Waxman, S. (1973). *Proc. Natl. Acad. Sci. U.S.A.* **70**, 11–14.
Prosser, J., Bradley, M. L., Muir, P. D., Vincent, P. C., and Gunz, F. W. (1978). *Leuk. Res.* **2**, 151–161.

Robins, D. M., Ripley, S., Henderson, A. S., and Axel, R. (1981a). *Cell* **23**, 29–39.

Robins, D. M., Axel, R., and Henderson, A. S. (1981b). *J. Mol. Appl. Genet.* **1**, 191–203.

Rogers, A. W. (1973). "Techniques in Autoradiography." Elsevier, Amsterdam.

Rudkin, G. T., and Stollar, B. D. (1977). *Nature (London)* **265**, 472–473.

Sambrook, J., Botchan, M., Gallimore, P., Ozanne, B., Petersson, U., Williams, J., and Sharp, P. A. (1975). *Cold Spring Harbor Symp. Quant. Biol.* **39**, 615–632.

Sanchez, O., and Yunis, J. J. (1974). *Chromosoma* **48**, 191–202.

Saunders, G. F. (1974). *In* "Advances in Biological and Medical Physics" (J. H. Lawrence and J. W. Gotman, eds.), Vol. 15, pp. 19–46. Academic Press, New York.

Saunders, G. F., Hsu, T. C., Getz, M., Simes, E. L., and Arrighi, F. E. (1972). *Nature, (London) New Biol.* **236**, 244–246.

Sawin, V. L., Skalka, A. M., and Wray, W. (1978). *Histochemistry* **59**, 1–8.

Schachat, F. M., and Hogness, P. S. (1973). *Cold Spring Harbor Symp. Quant. Biol.* **38**, 371–381.

Seabright, M. (1971). *Lancet* **2**, 971–972.

Sealy, L., Hartley, J., Donelson, J., Chalkley, R., Hutchison, N., and Hamkalo, B. (1981). *J. Mol. Biol.* **145**, 291–318.

Segal, S., Garner, M., Singer, M., and Rosenberg, M. (1976). *Cell* **9**, 247–257.

Shapiro, I. M., Moar, M. H., Ohno, S., and Klein, G. (1978). *Exp. Cell Res.* **115**, 411–414.

Shoyab, M., Markham, P. D., and Baluda, M. A. (1974). *J. Virol.* **14**, 225–230.

Singh, L., Purdom, I. F., and Jones, K. W. (1977). *Chromosoma* **60**, 377–389.

Spiegelman, G., Haber, J. E., and Halvorson, H. O. (1973). *Biochemistry* **12**, 1234–1242.

Stanulis, B. M., Sheldon, S., Grove, G. L., and Cristofalo, V. J. (1979). *J. Histochem. Cytochem.* **27**, 1303–1307.

Steffensen, D. M. (1977). *In* "Molecular Structure of Human Chromosomes" (J. J. Yunis, ed.), pp. 59–88. Academic Press, New York.

Steffensen, D. M., and Wimber, D. E. (1972). *In* "Nucleic Acid Hybridization in the Study of Cell Differentiation" (H. Ursprung, ed.), pp. 47–63. Springer-Verlag, Berlin and New York.

Steffensen, D. M., Prensky, W., and Dufy, P. (1974). *Cytogenet. Cell Genet.* **13**, 153–154.

Steffensen, D. M., Prensky, W., Mutton, D., and Hamerton, J. L. (1975). *Cytogenet Cell Genet.* **14**, 434–441.

Stuart, W. D., and Potter, D. L. (1978). *Exp. Cell Res.* **113**, 219–222.

Szabo, P., Elder, R., and Uhlenbeck, O. (1975). *Nucleic Acids Res.* **5**, 647–653.

Szabo, P., Elder, R., Steffensen, D. M., and Uhlenbeck, O. C. (1977). *J. Mol. Biol.* **115**, 539–563.

Szabo, P., Yu, L. C., Borun, T., Varricchio, F., Siniscalco, M., and Prensky, W. (1978). *Cytogenet. Cell Genet.* **22**, 359–363.

Tereba, A., Lai, M. M. C., and Murti, K. G. (1979). *Proc. Natl. Acad. Sci. U.S.A.* **76**, 6486–6490.

Thomas, M., White, R. L., and Davis, R. W. (1976). *Proc. Natl. Acad. Sci. U.S.A.* **73**, 2294–2298.

Thurston, J. M., and Joftes, D. L. (1963). *Stain Technol.* **38**, 231–235.

Warburton, D., and Henderson, A. S. (1979). *Cytogenet. Cell Genet.* **24**, 168–175.

Warburton, D., Henderson, A. S., Shapiro, L. R., and Hsu, L. Y. F. (1973). *Am. J. Hum. Genet.* **25**, 439–445.

Warburton, D., Henderson, A. S., and Atwood, K. C. (1975). *Chromosoma* **51**, 35–40.

Warburton, D., Yu, M. T., Atwood, K. C., and Henderson, A. S. (1976a). *Cytogenet. Cell Genet.* **16**, 440–442.

Warburton, D., Atwood, K. C., and Henderson, A. S. (1976b). *Cytogenet. Cell Genet.* **17**, 221–230.

Warburton, D., Naylor, A. F., Henderson, A. S., and Atwood, K. C. (1978). *Cytogenet. Cell Genet.* **22**, 714–717.

Willey, A. M., and Yunis, J. J. (1975). *Exp. Cell Res.* **91,** 223–232.

Wolgemuth-Jarashow, D. J., Jagiello, G. M., Atwood, K. C., and Henderson, A. S. (1976). *Cytogenet. Cell Genet.* **17,** 137–143.

Wolgemuth-Jarashow, D. J., Jagiello, G. M., and Henderson, A. S. (1977). *Hum. Genet.* **36,** 63–68.

Wolgemuth, D., Jagiello, G., and Henderson, A. S. (1979). *Exp. Cell Res.* **118,** 181–190.

Wolgemuth, D., Jagiello, G. M., and Henderson, A. S. (1980). *Dev. Biol.* **78,** 598–604.

Woodcock, C. L. F., D'Amico-Martel, A., McInnis, C. J., and Annunziato, A. T. (1979). *J. Microsc.* **117,** 417–423.

Yasmineh, W. G., and Yunis, J. J. (1975). *Exp. Cell Res.* **90,** 191–200.

Young, B. D., Hell, A., and Birnie, G. D. (1976). *Biochim. Biophys. Acta* **454,** 539–548.

Yu, L. C., Szabo, P., Borum, T. W., and Prensky, W. (1977). *Cold Spring Harbor Symp. Quant. Biol.* **42,** 1101–1105.

Yunis, J. J., and Tsai, M. Y. (1978). *Cytogenet. Cell Genet.* **22,** 364–367.

Yunis, J. J., Kuo, M. T., and Saunders, G. F. (1977). *Chromosoma* **61,** 335–344.

zur Hausen, H., Diehl, V., Wolf, H., and Schulte-Holthausen, H. (1972). *In* "Molecular Studies in Neoplasia," pp. 516–530. Williams & Wilkins, Baltimore, Maryland.

INTERNATIONAL REVIEW OF CYTOLOGY, VOL. 76

Organization and Expression of Viral Genes in Adenovirus-Transformed Cells

S. J. FLINT

Department of Biochemical Sciences, Princeton University, Princeton, New Jersey

I. Introduction

The human adenoviruses, first isolated during the winter of 1952–1953 (Rowe *et al.*, 1953; Hilleman and Werner, 1954), comprise linear double-stranded DNA genomes, $20–23 \times 10^6$ daltons (Green *et al.*, 1967), packaged in an icosahedral capsid that contains no lipid (Horne *et al.*, 1959; Green and Piña, 1963). At least 33 human adenoviruses have been isolated and are responsible for relatively mild disease in the human population, infections of the upper respiratory tract for example (see Tooze, 1980). Fortunately, perhaps, for they are very widespread among human populations. Interest in this group of viruses was considerably heightened when Trentin and his colleagues reported that adenovirus type 12 (Ad12) induced tumors when inoculated into newborn hamsters (Trentin *et al.*, 1962). Since that observation, human adenoviruses have been the objects of intense scrutiny: much has been learned about their molecular organization and interactions with permissive and semi- or nonpermissive host cells.

The original observation that Ad12 possessed tumorigenic potential has been repeatedly confirmed and extended to several other human adenoviruses as well as adenoviruses isolated from other species including those of simian, bovine, canine, and avian origin (Huebner *et al.*, 1962; Yabe *et al.*, 1962; Girardi *et al.*, 1964; Rabson *et al.*, 1964; Larsson *et al.*, 1965; Pereira *et al.*, 1965; Hull *et al.*, 1965; Darbyshire, 1966; Sarma *et al.*, 1965, 1967; Rapoza *et al.*, 1967; Anderson *et al.*, 1969a,b,c). On the other hand, not all human adenoviruses are tumorigenic when inoculated into newborn rodents, a property that has provided one criterion for their classification. Members of group A, including serotypes 12, 18, and 31, induce tumors rapidly and at high frequency. Serotypes 3 and 7 fall into group B whose members cause tumors less frequently and with a much longer latent period, 9 to 23 months by contrast to the less than 3 months observed with group A viruses. Finally, quite a large number of human adenovirus serotypes, including types 1, 2, 5, 6, 9, and 10, group C serotypes, manifest no tumorigenic potential (Huebner *et al.*, 1965). Interestingly, this classification scheme based upon complex biological properties enhibited by these viruses correlates exactly with that made using the criterion of DNA sequence homology (Green and Piña, 1965). The tumorigenic potential of human adenoviruses immediately raises the question of whether they play a role in human neoplastic disease. No evidence for the persistence of adenoviral genetic information in any of the very many human tumors tested has ever been obtained, despite the use of methods that could detect even very small amounts of viral nucleic acid sequences (Mackey *et al.*, 1976; Green *et al.*, 1979a).

Although adenoviruses display differences in their ability to induce tumors in animals, all that have been tested nevertheless transform cells in culture. Transformation by adenoviruses is the outcome of a rare interaction between the virus and cells that either fail completely to support replication or are semipermissive for virus growth, cells that are usually of rodent origin. The factors that govern whether an adenovirus infection is fully productive, incomplete, or abortive remain mysterious, but the outcome is influenced by both the cell type and the virus serotype. Thus the human serotypes 2 and 5 can replicate quite well in hamster, rat, or mouse cells although the yields of virus are decreased by several orders of magnitude compared to those obtained from fully permissive human cells (Takahashi, 1972; Williams, 1973; Gallimore, 1974; Zucker and Flint, unpublished observations). On the other hand, Ad12 infection of hamster cells is completely abortive: no progeny virions are synthesized for the infection never progresses beyond the early phase (Doerfler, 1968, 1969; zur Hausen and Sokol, 1969; Doerfler and Lundholm, 1970). During such abortive infections, adenoviruses appear to induce such severe chromosomal damage that the majority of infected cells do not survive (zur Hausen, 1967, 1968a,b; Cooper *et al.*, 1968; Strohl *et al.*, 1969a,b). Similarly, most semipermissive cells are eventually killed when infected by adenoviruses, the normal result of virus production. It is therefore not surprising that the appearance of a transformed cell following adenovirus infection is an infrequent event: the transformation efficiency displayed by these viruses is low, one focus-forming unit in fibroblasts of rat, mouse, or rabbit origin containing 10^4 to 10^6 plaque-forming units (Rabson *et al.*, 1964; Pope and Rowe, 1964; Freeman *et al.*, 1967; McAllister *et al.*, 1969a,b; Levinthal and Peterson, 1965; Gallimore, 1974; Zucker and Flint, unpublished observations). It is worthy of note that although transformed human cells have never been obtained following infection by adenoviruses or intact viral

DNA, human cells are transformed when exposed to virus inactivated by exposure to ultraviolet light (Graham *et al.*, 1975). Thus it seems clear that human cells represent one extreme end of a spectrum ranging from completely nonpermissive to fully permissive and are normally destroyed so efficiently by adenovirus infection that there exists no opportunity for potential transformants to survive, unless functions essential for productive infection are destroyed.

After a culture of rodent fibroblasts has been infected by a human adenovirus, the majority of cells, as discussed previously, are killed. The few surviving transformants continue to grow to form dense foci which are usually visible within 2–3 weeks of infection. Such cells possess altered growth properties for they overgrow one another to form these foci, the property that permits their identification and selection. Adenovirus-transformed cells also exhibit an altered morphology, being considerably smaller and less flattened than their fibroblastic parents, and express virus-specific proteins, referred to as T-antigens. They also possess many of the other properties typical of mammalian cells transformed by viruses, including, in many cases, tumorigenicity in nude mice (see Tooze, 1980, for a review).

II. Retention and Expression of Viral Genetic Information in Adenovirus-Transformed Cells

A. Nature and Organization of Integrated Viral DNA Sequences

Once it was established that adenoviruses transform cells both in animals and in culture, it became of paramount importance to identify the viral genes responsible for transformation and elucidate their mode of action. The former goal has been attained, the result of much elegant work, but the latter has not: the polypeptides encoded by the viral genes whose products participate in transformation have been identified, but little of their molecular function(s) and nothing of their interactions with cellular components that result in the phenotype typical of transformed cells are understood.

The initial investigations of adenoviral DNA sequences present in transformed cell DNA employed reassociation kinetics (Gelb *et al.*, 1971): when ^{32}P-labeled viral DNA is denatured and allowed to reanneal in the presence of nontransformed cell or transformed cell, high-molecular-weight DNA, any additional viral DNA sequences present in the latter reaction accelerate its rate compared to the former. This increase in renaturation rate can be used to estimate the concentration of DNA sequences homologous to the labeled probe DNA present in transformed cell DNA. Moreover, when this method is applied using sets of restriction endonuclease fragments of the viral DNA as probes, it also becomes possible to map which regions of the viral genome are integrated and often to make some deductions about their arrangements relative to one another. In the adenovirus system, this approach was first applied to a series of rat and hamster cell lines established after transformation of embryonic fibroblast, brain, or muscle cells by Ad2 or Ad5 (Williams, 1973; Gallimore, 1974; Sambrook *et al.*, 1975; Flint *et al.*, 1976). This comprehensive series of experiments established a number of general points about the nature of integrated type C adenoviral sequences that have been confirmed by more recent investigations (Johanssen *et al.*, 1977, 1978; Frolova *et al.*, 1978; Frolova and Zalmazon, 1978; Visser *et al.*, 1979; Dorsch-Häsler *et al.*, 1980).

The most striking conclusion to emerge from the studies of Sambrook and his colleagues is that the only adenoviral sequences common to the 12 or so lines examined comprise the left-hand 12–14% of the viral genome. It must therefore be inferred that this region includes all functions necessary for the maintainence of transformed cell phenotypes exhibited by these cell lines. These results alone, of course, do not exclude the possibility that additional viral functions may be required to mediate the early events of the virus–cell interaction that ultimately lead to transformation, usually referred to as initiation of transformation. A second important finding of these studies is that none of the lines examined contained a complete copy of the viral genome, although several do possess many sequences homologous to regions other than the left-hand 12%, as illustrated in Fig. 1. This observation certainly explains previous failures to rescue infectious adenoviruses from transformed cell lines (Landau *et al.*, 1966; Larsson *et al.*, 1966).

In this respect, an interesting difference appears to have emerged between group C adenovirus transformants and cells transformed by Ad12, a member of group A. Although a few Ad12 hamster tumor cell lines possess only a part of the Ad12 genome (see, for example, Lee and Mak, 1977), the great majority of transformed rat or hamster cell lines obtained after infection by Ad12 virus retain sequences complementary to all regions of the viral genome (Green *et al.*, 1976; Mak *et al.*, 1979; Ibelgaufts *et al.*, 1980). A probable explanation for this quite considerable difference lies in the fact, cited in the previous section, that rodent cells are completely nonpermissive for Ad12, but really quite permissive for type C adenoviruses. The great majority of rodent cells in a culture infected by a type C adenovirus are therefore likely to succumb to successive cycles of productive virus growth; only those rare cells that receive defective particles with incomplete genomes that therefore cannot mount a complete infection would survive to become transformed. In other words, the permissive nature of rodent cells for type C adenoviruses might select against the survival of potential transformants that contain the entire adenoviral genome. Rodent cells cannot support Ad12 replication (see Section I) and thus there exists a greater possibility that cells that received the normal viral genome will survive and become transformed. This hypothesis is lent credence by the results of transformation by the Ad5 early

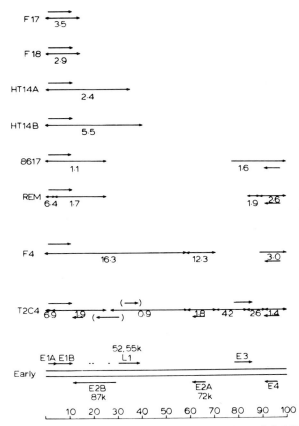

FIG. 1. Integration and expression in cytoplasmic mRNA sequences of viral DNA sequences in type C adenoviral-transformed rodent cell lines. The solid horizontal line at the bottom of the figure represents the adenoviral genome as 100 map units. The pair of horizontal lines immediately above depict the viral genome. Early mRNA sequences are shown by the horizontal arrows, →, drawn in the direction of transcription above or below the regions of the genome to which the mRNA is complementary. In the other parts of the figure, viral DNA sequences integrated in the transformed rodent cell lines listed on the left are depicted by the solid, double-headed arrows, ↔. The number below these arrows give the concentration of the DNA sequences in copies/diploid quantity of cell DNA. Transformed cell cytoplasmic RNA sequences are depicted in the same fashion as early mRNA. This figure is based on data of Sambrook *et al.* (1975), Flint *et al.* (1975, 1976), and Binger *et al.* (1982).

mutant H5ts125 that carries a lesion in early region 2 (Ginsberg *et al.*, 1975; Grodzicker *et al.*, 1977) whose product is a single-stranded DNA binding protein (Levine *et al.*, 1975) essential for adenoviral DNA replication (Ginsberg *et al.*, 1975; van der Vliet *et al.*, 1977). At a permissive temperature H5ts125 transforms rat embryonic cells at a frequency similar to wild-type Ad5, but at a nonpermissive temperature appears to transform more efficiently (Ginsberg *et al.*, 1975; Meyer and Ginsberg, 1977). More importantly, however, cell lines established after transformation at a permissive temperature are like those transformed by wild-type adenoviruses and usually contain only sequences comprising the left-hand end of the viral genome, whereas lines established after transformation at a nonpermissive temperature contain the entire viral genome (Meyer and Ginsberg, 1977; Dorsch-Häsler *et al.*, 1980). Thus, when a full productive infection is inhibited by the H5ts125 mutation, complete copies of the viral genome can become integrated, as the hypothesis outlined above would suggest.

The implication of this explanation, that adenoviral type C transformants arise from those rare cells infected by defective virus particles, has not been tested directed and, indeed, is a point difficult to prove. However, defective adenoviruses do retain their ability to transform cells in culture when they are no longer able to induce productive infection (Schaller and Yohn, 1974). In addition, very similar patterns of integrated viral DNA sequences are observed among cell lines independently transformed following infection by the same virus stock (Sambrook *et al.*, 1975; Flint *et al.*, 1976), an observation that can be interpreted as transformation by defective genomes of defined structure present in the stock.

A second approach to the identification of transforming functions relies on the ability to transfect purified, deproteinized DNA into cells in culture. Employing this technique, van der Eb and his colleagues have demonstrated that adenovirus-transforming functions lie at the left-end of the viral genome, (Graham *et al.*, 1975), in complete agreement with the conclusion based on reassociation kinetics. In addition, the use of small restriction endonuclease fragments spanning this region has established that the left-most 4.5%, that is 0 to 4.5 units of the Ad2 or Ad5 genome, is sufficient to induce the appearance of transformed cell foci. Such foci, however, are often difficult to establish as cell lines and do not display a fully transformed phenotype (van der Eb *et al.*, 1979; Houewling *et al.*, 1980). By contrast, fragments that comprise at a minimum the left-hand 8% of the viral genome induce completely typical transformation. These results therefore suggest that functions required for initiation of transformation are encoded to the left of 4.5 units, whereas sequences to the right specify functions involved in maintainence, a conclusion in agreement with the results of transformation studies using mutants bearing lesions that lie within these regions (Graham *et al.*, 1978; Jones and Shenk, 1979). Similar results have been obtained using restriction endonuclease fragments of Ad7 or Ad12, members of groups B and A, respectively (Dijkema *et al.*, 1979; Sawada *et al.*, 1979; Seikikawa *et al.*, 1978; Shiroki *et al.*, 1977, 1980).

The organization of integrated viral DNA sequences has been examined by the blotting technique (Southern, 1975) and more recently by cloning integrated viral DNA sequences into prokaryotic vectors, a procedure that permits refined restriction endonuclease mapping as well as sequencing. It is clear from comparisons among the patterns of integration of adenoviral DNA sequences in different transformed cell lines that there exists no preferred location in the cellular genome at which viral DNA sequences integrate: this is true of both type C and type A adenovirus-transformed cells (Sambrook *et al.*, 1979; Visser *et al.*, 1979; Sawada *et al.*, 1979; Groneberg *et al.*, 1977; Sutter *et al.*, 1978; Doefler *et al.*, 1979), although repetitive sequences have been observed as the sites of integration of adenoviral DNA sequences in some transformed cells (Sutter and Doefler, 1978; Stabel *et al.*, 1980). In many type C adenovirus-transformed cell lines, the pattern of integration appears to be quite simple, comprising colinear segments of subgenomic regions. In other cell lines, substantial rearrangements of viral DNA sequences have been observed. These include the junction of sequences that normally lie at opposite ends of the adenoviral genome, sometimes in an orientation inverted with respect to their normal orientation in the genome (Sambrook *et al.*, 1979; Visser *et al.*, 1979). Tandem repetitions of integrated adenoviral sequences have also been noted in type C adenovirus-transformed cells (Sambrook *et al.*, 1979).

Many lines of Ad12-transformed cells that have been examined contain a complete, colinear copy of the Ad12 genome, that is, sequences at or near the normal left and right-hand termini of free Ad12 DNA are linked to cellular DNA sequences, as well as subgenomic fragments integrated at separate sites (Sutter *et al.*, 1978; Doefler *et al.*, 1979; Ibelgaufts *et al.*, 1980). In some cases, integrated Ad12 sequences, probably with adjacent cellular sequences, are reiterated (Sutter *et al.*, 1978; Stabel *et al.*, 1980).

Some of the specific patterns of integrated patterns described have been interpreted in terms of specific DNA structures that serve as substrates for integration; junction of left and right end genomic sequences, for example, suggest circular intermediates (Sambrook *et al.*, 1979; Visser *et al.*, 1979). If such interpretations be correct, then the variable patterns observed would imply that a variety of structures must be suitable integration intermediates or that integrated sequences are subject to extensive rearrangement during passage of the transformed cells. The existence of tandem reiterations provides support for the latter view, but little is known of the original integration events, so rare in infected cell populations that they have been impossible to investigate. Nor has much attention been paid to the history of integrated adenoviral DNA sequences during passage of transformed cells.

B. Expression of Viral Genetic Information in Adenovirus-Transformed Cells

It is clear from the results discussed in Section II,A that sequences comprising approximately the left-hand 10–12% of the human adenoviral genome are suffi-

cient to induce and maintain the transformed cell phenotype. It would therefore
seem likely that this set of viral DNA sequences is expressed in transformed
cells, an expectation fully confirmed by hybridization analysis of viral mRNA
sequences present in cytoplasmic fractions of adenovirus type 2, 5, and 12
transformed cell lines and by the hybridization-selection and translation of such
RNA *in vitro* (Flint *et al.*, 1975, 1976; Lewis *et al.*, 1976, 1979; Chinnadurai *et
al.*, 1978; Johansson *et al.*, 1978; Frolova *et al.*, 1978; Frolova and Zalmanzon,
1978; Wilson *et al.*, 1978; Wold and Green, 1979; Mak *et al.*, 1979; Fujinaga *et
al.*, 1979; Shiroki *et al.*, 1980; Sawada and Fujinaga, 1980; Yoshida and
Fujinaga, 1980; Jochemsen *et al.*, 1981; Lupker *et al.*, 1981). This segment of
the adenoviral genome includes early regions E1A and E1B, expressed prior to
the onset of viral DNA synthesis. The mRNA species and protein products
encoded by these two regions in the type C adenoviral genome have been exhaus-
tively analyzed, indeed mapped down to the last nucleotide (Perricuadet *et al.*,
1979; 1980a; van Ormondt *et al.*, 1980a,b). The products synthesized during
productive infection are illustrated in Fig. 2: both E1A and 1B encode multiple
mRNA species that share some, but not all, of their sequences, the result of
differential splicing of the primary products of transcription of these two independent transcriptional units. In the case of E1A, such splicing events do not alter
the reading frames for translation used in the different mRNA species. Thus the
E1A mRNA species encode partially related polypeptides. Regions E1A and E1B
of Ad12 exhibit a similar organization to that shown in Fig. 2 for Ad2 or Ad5,
although differences in the temporal patterns of expression of individual E1B

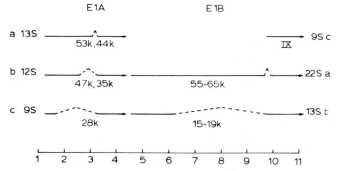

FIG. 2.   Type C adenoviral early mRNA species complementary to regions E1A and E1B. The
relevant section of the type C adenoviral genome is represented by the horizontal solid line divided
into map units at the bottom of the figure. Sequences represented in mRNA (Berk and Sharp, 1978;
Chow *et al.*, 1979) are shown by the horizontal arrows drawn in the direction of transcription: caret
symbols indicate sequences removed during posttranscriptional splicing. The species are named
according to their observed sedimentation coefficients (Spector *et al.*, 1978) or according to the
nomenclature of Chow *et al.* (1979). The polypeptides encoded by the individual mRNA species
(Halbert *et al.*, 1979; Esche *et al.*, 1980) are listed below each mRNA.

mRNA species during productive Ad12 infection may exist (Perricaudet *et al.*, 1980b; Sawada and Fujinaga, 1980).

Viral mRNA species unique to transformed cell lines have been observed in some circumstances. A striking example is provided by the Ad2-transformed cell line F4 which contains a novel viral DNA segment comprising inverted sequences from the right-hand end of the free viral genome linked to the normal left-hand end (see Section II,A). Inspection of Fig. 1 reveals that this inversion brings early regions E4 and E1A into the same orientation, with respect to direction of transcription as well as joining their sequences. The cytoplasm of F4 cells does contain unusual mRNA species that include sequences derived from both E4 and E1A apparently transcribed from the E4 promoter site and utilizing both E4 and E1A splicing signals (Sambrook *et al.*, 1979). Chimeric mRNA species, probably derived either from readthrough transcripts of tandemly integrated sets of E1A and E1B sequences or from transcription initiated at the E1B promoter site into cellular DNA have also been observed to be synthesized in cell lines established after Ad7 and Ad12 transformation (Sawada and Fujinaga, 1980; Yoshida and Fujinaga, 1980). Such examples are, however, relatively rare: the majority of adenoviral cell lines that have been examined appear to synthesize 1A and E1B mRNA species and polypeptides identical to those made during the early phase of productive infection (Chinnadarai *et al.*, 1976; Lewis *et al.*, 1976; Flint, 1977a; Lassam *et al.*, 1978; Spector *et al.*, 1978; Sambrook *et al.*, 1979; Beltz and Flint, 1979; Green *et al.*, 1979b; van der Eb *et al.*, 1979; Wold and Green 1979; Ross *et al.*, 1980; Yoshida and Fujinaga 1980; Sawada and Fujinaga, 1980; Jochemsen *et al.*, 1981).

The great similarity between expression of regions E1A and E1B as mRNA species in transformed and productively infected cells has led to the notion that viral promoter sites are recognized normally when they become integrated into cellular chromatin (see, for example, Sambrook *et al.*, 1979). Support for this idea also comes from the fact that in those type C transformed cell lines that contain viral DNA sequences in addition to those homologous to E1A and E1B, early sequences are expressed both as spliced mRNA species and polypeptides that appear identical to those made during productive infection (Lewis *et al.*, 1976; Flint, 1977a; Johansson *et al.*, 1978; Wold and Green, 1979; Wilson *et al.*, 1978; Sambrook *et al.*, 1979; Ross *et al.*, 1980; Flint, unpublished observations). As the 5' termini of early mRNA species correspond to the sites at which transcription is initiated (Fire *et al.*, 1981), these observation strongly imply that adenoviral early promoters are recognized when present in cellular chromatin. It is only fair to point out, however, that little work has been done to examine directly the transcription of adenoviral genetic information in transformed cells.

Of course, the arguments advanced in the previous paragraph are not meant to imply that adenoviral genes are always expressed in transformed cells in a manner identical to that employed during productive infection of human cells. Some differences that stem from rearrangements of viral DNA sequences or their relationship to cellular DNA sequences have been discussed. Potentially more significant deviations have also been noted in transformed cells. A major difference is the apparent failure to synthesize late mRNA sequences when the transformed cell line retains the major late promoter site, located near 16.45 units in the r-strand (Ziff and Evans, 1978), and large segments of the r-strand transcriptional unit, as in the case in lines such as HT14A, HT14B, F4, and T2C4 shown in Fig. 1 (Flint et al., 1975; 1976; Sambrook et al., 1979). Some late RNA sequences have been reported to present in some Ad12-transformed cell lines, but these have not been well defined (Doerfler et al., 1979). Whether this failure reflects inactivity of the promoter site itself, that is, no transcription of late sequences, or ineffective processing, perhaps because the complete late transcriptional unit is absent, is not established. Whatever the explanation, this difference is rendered more interesting by recent findings that the so-called late promoter, from which the majority of late mRNA sequences are transcribed (see Flint, 1982, for a review) is in fact active from the earliest periods during a productive infection to direct the synthesis of one specific late mRNA species (Chow et al., 1979; Shaw and Ziff, 1980; Nevins and Wilson, 1981; Akusjärvi and Persson, 1981). Thus it cannot simply be argued that activity of the late promoter site depends upon replication of free viral DNA, an event that does not occur in transformed cells. It is therefore of considerable interest to determine whether the late promoter site, when present in transformed cells like F4 or T2C4, is indeed inactive, a situation that might emphasize the importance of the chromosomal structure in which viral DNA sequences are integrated, or whether expression of late mRNA species is restricted by incomplete or inappropriate posttranscriptional processing. Cell lines, established after transformation by H5ts125 or Ad12, that contain integrated, colinear copies of the complete genome would be particularly good candidates for such experiments.

Another interesting difference between expression of integrated and "free" adenoviral genes concerns their regulation. During a productive infection, the various E1A mRNA species shown in Fig. 2 display characteristic patterns of temporal expression (Spector et al., 1978; Chow et al., 1979; Esche et al., 1980). Thus at the earliest times after infection species E1A-a (see Fig. 2) is predominant, whereas synthesis of species E1A-b begins at somewhat later times, and this mRNA continues to accumulate during the early part of the late phase. The human transformed cell line 293 contains and expresses both region E1A and E1B (Lassam et al., 1978; Aeillo et al., 1979). When this cell line is infected by mutants of Ad5 deleted in E1A (Jones and Shenk, 1979) the transformed cell complements the defective virus and a complete productive infection ensues. However, the E1A mRNA species, which must be synthesized from the integrated E1A DNA sequences, do not display such temporal regulation of expression (Spector et al., 1980). In a productive infection, regulation of synthesis of E1A mRNA species must occur by modulation of splicing, for, as illustrated in Fig. 2, the E1A mRNA species differ only in the set of internal sequences removed during splicing. It is not yet understood how such modulation is achieved or how it is linked to the unfolding of productive infection. A trivial explanation, that modulation of splicing of E1A transcripts requires an adenoviral gene product not made in 293 cells can be eliminated by the experiment performed by Spector and colleagues (1980), for the Ad5 mutant deleted in E1A provides all functions except those specified by E1A itself. Although the mechanisms underlying these observations are not yet understood, the differential regulation of E1A expression in 293 transformed cells once again points to a significant influence of the cellular milieu upon adenoviral gene expression.

In summary, then, it seems reasonable to conclude that adenoviral early genes integrated into cellular chromatin are frequently expressed as mRNA using promoter and splice site signals supplied by the viral DNA. On the other hand, it is clear that the mere presence of an adenoviral promoter site in the cellular genome does not guarantee its expression and subtle differences between regulation of integrated and free adenoviral genes have been reported. In this context, it becomes of considerable interest to examine directly the influence of the chromosomal organization of the regions into which viral DNA sequences become integrated upon their expression.

III. Chromosomal Organization of Integrated Viral Genes in Adenovirus-Transformed Cells

Although actively transcribed eukaryotic genes are clearly organized into structures that resemble nucleosomes (see, for example, Lacy and Axel, 1975; Foe *et al.*, 1976; Pruitt and Grainger, 1981) in nuclei or chromatin, they exhibit preferential sensitivity to digestion by DNase I: thus, globin gene sequences but not the ovalbumin gene in chicken red blood cells are digested under mild conditions to fragments too small to form stable hybrids with globin DNA probes (Weintraub and Groudine, 1976), while the converse is true in hen oviduct synthesizing ovalbumin (Garel and Axel, 1976). In these initial experiments, performed before the advent of molecular cloning, the only probes available were cDNA copies of the mRNA species of interest. These, while permitting analysis of sequences actually encoding mRNA in a gene cannot detect sequences of that gene that are transcribed but removed during posttranscriptional processing, or those flanking the transcriptional unit. Thus, it is not possible using cDNA to correlate DNase I sensitivity with the transcriptional unit or to measure how far a DNase I-sensitive domain might extend beyond the borders of a transcriptional unit. Adenovirus-transformed cells that contain defined segments of the viral genome, only some sequences of which are expressed as mRNA, have provided a convenient solution to that limitation (Flint and Weintraub, 1977; Frolova *et al.*, 1978; Frolova and Zalmanzon, 1978) as, more recently, has the use of cloned segments of genomic DNA spanning both genes expressed in specific tissues, such as the clusters of chicken α and β globin genes, and extensive 5' and 3' flanking regions (Stalder *et al.*, 1980a,b).

The two lines of Ad5-transformed cells originally chosen for study, HT14A and HT14B contain, as illustrated in Fig. 1, the left-hand 35 and 40%, respectively, of the viral genome, probably in colinear arrays in each case, at concentrations of 2.4 and 5.5 copies/diploid amount of cell DNA, respectively (Sambrook *et al.*, 1975). Of this information, only that comprising to E1A and E1B regions is expressed as mRNA (Flint *et al.*, 1976; Binger *et al.*, 1982). Thus, in these cell lines, viral DNA sequences that are expressed are linked to those that are not and both expressed and nonexpressed regions can be readily assayed with defined restriction endonuclease fragments of adenoviral DNA. In order to determine the extent of DNase I-sensitive of integrated viral sequences in these cell lines, nuclei isolated from HT14A or HT14B cells were subjected to mild DNase I digestion and the purified DNA subsequently allowed to reanneal with the labeled DNA of small restriction endonuclease fragments homologous to either the expressed or nonexpressed regions. All viral sequences appear, by the criterion of resistance to *Staphylococcal* nuclease digestion, to be organized in nucleosomes (Flint and Weintraub, 1977). It was therefore possible on the basis of the results observed following DNase I digestion to provide a map, albeit one of low resolution, of the viral sequences sensitive to DNase I. In the case of HT14A cells, a clear-cut pattern was observed: within the region spanning approximately 0.5 to 11.5 units in the Ad5 genome, which includes regions E1A and E1B (see Fig. 2), all sequences are highly sensitive to DNase I. As the site at which transcription of E1A and E1B initiates corresponds to the 5' ends of these mRNA species at 0.8 and 4.6 units, respectively (Baker and Ziff, 1981), these data indicate that there exists a very good correlation between DNase I sensitivity of chromatin and transcriptional activity. Moreover, viral DNA segments that are not transcribed in HT14A cells, such as those to the right of 18.8 units, are resistant to digestion under the same conditions. Interestingly, additional sequences flanking the 5' and 3' ends of the mRNA coding region are sensitive to DNase I. However, the data obtained could not distinguish between the same sensitivity to digestion as the mRNA coding sequences of only a subset of the sequences detected by these probes or a lower degree of sensitivity of all sequences detected. In either case, it is clear that DNase I sensitivity extends some distance beyond the borders of the viral transcriptional units themselves.

These conclusions have been elegantly confirmed and extended by studies of the chicken erythrocyte α and β globin genes using blotting and cloned globin-region DNA fragments to probe the chromosomal organization of these genes (Stalder *et al.*, 1980a,b; Weintraub *et al.*, 1981). It is clear that actively transcribed genes exhibit preferential DNase I sensitivity and that such regions are bordered by chromosomal regions that exhibit intermediate sensitivity to DNase I digestion. Sharp boundaries separate such sensitive domains and regions that are insensitive to DNase I digestion. Chromosomal domains that are preferentially sensitive to DNase I digestion are associated with two specific nonhistone proteins, HMG 14 and 17, and the DNase I sensitivity of actively transcribed genes is dependent upon this association (Weisbrod and Weintraub, 1980). Globin gene sequences that exhibit DNase I sensitivity possess the ability to bind HMG proteins *in vitro* (Weisbrod and Weintraub, 1981; Sandeen *et al.*, 1981), but the structural basis for this distinctive property of transcribed DNA sequences is not yet completely understood. It is, however, noteworthy that transcribed and preferentially DNase I-sensitive sequences are generally undermethylated compared to genes that are not expressed in the same tissue (Bird *et al.*, 1979; Waalwyck and Flavell, 1978; McGhee and Ginder, 1979; Mandel and Chambon, 1979; Shen and Maniatis, 1980; van der Ploeg and Glavell, 1980; Weintraub *et al.*, 1981). In some cases, the correlation between hypomethylation and transcription is far from clear-cut, but absolute correlations between gene activity, DNase I sensitivity and hypomethylation have been observed in studies of α and β globin gene switching during chicken embryo development (Weintraub *et al.*, 1981; Groudine and Weintraub, 1981).

Of the adenoviral DNA sequences present in the second line examined, HT14B, some half display a pattern of DNase I sensitivity similar to that described for HT14A. The other half, however, appear to be completely resistant to DNase I digestion (Flint and Weintraub, 1977), suggesting that these copies became integrated into chromosomal sites at which they cannot be expressed. It should, however, be noted that small rearrangements of the viral genome in some copies of the integrated adenoviral DNA sequences in HT14B cells, sufficient, for example, to deplete a promoter site, cannot be formally excluded at this time. This explanation, however, seems unlikely for two such changes would be required to inactivate both the E1A and E1B promoter sites and any major change would have been detected. Furthermore, similar observations have been made using other lines of transformed cells established after infection of rat embryo fibroblasts by Ad5 (Frolova *et al.*, 1978; Frolova and Zalmanzon, 1978). In the cell lines examined, a good correlation exists among DNase I sensitivity and transcription of integrated viral DNA sequences, but several segments that are transcribed and present in multiple copies exhibit only partial sensitivity to DNase I digestion. Thus it seems likely that adenoviral DNA sequences do integrate at sites in the host genome whose chromosomal organization precludes their expression. Such integrated sequences are presumably detected only because other sets of viral transforming sequences are integrated at additional sites, from which they are expressed to induce and/or maintain the transformed phenotype. It would be of interest to learn whether copies of integrated viral sequences that are not expressed are lost over time, but, as mentioned previously, there are no data available to relate patterns of integrated adenoviral DNA sequences to the passage history of the cell line.

As we have seen, the conclusions made about the chromosomal organization of integrated adenoviral DNA sequences have been confirmed and elaborated for normal cellular genes using cloned DNA probes. Conversely, everything that is known of the packaging of integrated viral genes is completely consistent with the notions that they are packaged in chromatin as any cellular gene and that the nature of their chromosomal organization can influence their expression. Moreover, the evidence available, although not complete, suggests that the majority of integrated viral genes that have been examined are expressed from viral promoter sites, although these may be present in novel arrangements compared to the free viral genome. It is not at all surprising that viral promoter sites can be recognized in a chromosomal context for when present in adenoviral genomes they are transcribed by cellular RNA polymerase form II (see Flint, 1977b; Tooze, 1980, for reviews). The similarity between the chromosomal structure of integrated viral genes that are expressed in transformed cells and their normal cellular counterparts extends to methylation. In an elegant study, Sutter and Doerfler (1980) have shown that in two lines of Ad12-transformed hamster cells early viral genes that are transcribed are hypomethylated: In two other lines of Ad12-transformed rat brain cells, some sequences that specify late gene products are expressed and these too are undermethylated, although they are extensively methylated in the two hamster lines in which they are not transcribed. Thus, there appears to be a perfect correlation between transcriptional activity and hypomethylation in these Ad12-transformed cells, similar to that observed with chicken globin genes. Unfortunately DNase I sensitivity of the viral DNA sequences was not examined in these experiments.

It would also be of considerable interest to compare the structure of integrated viral genes as revealed by these assays to that exhibited when they are expressed from an intact viral genome during lytic infection. Unfortunately, little is known of the nucleoprotein structure and organization of nonintegrated viral genome in productively infected cells. Digestion of such intracellular DNA, and virion DNA, with micrococcal nuclease has provided evidence for some kind of nucleosome-like structure (Corden et al., 1976; Tate and Philipson, 1978; Sergent et al., 1979; Brown and Weber, 1980), but the details of this organization have not been elucidated. Nor has the identity of the proteins associated with intracellular viral DNA been clearly established: within virions, adenoviral DNA is packaged by two core proteins, VII and V (see Tooze, 1980), but it has been suggested that these may be replaced by histones in infected cells (Tate and Philipson, 1978). Much further work is clearly required to clarify these issues.

It may well be that significant differences between free and integrated homologous sequences will emerge. It is, for example, established that genomic adenoviral DNA is never methylated during productive infection (Vardinon et *al.*, 1980) even at high multiplicities of infection at early times after infection, when it is unlikely that all copies of a given early gene are transcribed (Vayda and Flint, unpublished observations; see Flint, 1982, for a review). It is also now clear that considerably more of the adenoviral genome is actually expressed in the absence of viral DNA replication than was originally supposed, including sequences transcribed from the major late promoter site at 16.45 units in the r-strand (Chow *et al.*, 1979; Shaw and Ziff, 1980; Nevins and Wilson, 1981; Akusjärvi and Persson, 1981). It is' not, however, correct to suppose that the entire adenoviral genome is a permanently active unit of expression, for transcription of the major late unit is clearly limited during the early phase of infection (Shaw and Ziff, 1980; Nevins and Wilson, 1981; Akusjärvi and Persson, 1981). Thus, there must be some other explanation for the lack of methylation of free adenoviral DNA sequences. It therefore might be speculated that the "chromosomal" organization of nonintegrated adenoviral genomes in productively infected cells is built by viral proteins and that regulation of viral gene expression as productive infection unfolds is a response to signals unique to the viral DNA, in other words, that adenoviruses have evolved mechanisms for organization and regulation of expression of their genetic information that do not employ the components, or signals, that mediate cellular gene expression. Whether this is so remains to be tested experimentally, but it is an especially interesting possibility in view of the fact that when integrated, adenoviral DNA sequences behave like any cellular gene. Thus, they possess the potential to become methylated when not expressed, or methylated and packaged into a DNase I-sensitive conformation associated with transcription. It may therefore be that viral products made during productive infection, perhaps in combination with signals present in an intact viral genome, possess the ability to override the signals recognized when segments of the viral genome become integrated.

Of course these ideas are speculative at the present time, but they serve to illustrate the belief that analysis of the structure, chromosomal organization, and expression of adenoviral genetic information is transformed compared to productively infected cells will continue to provide fascinating insights into mechanism(s) of regulation of gene expression in eukaryotic cells.

ACKNOWLEDGMENTS

Unpublished work by the author cited in this article was supported by a grant from the American Cancer Society.

REFERENCES

Aeillo, L., Guilfoyle, K., Huebner, K., and Weinmann, R. (1979). *Virology* **94**, 460–469.
Akusjärvi, G., and Persson, H. (1981). *Nature (London)* **292**, 420–426.
Anderson, J., Yates, V. J., Jasty, V., and Manani, L. O. (1969a). *J. Natl. Cancer Inst.* **42**, 1–5.

Anderson, J., Yates, V. J., Jasty, V., and Manani, L. O. (1969b). *J. Natl. Cancer Inst.* **43**, 65–70.

Anderson, J., Yates, V. J., Jasty, V., and Manani, L. O. (1969c). *J. Natl. Cancer Inst.* **43**, 575–580.

Baker, C. C., and Ziff, E. (1981). *J. Mol. Biol.* **148**, 189–222.

Beltz, G. A., and Flint, S. J. (1979). *J. Mol. Biol.* **131**, 353–373.

Berk, A. J., and Sharp, P. A. (1978). *Cell* **14**, 694–712.

Bird, A. P., Taggart, M. H., and Smith, B. A. (1979). *Cell* **17**, 889–901.

Brown, M., and Weber, J. (1980). *Virology* **107**, 306–310.

Chinnadurai, G., Rho, M. H., Horton, R. B., and Green, M. (1976). *J. Virol.* **20**, 255–263.

Chinnadurai, G., Fujinaga, K., Rho, M. H., van der Eb, A. J., and Green, M. (1978). *J. Virol.* **28**, 1011–1014.

Chow, L. T., Broker, T. R., and Lewis, J. B. (1979). *J. Mol. Biol.* **134**, 265–303.

Cooper, J. E. K., Yohn, D. S., and Stich, H. F. (1968). *Exp. Cell Res.* **53**, 225–240.

Corden, J., Engelking, H. M., and Pearson, G. D. (1976). *Proc. Natl. Acad. Sci. U.S.A.* **73**, 401–404.

Darbyshire, J. H. (1966). *Nature (London)* **211**, 102–104.

Dijkema, R., Dekker, B. M. M., van der Feltz, J. M., and van der Eb, A. J. (1979). *J. Virol.* **32**, 943–950.

Doerfler, W. (1968). The fate of the DNA of adenovirus 12 in BHK cells. *Proc. Natl. Acad. Sci. U.S.A.* **60**, 636–643.

Doerfler, W. (1969). *Virology* **38**, 587–606.

Doerfler, W., and Lundholm, U. (1970). *Virology* **40**, 754–756.

Doerfler, W., Stabel, S., Ibelgaufts, H., Sutter, D., Neumann, R., Groneberg, J., Scheidetmann, K. H., Deuring, R., and Winterhoff, U. (1979). *Cold Spring Harbor Symp. Quant. Biol.* **44**, 551–564.

Dorsch-Häsler, K., Fisher, P. B., Weinstein, I. B., and Ginsberg, H. S. (1980). *J. Virol.* **34**, 305–314.

Eb, A. J., van der, van Ormondt, H., Schrier, P. J., Lupker, J. H., Jochemsen, H., van den Elsen, P. J., Deleys, R. J., Maat, J., van Beveren, C. P., Dijkema, R., and deWaard, A. (1979). *Cold Spring Harbor Symp. Quant. Biol.* **44**, 383–400.

Esche, H., Mathews, M. B., and Lewis, J. B. (1980). *J. Mol. Biol.* **142**, 399–417.

Fire, A., Baker, C. C., Manley, J. L., Ziff, E. B., and Sharp, P. A. (1981). *J. Virol.* **40**, 703–719.

Flint, S. J. (1977a). *J. Virol.* **23**, 44–52.

Flint, S. J. (1977b). *Cell* **10**, 153–166.

Flint, S. J. (1982). *BBA Rev. Cancer,* in press.

Flint, S. J., and Weintraub, H. (1977). *Cell* **12**, 783–794.

Flint, S. J., Gallimore, P. H., and Sharp, P. A. (1975). *J. Mol. Biol.* **96**, 47–68.

Flint, S. J., Sambrook, J. F., Williams, J., and Sharp, P. A. (1976). *Virology* **72**, 456–470.

Foe, V. E., Wilkinson, L. E., and Laird, C. D. (1976). *Cell* **9**, 131–146.

Freeman, A. E., Black, P. H., Vanderpool, A. E., Henry, P. H., Austin, J. B., and Huebner, R. J. (1967). *Proc. Natl. Acad. Sci. U.S.A.* **58**, 1205–1212.

Frolova, E. I., and Zalmanzon, E. D. (1978). *Virology* **89**, 347–359.

Frolova, E. I., Zalmanzon, E. S., Lukanidin, E. M., and Georgiev, G. P. (1978). *Nucleic Acids Res.* **5**, 1–11.

Fujinaga, K., Sawada, Y., Uemizu, Y., Yamashita, T., Shimojo, H., Shiroki, K., Sugisaki, H., Sugimoto, K., and Takanami, M. (1979). *Cold Spring Harbor Symp. Quant. Biol.* **44**, 519–532.

Gallimore, P. H. (1974). *J. Gen. Virol.* **25**, 263–273.

Garel, A., and Axel, R. (1976). *Proc. Natl. Acad. Sci. U.S.A.* **73**, 3966–3970.

Gelb, L. D., Kohne, D. E., and Martin, M. A. (1971). *J. Mol. Biol.* **57**, 129–145.

Ginsberg, H. S., Ensinger, M. J., Kaufman, R. S., Mayer, A. J., and Lundholm, U. (1975). *Cold Spring Harbor Symp. Quant. Biol.* **39**, 419–426.

Girardi, A. S., Hilleman, M. R., and Zwickey, R. E. (1964). *Proc. Soc. Exp. Biol. Med.* **115,** 1141–1150.

Graham, F. L., and van der Eb., A. J. (1973). *Virology* **54,** 536–539.

Graham, F. L., Abrahams, P. C., Mulder, C., Heinjneker, H. L., Warnaar, S. O., deVries, F. A. J., Fiers, W., and van der Eb, A. J. (1975). *Cold Spring Harbor Symp. Quant. Biol.* **39,** 637–650.

Graham, F. L., Harrison, T., and Williams, J. (1978). *Virology* **86,** 10–21.

Green, M., and Piña, M. (1963). *Proc. Natl. Acad. Sci. U.S.A.* **50,** 44–46.

Green, M., Piña, M., Kimes, R., Wensink, P. C., MacHaltie, L., and Thomas, C. A. (1967). *Biochemistry* **57,** 1302–1309.

Green, M. R., Chinnadurai, G., Mackey, J. R., and Green, M. (1976). *Cell* **7,** 419–428.

Green, M., Wold, W. S. M., Mackey, J. K., and Rigden, P. (1979a). *Proc. Natl. Acad. Sci. U.S.A.* **76,** 6606–6610.

Green, M., Wold, W. S. M., Brackmann, K., and Cartas, M. A. (1979b). *Cold Spring Harbor Symp. Quant. Biol.* **44,** 457–469.

Grodzicker, T., Anderson, C., Sambrook, J., and Mathews, M. B. (1977). *Virology* **80,** 111–126.

Groneberg, J., Chardonnet, Y., and Doerfler, W. (1977). *Cell* **10,** 101–111.

Groudine, M., and Weintraub, H. (1981). *Cell* **24,** 393–402.

Halbert, D. N., Spector, D. J., and Raskas, H. J. (1979). *J. Virol.* **31,** 621–621.

Hausen, H., zur. (1967). *J. Virol.* **1,** 1174–1185.

Hausen, H., zur. (1968a). *J. Virol.* **2,** 218–223.

Hausen, H., zur. (1968b). *J. Virol.* **2,** 918–924.

Hausen, H., zur, and Sokol, F. (1969). *J. Virol.* **4,** 255–263.

Hilleman, M. R., and Werner, J. H. (1954). *Proc. Soc. Exp. Biol. Med.* **85,** 183–188.

Hoeuwling, A., van den Elsen, P. J., and van der Eb, A. J. (1980). *Virology* **105,** 537–550.

Horne, R. W., Brenner, S., Waterson, A. P., and Wildy, P. (1959). *J. Mol. Biol.* **1,** 84–86.

Huebner, R. J., Rowe, W. P., and Lane, W. T. (1962). *Proc. Natl. Acad. Sci. U.S.A.* **48,** 2051–2058.

Huebner, R. J., Casey, H. J., Chanock, R. M., and Schell, K. (1965). *Proc. Natl. Acad. Sci. U.S.A.* **54,** 381–388.

Hull, R. N., Johnson, I. S., Culbertson, C. G., Reimer, C. B., and Wright, H. F. (1965). *Science* **150,** 1044–1046.

Ibelgaufts, H., Doerfler, W., Scheidtman, H., and Wechsler, W. (1980). *J. Virol.* **33,** 423–437.

Jochemsen, H., Hertoghs, J. J. L., Lupker, J. H., Davis, A., and van der Eb, A. J. (1981). *J. Virol.* **37,** 530–534.

Johansson, K., Pettersson, U., Philipson, L., and Tibbetts, C. (1977). *J. Virol.* **23,** 29–35.

Johansson, K., Persson, H., Lewis, A. M., Pettersson, U., Tibbetts, C., and Philipson, L. (1978). *J. Virol.* **27,** 628–639.

Jones, N. C., and Shenk, T. (1979). *Cell* **13,** 181–188.

Lacy, E., and Axel, R. (1975). *Proc. Natl. Acad. Sci. U.S.A.* **72,** 3978–3982.

Landau, B. J., Larsson, V. M., Devers, G. A., and Hilleman, M. R. (1966). *Proc. Soc. Exp. Biol. Med.* **122,** 1174–1182.

Larsson, V. M., Girardi, A. J., Hilleman, M. R., and Zwickey, R. E. (1965). *Proc. Soc. Exp. Biol. Med.* **118,** 15–24.

Larsson, V. M., Gosnell, P. A., and Hilleman, M. R. (1966). *Proc. Soc. Exp. Biol. Med.* **122,** 1182–1191.

Lassam, N. T., Bayley, S. T., and Graham, F. L. (1978). *Virology* **87,** 463–467.

Lee, K. C., and Mak, S. (1977). *J. Virol.* **24,** 408–411.

Levine, A. J., van der Vliet, P. C., Rosenwirth, B., Rabak, J., Frenkel, G., and Ensinger, M. (1975). *Cold Spring Harbor Symp. Quant. Biol.* **39,** 559–566.

Levinthal, J. D., and Petersen, W. (1965). *Fed. Proc. Fed. Am. Soc. Exp. Biol.* **24,** 174–177.

Lewis, J. B., Atkins, J. F., Baum, P. R., Solem, R., Gesteland, R. F., and Anderson, C. W. (1976). *Cell* **7**, 141–151.

Lewis, J. B., Esche, H., Smart, J. E., Stillman, B. W., Harter, M. L., and Mathews, M. B. (1979). *Cold Spring Harbor Symp. Quant. Biol.* **44**, 493–508.

Lupker, J. H., Davis, A., Jochemsen, H., and van der Eb, A. J. (1981). *J. Virol.* **37**, 524–529.

MacAllister, R. M., Nicolson, M. O., Lewis, A. M., MacPherson, I., and Huebner, R. J. (1969a). *J. Gen. Virol.* **4**, 29–36.

MacAllister, R. M., Nicolson, M. O., Reed, G., Kern, J., Gilden, R. V., and Huebner, R. J. (1969b). *J. Natl. Cancer Inst.* **43**, 917–923.

McGhee, J., and Ginder, G. D. (1979). *Nature (London)* **280**, 419–420.

Mackey, J. K., Rigden, P. M., and Green, M. (1976). *Proc. Natl. Acad. Sci. U.S.A.* **73**, 4657–4661.

Mak, S., Mak, I., Smiley, J. R., and Graham, F. L. (1979). *Virology* **98**, 456–460.

Mandel, J., and Chambon, P. (1979). *Nucleic Acids. Res.* **7**, 2081–2103.

Meyer, A. J., and Ginsberg, H. S. (1977). *Proc. Natl. Acad. Sci. U.S.A.* **74**, 785–788.

Nevins, J. R., and Wilson, M. C. (1981). *Nature (London)* **290**, 113–118.

Ormondt, H., van. Maat, J., and Dijkema, R. (1980a). *Gene* **12**, 63–76.

Ormondt, H., van, Maat, J., and van Beveren, C. P. (1980b). *Gene* **11**, 299, 309.

Pereira, M. S., Periera, H. G., and Clarke, S. K. R. (1965). *Lancet* **i**, 21–23.

Perricaudet, M., Akusjärvi, G., Virtonen, A., and Pettersson, U. (1979). *Nature(London)* **281**, 694–696.

Perricaudet, M., Akusjärvi, G., Virtonen, A., and Pettersson, U. (1980a). *Nature(London)* **288**, 174–176.

Perricaudet, M., LeMoullec, J. M., and Pettersson, U. (1980b). *Proc. Natl. Acad. Sci. U.S.A.* **77**, 3778–3782.

Ploeg, L. H. T., van der, and Flavell, R. A. (1980). *Cell* **19**, 947–958.

Pope, J. H., and Rowe, W. P. (1964). *Proc. Soc. Exp. Biol. Med.* **120**, 577–587.

Pruitt, S. C., and Grainger, R. M. (1981). *Cell* **23**, 711–720.

Rabson, A. S., Kirchstein, R. L., and Paul, F. J. (1964). *Natl. Cancer Inst. J.* **32**, 77–87.

Rapoza, N. P., Merkow, L. P., and Slifkin, M. (1967). *Cancer Res.* **27**, 1887–1894.

Ross, S., Flint, S. J., and Levine, A. J. (1980). *Virology* **100**, 419–432.

Rowe, W. P., Huebner, R. J., Gilmore, L. K., Parrott, R. H., and Ward, T. G. (1953). *Proc. Soc. Exp. Biol. Med.* **84**, 570–573.

Sambrook, J., Botchan, M., Gallimore, P., Ozanne, B., Pettersson, U., Williams, J., and Sharp, P. A. (1975). *Cold Spring Harbor Symp. Quant. Biol.* **39**, 615–632.

Sambrook, J., Greene, R., Stringer, J., Mitchison, T., Hu, S.-L., and Botcham, M. (1979). *Cold Spring Harbor Symp. Quant. Biol.* **44**, 569–584.

Sandeen, G., Wood, W., and Felsenfeld, G. (1981). *Nucleic Acid Res.* **8**, 3757–3778.

Sarma, P. S., Huebner, R. J., and Lane, W. T. (1965). *Science* **149**, 1108.

Sarma, P. S., Vass, W., Huebner, R. J., Igel, H., Lane, W. T., and Turner, H. C. (1967). *Nature (London)* **215**, 293–294.

Sawada, Y., and Fujinaga, K. (1980). *J. Virol.* **36**, 639–651.

Sawada, Y., Ojima, S., Shimojo, H., Shiroki, K., and Fujinaga, K. (1979). *J. Virol.* **32**, 379–385.

Schaller, J. D., and Yohn, D. S. (1974). *J. Virol.* **14**, 392–401.

Seikikawa, K., Shiroki, K., Shimojo, H., Ojima, S., and Fujinaga, K. (1978). *Virology* **88**, 1–7.

Sergent, A., Tigges, M. A., and Raskas, H. J. (1979). *J. Virol.* **29**, 888–898.

Shaw, A. R., and Ziff, E. (1980). *Cell* **22**, 905–916.

Shen, S. T., and Maniatis, T. (1980). *Proc. Natl. Acad. Sci. U.S.A.* **77**, 6634–6638.

Shiroki, K., Handa, H., Shimojo, H., Yano, S., Ojima, S., and Fujinaga, K. (1977). *Virology* **82**, 462–471.

Shiroki, K., Segawa, K., and Shimojo, H. (1980). *Proc. Natl. Acad. Sci. U.S.A.* **77**, 2274–2278.

Southern, E. (1975). *J. Mol. Biol.* **98**, 504–517.

Spector, D. J., McGrogan, M., and Raskas, H. J. (1978). *J. Mol. Biol.* **126**, 395–414.

Spector, D. J., Halbert, D. N., and Raskas, H. J. (1980). *J. Virol.* **36**, 860–871.

Stabel S., Doerfler, W., and Fris, R. R. (1980). *J. Virol.* **36**, 22–40.

Stalder, J., Groudine, M., Dodgson, J. B., Engel, J. D., and Weintraub, H. (1980a). *Cell* **19**, 973–980.

Stalder, J., Larsen, A., Engel, J. D., Dolan, M., Groudine, M., and Weintraub, H. (1980b). *Cell* **20**, 451–460.

Strohl, W. (1969a). *Virology* **39**, 642–652.

Strohl, W. (1969b). *Virology* **39**, 653–665.

Sutter, D., and Doerfler, W. (1980). *Proc. Natl. Acad. Sci. U.S.A.* **77**, 253–256.

Sutter, D., Westphal, M., and Doerfler, W. (1978). *Cell* **14**, 569–585.

Takahashi, M. (1972). *Virology* **49**, 815–817.

Tate, V., and Philipson, L. (1978). *Nucleic Acids. Res.* **6**, 2769–2785.

Tooze, J., ed. (1980). "The Molecular Biology of Tumor Virus" (2nd Ed. DNA Tumor Viruses). Cold Spring Harbor Lab., Cold Spring Harbor, New York.

Trentin, J. J., Yabe, Y., and Taylor, G. (1962). *Science* **137**, 835–841.

van der Vliet, P. C., Landberg, J., and Jansz, H. S. (1977). *Virology* **80**, 98–110.

Vardimon, L., Neuman, R., Kuhlman, I., Sutter, D., and Doerfler, W. (1980). *Nucleic Acids. Res.* **8**, 2461–2473.

Visser, L., van Maarschalkerweerd, M. W., Rozjin, T. H., Wassenaar, A. D. C., Reemst, A. M. C. B., and Sussenbach, J. S. (1979). *Cold Spring Harbor Symp. Quant. Biol.* **44**, 541–550.

Waalwyck, C., and Flavell, R. A. (1978). *Nucleic Acids Res.* **5**, 4631–4641.

Weintraub, H., and Groudine, M. (1976). *Science* **93**, 848–858.

Weintraub, H., Larsen, A., and Groudine, M. (1981). *Cell* **24**, 333–344.

Weisbrod, S., and Weintraub, H. (1980). *Cell* **19**, 289–301.

Weisbrod, S., and Weintraub, H. (1981). *Cell* **23**, 391–400.

Williams, J. F. (1973). *Nature (London)* **243**, 162–163.

Wilson, M. C., Sawicki, S. G., Salditt-Georgieff, M., and Darnell, J. E. (1978). *J. Virol.* **25**, 97–103.

Wold, W. S. M., and Green, M. (1979). *J. Virol.* **30**, 297–310.

Yabe, Y., Trentin, J. J., and Taylor, G. (1962). *Proc. Soc. Exp. Biol. Med.* **111**, 343–344.

Yoshida, K., and Fujinaga, K. (1980). *J. Virol.* **36**, 337–352.

Ziff, E., and Evans, R. (1978). *Cell* **15**, 1463–1475.

INTERNATIONAL REVIEW OF CYTOLOGY, VOL. 76

Highly Repeated Sequences in Mammalian Genomes

MAXINE F. SINGER

Laboratory of Biochemistry, National Cancer Institute, National Institutes of Health, Bethesda, Maryland

I. Introduction

The discovery of restriction endonucleases and the development of molecular cloning and DNA sequencing techniques have revolutionized the study of the structure of complex genomes. The unexpected insights generated by these new techniques have created profound changes in our understanding of the way genetic information is organized and expressed. Consequently, many new concepts regarding the highly repeated DNA sequences in eukaryotic genomes are rapidly developing. Perhaps the only generalization that can safely be made at this time is that both the repeated sequences and their genomic organization are much more complex than previously suspected. This article attempts to summarize the emerging picture regarding the structure and organization of mammalian highly repeated sequences at the molecular level. Both the tandemly repeated sequences widely termed "satellites" and those segments that are interspersed among other genomic DNA sequences will be described. As a general rule of thumb, any sequence repeated more than 10^4 times is included. I have chosen to concentrate on mammals because of my own interests and because other recent reviews emphasize plants (Bedbrook and Gerlach, 1980; Flavell, 1980) and invertebrates including crabs (Christie and Skinner, 1980), *Drosophila* (Appels and Peacock, 1978; John and Miklos, 1979; Brutlag, 1980; Hilliker *et al.*, 1980; Spradling and Rubin, 1981), and sea urchins (Moore *et al.*, 1980; Posakony *et al.*, 1981). The available information is for the most part structural. Unfortunately, the function, if any, of most of these sequences remains an enigma.

Most articles covered here were published prior to the end of July 1981 although I had, through the courtesy of colleagues, preprints of other papers. My own interests dictated selection of the material. The interesting matter of the organization of highly repeated sequences within chromatin is discussed extensively in several recent reviews (Zachau and Igo-Kemenes, 1981; Kornberg, 1981; Igo-Kemenes *et al.*, 1982) and is not included here. Repeated sequences that are known to be genes (e.g., histone and ribosomal RNA genes) have also been reviewed recently (Long and Dawid, 1980). There are available several summaries of current thinking about the evolution of highly repeated sequences (Brutlag, 1980; Bostock, 1980; Dover, 1981). My own bias is that while relevant structural information is rapidly accumulating, we are still largely ignorant of overriding principles. In particular, with the question of function unresolved, discussions of evolution are perforce carried out in the absence of any sense of selective pressure, a critical defect.

Throughout this article the words "repeated" and "reiterated" are used interchangeably. Neither one implies that the many copies of a given family of sequences are identical. Unless demonstrated otherwise, all "repeats" are assumed to be members of a set of closely similar but somewhat variant DNA segments; "closely similar" means that they hybridize to one another under stringent conditions (0.45 M NaCl, greater than 65°C). The term "consensus sequence" will be used to describe a nucleotide sequence representing the most abundant nucleotide at each position in the repeat units comprising a set.

II. Perspectives on Methods

In the past, two methods dominated analysis of repeated DNA sequences: measurement of DNA renaturation kinetics (Britten *et al.*, 1974) and isopycnic centrifugation in gradients of CsCl and $CsSO_4$ (Szybalski, 1968). Gross physical separation of highly repeated, middle repeated, and unique sequences was achieved by taking advantage of (1) the dependence of renaturation rate on the concentration of a sequence, (2) the ability of hydroxyapatite to separate single-

and double-stranded DNA, and (3) the separation of some satellite sequences from bulk DNA by virtue of a unique density. Together, these methods demonstrated that (1) eukaryote genomes can be divided up into classes of DNA sequences according to the reiteration frequency (Britten and Kohne, 1968), (2) many very highly repeated sequences are in long tandem arrays (satellites), and (3) some repetitive sequences are dispersed throughout major portions of genomes amid either other repeated sequences or sequences present only once per genome, that is, "unique" sequences.

Even before molecular cloning came into wide use, it was apparent that these useful characterizations masked much greater complexity. For example, the mixture of sequences in the "middle repeated" category includes segments repeated anywhere from two times to tens or hundreds of thousands of times. Some of these sequences proved to be functional genes—either identical multicopy genes such as those for ribosomal RNA and histones, or closely related genes encoding similar but distinctive gene products such as the families of actin, β-globin, or immunoglobulin genes (see review by Long and Dawid, 1980). Furthermore, because divergence among the members of a set of repeated sequences decreases the rate at which they reassociate, middle repeated sequences can appear in the single copy class and the copy number of highly repeated sequences is easily underestimated. Similarly, although many organisms show one or more discrete satellite DNA fractions apart from the main band upon isopycnic centrifugation, others yield no such fractions. Centrifugation in the presence of heavy metals or various antibiotics or dyes, tease "cryptic satellites" out of the DNA of many organisms, but others yield no satellite even though renaturation kinetics shows the presence of highly repeated sequences.

New methods have now begun to unravel some of the complexities and it is the application of these methods that is the basis for the observations summarized in this article. Restriction endonucleases cleave DNA molecules after recognition of specific short nucleotide sequences. The recognition and cleavage sites of some of the enzymes referred to in this article are shown in Table I. Complete current lists of known restriction endonucleases are available (Roberts, 1980). Digestion of DNA with a specific enzyme reproducibly divides it up into a set of fragments whose sizes depend on the spacing between recognition sites. The mixture of fragments can be conveniently separated according to size by electrophoresis on semisolid supports such as agarose or polyacrylamide (Southern, 1979). When the gels are stained to mark DNA the mixture appears as a continuous smear of fragments of all possible sizes. However sequences that are repeated sufficiently to represent at least 0.5% of the genome often appear as distinct bands against the background smear. The entire collection of fragments on the gel can be transferred without disturbing their distribution by blotting after denaturation onto sheets of nitrocellulose (Southern, 1975b, 1979) or diazotized paper (Alwine *et al.*, 1979). Thereafter, the sheets can be incubated in the presence of radioactive

MAXINE F. SINGER

TABLE I
RECOGNITION AND CLEAVAGE SITES OF SOME RESTRICTION
ENDONUCLEASES[a]

Enzyme	Sites
Sau96I	5'-G'G N C C – –C C N G'G –5'
AvaII	A 5'-G'G (T)C C – –C C (T)G'G –5' A
TaqI	5'-T' C G A – –A G C' T –5'
Sau3A	5'-'G A T C –
MboI	–C T A G'–5'
EcoRI	5'-G'A A T T C– –C T T A A'G–5'
AluI	5'-A G'C T – –T C'G A –5'
HindIII	5'-A'A G C T T– –T T C G A'A–5'
BamHI	5'-G'G A T C C– –C C T A G'G–5'

[a] Taken from Roberts (1980) where a complete list can be found. The ' indicates the site of cleavage within the recognition site. Enzymes with identical specificity (Sau3A and MboI) are called isoschizomers.

DNA or RNA fragments to permit hybridization of the probe with homologous sequences on the sheet. Exposure of the sheet to X-ray film provides an autoradiogram on which a darkened band reveals the DNA fragments homologous to the probe. These techniques are sufficiently sensitive to reveal even unique DNA sequences amid fragments generated from an entire mammalian genome.

Molecular cloning, usually in *E. coli* host vector systems, provides ways to purify and amplify DNA segments generated after restriction endonuclease digestion or after random generation of genomic fragments by shearing or nonspecific nuclease digestion (Morrow, 1979). Hybridization techniques with radioactive probes permit identification of the clones of interest and primary nucleotide sequence is readily determined (Maxam and Gilbert, 1980; A. J. H. Smith, 1980). In this way, single molecular species corresponding to the individual members of a repeated set are available as are segments corresponding to a tandem array or a dispersed repeated sequence enclosed within its natural neighbors. One very important consequence of these methods is the availability of pure DNA segments for use as probes in the analysis of genome organization by the blotting procedures described above. Uncloned probes are likely to be contaminated by extraneous DNA sequences. Because of the very high sensitivity of the blotting methods even minor contaminants lead to misleading data. A word of caution: molecular cloning is not without problems. Many investigators have noted that deletions often occur when tandemly repeated arrays are cloned and amplified in *E. coli* (e.g., Brutlag *et al.*, 1977; Carlson and Brutlag, 1977; Sadler *et al.*, 1980; Sakano *et al.*, 1980).

III. Satellites

A. CHANGING CONCEPTS

The characteristic organizational feature of satellites and cryptic satellites is the tandem repetition of a unit DNA sequence. In accordance with a previous suggestion (Pech *et al.*, 1979b), the term satellite will be used to describe such DNA regions regardless of whether or not they are separable as classical satellites by isopycnic centrifugation. This usage, while not completely accurate, conforms with widespread practice. Satellites comprise anywhere from a few percent (human) to over 50% (Kangaroo rat) of mammalian genomes. Common characteristics of satellites include (1) association with heterochromatin, (2) lack of measurable transcription (but see below), (3) replication late in S-phase, and (4) underreplication in polytene chromosomes (see Brutlag, 1980, for review).

At one time it was believed that all satellites contain simple redundant repeats of short oligonucleotide segments. However molecular analysis has demonstrated that repeat units vary from a few to several thousand base pairs and that enormous complexity resides within these sequences. Satellite arrays frequently resist separation by isopycnic centrifugation and remain instead within the main density fraction of genomic DNA. This is true even when some satellite is separable by centrifugation; additional satellite may remain sequestered in the main band. Furthermore, neither the purity nor the unique character of an isopycnic satellite fraction can be assumed. Sequences present in one satellite band may also occur in others and in main band. Also several nonhomologous repeat units may reside in one density satellite fraction, either linked in one molecule or on separate molecules.

Regardless of whether separated satellite or total DNA is used, both the repeat unit and its tandem organization can be revealed by restriction endonuclease digestion. When a site for a particular restriction endonuclease occurs within a typical repeat unit, digestion converts a tandem array to a set of DNA fragments of repeat unit length (a type "A" digestion; Hörz and Zachau, 1977). Partial digests generate "ladders" of fragments that are demonstrable upon electrophoresis. The "ladder" fragments are integral multiples of the basic repeat unit in length, and provide evidence for the tandem organization of the satellite.

After exhaustive digestion, a "ladder" of resistant multiples usually remains. The resistant ladder arises because of sequence alteration at the canonical restriction endonuclease site in occasional copies of the repeat unit. Such alterations may be (1) randomly dispersed among the copies of the unit repeat, or (2) clustered within neighboring repeats, or (3) occur at regular intervals along an array. All three possibilities occur. The latter two nonrandom arrangements (and combinations of the two) define distinctive subarrays of a satellite and these are here called satellite domains. Restriction endonuclease sites that are missing from the typical unit repeat sequence may occur occasionally within an array because of sequence variation and yield distinctive digestion patterns with a preponderance of very long fragments (a type "B" digestion). Again the same 3 distributions occur and may define specific satellite domains. The absence of an A-type site or the presence of a B-type site at a fixed frequency within an array of repeats is seen as a long repeat length superimposed on a shorter canonical reiterated unit.

The primary nucleotide sequence of a repeat unit can be determined using an entire set of monomeric units generated by type A digestion or a molecularly cloned member of the set. Unambiguous consensus sequences are frequently obtained with uncloned sets, indicating that variant bases occur in fewer than 15% of the copies at any given position. The 15% limit reflects the limits of detectability in the DNA sequencing procedures. In the typical case one restriction endonuclease is used to generate monomer units; any repeat units lacking the site appear in the higher ladder and are not represented in the consensus. Subsequent steps in the preparation of fragments for sequencing often utilize additional restriction sites in the repeat unit to generate subunit size sections. At each such cleavage repeat units lacking the site in question will be discarded with undigested material and will not be represented in the consensus. Thus, the existence of subsets of a given monomer bias the sequence data on uncloned fragments.

Some organisms have multiple distinguishable satellites. Sequence analysis frequently shows that these may be related to one another in spite of different densities and restriction endonuclease cleavage patterns or inability to cross hybridize. Other organisms appear to have one predominant satellite. In these cases, molecular analysis may show the presence of domains. It is possible that domains and different satellites are organizationally equivalent, representing localized amplifications of particular variants of a basic sequence. There is growing evidence that different domains or distinguishable satellites may tend to be localized to specific chromosomes. Both the domains of a relatively homogeneous satellite and the several satellites in some organisms can all be seen as the product of alternating cycles of mutation (including single base pair changes, deletions, and insertions) and amplification and deletion. This scheme,

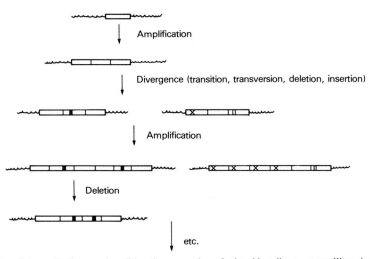

FIG. 1. Schematic diagram describing the generation of related but divergent satellites, including domains, by alternating cycles of mutation and amplification (after Southern, 1970).

first suggested by Southern (1970), is outlined in Fig. 1; it is consistent with the data on various satellites, as reviewed below.

B. RODENTS

1. *Mouse*

The single density satellite of *Mus musculus* makes up 5 to 10% of the genome (Kit, 1961) and is localized at centromeric heterochromatin in all mouse chromosomes except the Y chromosome (Pardue and Gall, 1970). Most (60–70%) of isolated mouse satellite is digested to a set of 234-bp-long monomeric units by *Sau*96I (Hörz and Altenburger, 1981), *Eco*RII (Southern, 1975a; Hörz and Zachau, 1977), and *Ava*II (Dover, 1978). The set yields an unambiguous consensus sequence (Hörz and Altenburger, 1981; Manuelidis, 1981a) (Fig. 2a). The asymmetric distribution of A and T residues on the two DNA strands of the repeat unit may account for anomalous prior estimates of both the size of the repeat unit (by gel electrophoresis) and of the GC content (by density gradient centrifugation or melting temperature determination). The 234-bp segment comprises four related internal tandem repeats—58, 60, 58 and 58 bp in length, respectively. And each of these can be further divided into two related but variant 28- and 30-bp-long segments. Thus the 234-bp repeat encloses 8 shorter tandem repeats. Hörz and Altenburger (1981) deduced a common progenitor sequence

a

```
          10         20        ,30         40         50          60
 [GGACCTGGAA TATGGCGAGA AAACTGAAAA TCACGGAAAA TGAGAAATAC ACACTTTAGG

          70         80        ,90        100        110         120
  ACGTGAAATA TGGCGAGGAA AACTGAAAAA GGTGGAAAAT TTAGAAATGT CCACTGTAGG

         130        140        ,150       160        170          180
  ACGTGGAATA TGGCAAGAAA ACTGAAAATC ATGGAAAATG AGAAACATCC ACTTGACGAC

         190        200         210       220        230
  TTGAAAAATG ACGAAATCAC TAAAAAACGT GAAAAATGAG AAATGCACAC TGAA ]
```

b

```
          10         20         30         40         50          60
 [GAATTCACAG AGAAACAGTG TTTCAGTTCG TTAAAACGTT GCTCTATCTT GAATAACAAG

          70         80         90        100        110         120
  CTTATTACAT GCGAATCCTA TTGGGAACCT ACTGAATTCA CCATGATACT TAGATTCCGT

         130        140        150        160        170         180
  TCCTCAAAAT GTTGCTCCAT ATTGAAAAGC AAACTCATAC AAGCAGGTCC CATTGGGAAC

         190        200        210        220        230         240
  TCACTGAATT CGCCTAGAAA TTTTGATTCC ATTCGTGAAA ATTTTTCTAT ATCCCGAACA

         250        260        270        280        290         300
  GTCCACTTAT TACTACTGCG GCCCACTGGG AACTAACCGA ATTCACCATG TTACTCAGAT

         310        320        330        340        350         360
  TCGGCTCACC AAATTTTGAT AAATCTTTAA AAGTACACAT ATTACAAGAG CAGGCTACTG

         370
  GGAACTAACT ]
```

FIG. 2. The consensus sequences of the major monomeric repeat units in (a) mouse satellite (Hörz and Altenburger, 1981) and (b) rat satellite I (Pech *et al.*, 1979b). Brackets enclose the long internal repeats and lines mark off the 8 shorter internal repeats in the mouse sequence. Only one DNA strand is shown and it is 5′ to 3′ reading from left to right.

from the structure of the 8 repeats and suggested that the satellite may have been constructed by alternating mutation and amplification starting with three similar nonanucleotides: GAAAAATGA, GAAAAAACT, GAAAAACGT.

Several variants of the 234-bp consensus sequence have been identified. Thus, approximately 5–10% of the 234-bp-long fragment set contains a single *Taq*I cleavage site that is missing in the consensus sequence (Hörz and Altenburger, 1981). Approximately 20% of the satellite is degraded to dimers (464 bp) after exhaustive cleavage with either *Sau*96I or *Eco*RII (Hörz and Zachau, 1977; Southern, 1975a) and thus variation at these (overlapping) restriction sites (residues 1–8 in Fig. 2a) is frequent; the dimer yields a consensus sequence almost identical to that of the monomer (Hörz and Altenburger, 1981). *Sau*96I digestion also yields small amounts of heretical fragments 0.25 × *n* × 234 bp in length (*n* is an integer); these structures have not been studied in detail. There is extensive methylation of C-G sequences in the satellite (Hörz and Altenburger, 1981). Further, some sequences that hybridize with mouse satellite are found in DNA from which satellite has been removed by density gradient centrifugation (Man-uelidis, 1980; Stambrook, 1981).

Individual mouse chromosomes may carry distinct satellite domains. DNA from a Chinese hamster/mouse hybrid cell line containing only the mouse X chromosome in 75% of the cells was studied by restriction endonuclease diges-tion and hybridization with ^{32}P-labeled density gradient purified satellite (Brown and Dover, 1980a). There is no cross-reaction with the Chinese hamster DNA. The data indicate that the X-chromosome contains mouse satellite domains whose basic organization is similar to the bulk of the satellite but that (1) contain fewer *Hinf*, *AluI*, and *Eco*RI sites, and (2) lack heretical size fragments.

Sequences homologous to the *M. musculus* satellite have been detected in other *Mus* species by hybridization with radioactive probes of *M. musculus* satellite purified by density gradient centrifugation; the data require confirmation with cloned probes. There was less *in situ* hybridization to *M. booduga* chromo-somes than to those of *M. musculus* although the distribution was similar; only one segment on the short arm of the X-chromosome in *M. dunni* showed signifi-cant hybridization (Sen and Sharma, 1980). Hybridization to restriction endonuc-lease fragments of total *M. spretus* DNA (Brown and Dover, 1980b) indicated sequences homologous to *M. musculus* satellite at about 1% the level even though *M. spretus* does not yield a classical density satellite. Almost all the homologous material in *M. spretus* was degraded by *AvaII*, as is *M. musculus* satellite. *AluI*, which degrades about 10% (type B) of *M. musculus* satellite, digested only about 2–3% of the homologous sequences in *M. spretus* to sepa-rable products suggesting different subsets of sequences in the two species. Restriction endonuclease digests suggest the presence of related satellite se-quences in two species of the field mouse genus *Apodemus* (Brown and Dover, 1979).

2. Rat

No substantial satellite, cryptic or otherwise, can be isolated from total *R. rattus* DNA by isopycnic centrifugation, although renaturation kinetics shows that almost 10% of the rat genome comprises highly repeated sequences (Bonner *et al.*, 1973). Digestion of total rat or Novikoff hepatoma ascites cell DNA with several restriction endonucleases yields ladders of DNA fragments of defined chain length, indicative of the presence of tandem repeated sequences (Philippsen *et al.*, 1974; Maio *et al.*, 1977; Fuke and Busch, 1979; Lapeyre and Becker, 1980; Sealy *et al.*, 1981). Some distinct differences in the pattern of *Eco*RI bands observed with rat liver nuclear and Novikoff hepatoma DNAs have been reported (Lapeyre and Becker, 1980). A predominant 370-bp-long *Hind*III band is degraded by *Eco*RI to yield bands about 93 bp long, which is the size of the smallest abundant band obtained by *Eco*RI digestion of total rat DNA. About 3% of the rat genome is accounted for by these fragments (Fuke *et al.*, 1979;

Pech *et al.*, 1979b) and the sequences may be concentrated in nucleolar DNA (Fuke *et al.*, 1979).

Approximately 1–3% of rat DNA is converted to fragments about 10 kbp in length by digestion with *Sau*3A (type "B" digestion) while most of the rest of the genome is simultaneously degraded to a large mixture of small fragments whose average chain length is 400 bp (Pech *et al.*, 1979b). The 10-kbp fraction was purified by preparative gel electrophoresis and named rat satellite I. Analysis of the purified satellite I (Pech *et al.*, 1979b) demonstrated a repeat length of 370 bp constructed of four related internal segments 93, 92, 93, and 92 bp in length, respectively (Fig. 2b). Each of the internal segments contains an *Eco*RI site (at residues 1–6, 94–99, 186–191, and 279–284 in Fig. 2b) and one contains a *Hin*dIII site (residues 58–63). The sequence data obtained with the set of *Hin*dIII monomers (Fig. 2b) revealed frequent variations at eight defined residues. These must represent alterations in at least 5–10% of the members of the set. Thus a substantial number of the members diverge in specific nonrandom ways. It is not known whether the different abundant variants are present in the same or different copies of the sequence. Domains have not been characterized, but they may occur since some portion of the sequence is left uncut by each of the restriction endonucleases with otherwise regular sites. The CpG sequences in the satellite are extensively methylated (Pech *et al.*, 1979b). Independent base sequence determination on *Eco*RI fragment sets 92 and 93 bp in length, respectively, and generated by cleavage of total rat DNA with *Eco*RI confirms the relatedness of the *Eco*RI fragments to rat satellite I (Lapeyre *et al.*, 1980; Sealy *et al.*, 1981).

As with *R. rattus* neither satellites nor cryptic satellites can be isolated from *R. norvegicus*, *R. sordidus*, or *R. villosissimus* by isopycnic centrifugation. Also, less than 5% of the three genomes are highly repeated segments (Miklos *et al.*, 1980). Analysis by restriction endonuclease digestion is consistent with the presence of very low amounts of satellite DNA in these three species.

3. *Other Rodents*

The oligomer 5′-TTAGGG-3′ is repeated frequently in a major satellite, Hs, of *Dipodomys ordii*, the kangaroo rat (Fry and Salser, 1977) and in guinea pig α-satellite (or satellite I) (Southern, 1970). The same oligomer probably occurs frequently in *Thomomys bottae*, the pocket gopher, *Ammospermophilus leucurus*, the antelope ground squirrel, and other species in the genus *Dipodomys* (Fry and Salser, 1977; Mazrimas and Hatch, 1977). Analysis of the total repetitive DNA of the guinea pig, *Cavia porcellus*, was made by isolating the fraction reassociating at a C_0t of about 7×10^{-2} and removing remaining single strands with S1 nuclease (Hubbell *et al.*, 1979). The data suggested that about 21% of the genome of *C. porcellus* is highly repeated while earlier estimates, based on density gradient isolation of satellites indicated about 10%. It is possible that the rapidly annealing fraction contains both satellites and interspersed repeated sequences as it is heterogeneous. The three *C. porcellus* satellites that are separable by isopycnic centrifugation were used to prepare ^{3}H-labeled cRNAs and the probes were hybridized *in situ,* to *C. porcellus* chromosomes (Duhamel-Maestracci *et al.,* 1979). All hybridization was to centromeric regions and some distributional specificity of the 3 satellites was observed; the Y chromosome hybridized to none of the three probes. Confirmation of the data with cloned probes is required.

C. Bovine

The eight different satellites that have been distinguished in calf thymus DNA (Macaya *et al.,* 1978; Kopecka *et al.,* 1978) are listed in Fig. 3 along with their buoyant densities in CsCl, their relative abundance in the genome, and, where known, a schematic diagram of the organization of the repeat units. The purification of these satellites depends on sophisticated use of differential density gradient centrifugation in Cs_2SO_4 in the presence of various additives (Macaya *et al.,* 1978; Streeck *et al.,* 1979). Together, the eight comprise over 23% of the genome although analysis of the kinetics of renaturation of calf thymus DNA suggested that less than 5% was in very rapidly renaturing components. This dramatic example emphasizes that satellite and "rapidly renaturing" are not synonymous. Recent analysis of bovine satellites on the molecular level illus-

NAME	BUOYANT DENSITY	% GENOME	ORGANIZATION
III	1.706	4.2	2350
	1.709	4.6	
	1.711a	1.7	1413
	1.711b	7.1	2600
I	1.715	5.1	1400
	1.720a	0.1	
	1.720b	0.1	46
	1.723	0.5	700

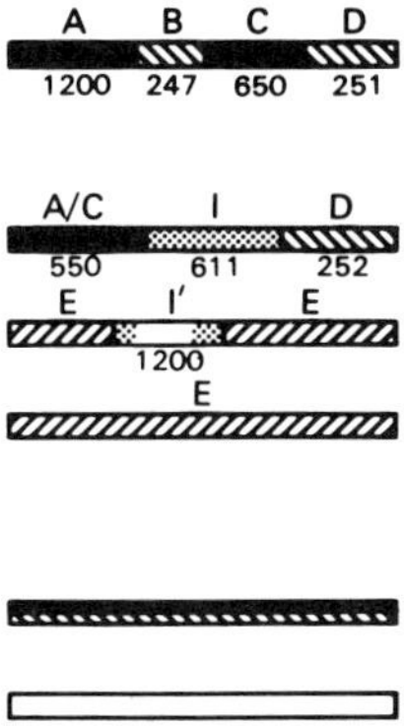

FIG. 3. Bovine satellites and their structural organization where known. See the text for references. The earlier "names" of the bovine satellites are given in the first column; the identity of II is uncertain. Closely related segments repeated in more than one satellite are indicated graphically and by letter. The numbers to the right are the sizes, in base pairs, of the overall repeating units. The sizes of different segments are indicated below. The drawings are not to scale.

trates how distinctions based on satellite density or even restriction endonuclease digestion can obscure striking similarities. Also, these studies show that the size of repeat units determined by restriction endonuclease digestion is not a reliable indicator of underlying structure. In one case, the 1.706 gm/cm³ bovine satellite, what initially seemed a complex satellite with a repeat unit of 2350 bp, has turned out to have true repeat unit about 0.01 times that size! And at least 5 of the 8 satellites are related in part to one another.

Digestion of purified 1.720b satellite with several different restriction endonucleases yields a series of fragment sets that are multiples of 46 bp in length [Streeck and Zachau, 1978 (note that the 1.720b satellite is erroneously termed 1.723 in that paper), Pöschl and Streeck, 1980]. The set of 46-bp fragments produced by *Alu*I gave an unambiguous consensus sequence (Fig. 4) and revealed an underlying periodicity of two related 23-bp repeats. Individual copies of the repeat units diverge from the consensus sequence. Whether or not particular divergent family members are collected in domains is not clear. The sequence contains several CG dinucleotides and these are probably frequently methylated.

Although quite distinct by density and restriction endonuclease patterns, the

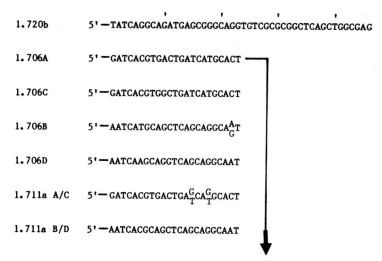

FIG. 4.   Sequence relations among the bovine satellites (see Fig. 3). The 46-bp consensus unit of the 1.720b satellite is from Pöschl and Streeck (1980); the 23-bp consensus units of 1.706A, B, C, and D are from Pech *et al.* (1979a); the 23-bp consensus units of 1.711a A/C and B/D are from Streeck (1981). Only one DNA strand is shown in each case. The data are somewhat simplified and the reader is referred to the original papers for details.

sequence of the 1.706 satellite (III) is related to that of the 1.720b satellite. The overall repeating unit of the 1.706 satellite (Streeck and Zachau, 1978; Streeck *et al.*, 1979; Pech *et al.*, 1979a) contains four distinct regions totaling 2350 bp. The distribution of restriction endonuclease sites and direct nucleotide sequencing of both cloned and uncloned fragments revealed that the four regions comprise two pairs of closely related alternating segments (A and C, B and D). The two kinds of segments (A/C, B/D) are themselves distantly related. The underlying unity is the tandem repetition of a 23-bp unit that is related to the repeat unit of the 1.720b satellite (Fig. 4). Furthermore, the 23-bp repeat unit of 1.706 A, for example, itself looks like a tandem repeat of a unit half its size (see Fig. 4).

The 1.711a satellite (Streeck, 1981) is also related to 1.720b and 1.706; an unambiguous consensus sequence was obtained with an uncloned fragment set. The long-range repeat length of 1.711a is 1413 bp and can be divided into 3 regions. Two are marked A/C and D in Fig. 3 to reflect their marked similarity to the corresponding segments in the 1.706 satellite, and the third region, which is unrelated to A/C or D and is not internally repetitive, is marked I. The I region has several features reminiscent of transposable elements (reviewed by Calos and Miller, 1980) including terminal direct and inverted repeats (Streeck, 1981). Indeed, the 1.711a repeat unit can be viewed as one portion of the 1.706 unit interrupted by the I region. Furthermore, I contains open reading frames suitable for translation into protein and a segment similar to the TATAAATA box that is one component of eukaryotic promotors.

More than 95% of the 1.715 satellite from either calf thymus or bovine kidney cells in tissue culture is degraded by *Eco*RI and by *Sal*I to a fragment set about 1400 base pairs in length (Botchan, 1974; Roizes, 1974, 1976; Philippsen *et al.*, 1974; Lipchitz and Axel, 1976; Gaillard *et al.*, 1981). Analysis of the 1400-bp monomer set with other restriction endonucleases established that the bulk of the set shares similar sequences but that variant forms exist (Roizes, 1976; Roizes *et al.*, 1980). At least four distinguishable domains may occur. Superimposed on this nonrandom division into domains other variations appear to reflect random base pair changes leading fortuitously to altered (gain or loss) restriction sites (Roizes *et al.*, 1980). The sequence of the 1.715 satellite monomer does not contain tandemly reiterated internal repeats like the 23 bp regularity found in satellites 1.720 and 1.711a but does contain multiple repeats of short segments of the consensus sequences of those satellites (Roizes *et al.*, 1980; Gaillard *et al.*, 1981; A. Plucienniczak, J. Skowronski, and J. Jaworski, personal communication). Comparison of restriction sites in the cloned and sequenced monomer with those in the bulk satellite suggests extensive methylation (Roizes *et al.*, 1980; Kaput and Sneider, 1979; Gaillard *et al.*, 1981). The extent of methylation appears to differ markedly in sperm and thymus DNA (Sturm and Taylor, 1981).

The 1.711b satellite is similar to 1.715 except that the 1400-bp repeat unit is interrupted about 60% of the time by a 1200-bp-long insertion (I' on Fig. 3). The arrangement of 1.711b repeat units with and without the I' element has no known pattern. I' shares sequence homologies with the I element in 1.711a (Streeck, 1981). The I' region of 1.711b has inverted terminal repeated sequences and is flanked by duplications of the 6 base pairs at the point of interruption of the 1400-bp repeats. In contrast, the I region in 1.711a is not surrounded by direct repeats of the target sequence.

D. PRIMATES

1. α-Satellites

At least one satellite in every primate that has been investigated is part of a kinship of sequences referred to as α-satellite (or alphoid) after the prototypical African green monkey (AGM) *(Cercopithecus aethiops)* satellite (Maio, 1971; Kurnit and Maio, 1974). The AGM α-satellite represents about 20% of the genome (Maio, 1971; Fittler, 1977; Singer, 1979), hybridizes to centromeres (Kurnit and Maio, 1973, 1974; Segal *et al.*, 1976), and comprises a substantial amount of nucleolar DNA (Kurnit and Maio, 1973). Both *in situ* hybridization to chromosomes (Segal *et al.*, 1976) and kinetic analysis of interspersion frequency (Singer, 1979) suggested that some copies of the satellite might be dispersed into other parts of the genome. However this suggestion has not been confirmed on a molecular basis and the kinetic data may reflect the interspersion of *Alu* sequences (see below) rather than α-satellite. Restriction endonuclease analysis and hybridization experiments suggest that related but nonidentical α-satellites are present in *Gorilla gorilla, Pan troglodytes, Cercocebus aterrimus, Macaca mulatta, Mandrillus sphinx, Papio cynocephalus, Cercopithecus pygerythrus, Colobus badius,* and *Cebus capuchinus* (Donehower and Gillespie, 1979; Singer and Donehower, 1979; Gillespie, 1977; Musich *et al.*, 1980; Maio *et al.*, 1981a). Typically, digestion of total or satellite DNA with one or another restriction endonuclease yields a ladder of fragments that are integral multiples of approximately 170 bp in length.

Sequence analysis confirms the presence of related α-satellites in AGM (Rosenberg *et al.*, 1978), baboon (*Papio papio*) (Donehower *et al.*, 1980), bonnet monkey (*Macaca radiata*) (Rubin *et al.*, 1980a), and human (*Homo sapiens*) (Manuelidis and Wu, 1978; Wu and Manuelidis, 1980) DNA (Fig. 5). The length of the basic repeat unit is 172 bp in the AGM and close to twice that in the three other species. The 2n units are composed of two different variations of the basic repeat unit (an a-b-a-b-a-b type structure, see legend to Fig. 5). No regular internal periodicity has been discerned although large portions of the sequences are alternating blocks of purines and pyrimidines (Rosenberg *et al.*, 1978) and

1   AGCTTTCTGAGAAACTGCTCTGTGTTCTGTTAATTCATCTCACAGAGTTACATCTTTCCCTTCAAGAAGC

2a AGCTTTCTGAGAAACTGCTTAGTGTTCTGTTAATTCATCTCACAGAGTTACATCTGTATTTCGTGGATCT

2b AGCTTTCTGAGAAACTTCTTTGTGTTCTGTGAAATCATCTCACAGAGTTACAGCTTTCCCCTCAAGAAGC

3a AGCTTTCTGAGAAACTTCTTTGTGTTCTGTGAAATCATCTCACAGAGTTACAGCTTCCCCCTCAAGAAGC

3b AGCTTTCTGAGAAACTGCTTAGTGTTCTGTTAATTCCTCTCGCAGAGTTACATCTGTATTTCGTGGATCT

4a AGAATTCTCAGTAACTTCCTTGTGTTGTGTGTATTCAACTCACAGAGTTGAACGATCCTTTACACAGAGC

4b ATGATTCTCAGAAACTCCTTTGTGATGTGTGCGTTCAACTCACAGAGTTTAACCTTTCTTTTCATAGAGC

1   CTTTCGCTAAGGCTGTTCTTGTGGAATTGGCAAAGGGATATTTGGAAGCCCATAGAGGGCTATGGTGAAA

2a CTTTGCTAGCCTTATTTCT GTGGAATCTGAGAACAGATATTTCGGATCCCTTTGAAGACTATAGGGCCA

2b CTTTCGCTAAGACAGTTCTTGTGGAATTGGCAAAGTGATATTTGGAAGCCCATAGAGGGCTATGGTGAAA

3a CCTTCGCTAAGACAGTTCTTGTGGAATTGGCAAAGTGATATTTGGAAGCCCATAGAGGGCTATGGTGAAA

3b CTTT GCTAGCCTTATTTCTGTGGAATCTGAGAACAGATATTTCGGATCCCTTTGAAGACTATAGGGC C

4a AGACTTGAAACACTCTTTTTGTGGAATTTGCAAGTGGAGATTTCAGCCGCTTT GAGGTCAATGGTAGAA

4b AGTTAGGAAACACTCTGTTTGTAAAGTCTGCAAGTGGATATTCAGACCTCTTT GAGGCCTTCGTTGGAA

1   AAGGAAATATCTTCCGTTCAAAACTGGAAAGA

2a AAGGAAATATCCTCCGATAACAAAGAGAAAGA

2b AAGGAAATATCCTCAGATGAAATCTGGAAAGA

3a AAGGAAATATCCTCAGATGAAATCTGGAAAGA

3b AAGGAAATATCCTCCGATAACAAAGAGAAAGA

4a TAGGAAATATCTTCCTATAGAAACTAGACAGA

4b ACGGGATT TCTTCATATTATG CTAGACAGA

FIG. 5.   Sequence relations among primate α-satellites. (1) African green monkey (Rosenberg *et al.*, 1978); consensus sequence of uncloned *Hin*dIII monomer (172 bp). The sequence is displayed from one *Hin*dIII site to the next. (2) Baboon (Donehower *et al.*, 1980); consensus sequence of the dimeric uncloned *Bam*HI monomer (343 bp). The two portions should be read a then b. The sequences are aligned with the African green monkey sequence. The *Bam*HI cleavage site is between residues 115/116 in a. (3) Bonnet monkey (Rubin *et al.*, 1980); consensus sequence of the dimeric uncloned *Hae*III monomer (343 bp). The two portions should be read a then b. The *Hae*III cleavage site is between residues 137–138 in b. (4) Human (Wu and Manuelidis, 1980); consensus sequence of the dimeric uncloned *Eco*RI monomer (340 bp). The two segments should be read a then b. The *Eco*RI cleavage site is between residues 2/3 in a. In each case only a single strand is shown; it reads 5' to 3', left to right.

there is a high frequency of 5'-GAAA and 5'-CTTT on the displayed strands (Donehower *et al.*, 1980; Rubin *et al.*, 1980a).

The sequences of cloned monomers of the AGM satellite indicate that the consensus sequence shown in Fig. 5 reflects a mixture of many versions that differ from one another at a few positions (Rosenberg *et al.*, 1978; Graf *et al.*, 1979; Graf, 1979; Thayer *et al.*, 1981). No cloned unit has a sequence identical to that of the consensus sequence. The same situation may exist in the α-satellites of other species since "ladder" patterns are always observed upon exhaustive type "A" digestion (Rubin *et al.*, 1980a; Wu and Manuelidis, 1980; Donehower *et al.*, 1980; Musich *et al.*, 1980; Maio *et al.*, 1981a). At least some variants of the AGM sequence are within satellite domains. For example, as many as 10% of the monomer units contain an *Eco*RI cleavage site between residues 31 and 32 [inspection of the consensus sequence (Fig. 5) shows that a single base pair change at residue 31 (T→ G) in the consensus sequence yields an *Eco*RI site]. Digestion with *Eco*RI yields a typical ladder pattern indicating a clustering of units with *Eco*RI sites (Fittler, 1977). Two long α-satellite segments containing frequent *Eco*RI cleavage sites were recently isolated by molecular cloning (McCutchan *et al.*, 1982). Furthermore, essentially all the α-satellite in a single AGM chromosome isolated within a mouse–monkey somatic cell hybrid contains cleavage sites for *Eco*RI; a large percentage of this α-satellite also has cleavage sites for *Hae*III (T.N.H. Lee and M. F. Singer, unpublished) although this site is present in less than 3% of total α-satellite (Fittler, 1977; Rosenberg *et al.*, 1978; Graf *et al.*, 1979; Thayer *et al.*, 1981). Heretical size α-satellite fragments are generated in small amounts by *Hin*dIII digestion (Fittler, 1977; McCutchan *et al.*, 1982). Some tandem arrays of AGM α-satellite are interrupted (McCutchan *et al.*, 1982) by dispersed repeated sequences such as *Alu*-SINE or *Kpn*-LINE (see below) family members (Grimaldi *et al.*, 1981; Grimaldi and Singer, 1982). The frequency at which heretical repeat units and interruptions occur is not known.

2. *Classical Primate Satellites*

By classical, I refer to fractions separable by density gradient centrifugation, cryptic or not. In the African green monkey the overwhelming mass of the major classical satellite is α-satellite; three or four additional classical satellites have been noted but not characterized (Kurnit and Maio, 1974; Fittler, 1977).

The classical human satellites present a complex picture compounded by the fact that similarly named preparations are not necessarily identical from laboratory to laboratory (Macaya *et al.*, 1977; Manuelidis, 1978; Miklos and John, 1979; Mitchell *et al.*, 1979). Macaya and co-authors (1977) and Miklos and John (1979) valiantly summarized the confusions. Recently some clarity has begun to emerge, but evaluation of the literature remains difficult. In particular, data on chromosomal location obtained by *in situ* hybridization with different satellites are problematic because of the cross-hybridization of the fractions (see below) and the use of uncloned probes; it will not be summarized here. The story of the complex interrelations among the classical calf satellites (see above) perhaps hints at what we may, in the long run, expect to learn. Table II summarizes one

TABLE II
CLASSICAL HUMAN SATELLITES[a]

	ρ (gm/cm^3)	Estimated percentage of genome
I	1.687	0.2–0.5
II	1.693	1–2
III	1.697	1–3
IV	1.700	0.5–2

[a] Data summarized from Macaya et al. (1977) and Mitchell et al. (1979).

overview of the properties of the major classical human satellites I, II, III, and IV (Corneo et al., 1968, 1970, 1971; Macaya et al., 1977; Mitchell et al., 1979). Together they represent only 2–5% of the genome although fractionation of human DNA on the basis of renaturation kinetics suggests that this may underestimate the total (Marx et al., 1976; Schmid and Deininger, 1975). The percentages of G·C base pairs estimated from the densities and melting temperatures are at variance with one another (Macaya et al., 1977; Mitchell et al., 1979). Furthermore, Mitchell et al. (1979) have presented data indicating that the density fractions contain related sequences. Trace amounts of ^{32}P-labeled III or IV reassociate with identical kinetics in the presence of an excess of unlabeled III and the melting curves of the resulting labeled duplexes are indistinguishable. Satellites I and II each share common sequences with III (and thus presumably IV); ^{32}P-labeled II hybridizes with III, as does ^{32}P-labeled I. The III-like sequences in I and II are probably not the same. It is not surprising then that all the satellites hybridized in situ to the same regions of the same human chromosomes. The classical satellites do not tell the whole story either; substantial satellite sequence may be buried in the main density band of DNA (Corneo et al., 1980).

Of all the classical human satellites, III has been best studied (assuming that what is called III is the same in each instance). It contains at least four different types of sequence as defined by restriction endonuclease products and the ability to cross hybridize. The extent to which the different sequence types are covalently linked together is not known.

One portion of satellite III is degraded to a ladder of bands $n \times 170$ base pairs in length by restriction endonuclease HaeIII (Manuelidis, 1976; Bostock et al., 1978; Mitchell et al., 1979, 1981). Similar bands are produced from satellite II (Mitchell et al., 1979). The repeat length of these units is immediately reminiscent of α-satellite type sequences and they may indeed be related. The uncloned 340-bp α fragment set (see Fig. 5) isolated from total human DNA after cleavage with EcoRI hybridizes with HaeIII ladders produced by digestion of total genomic human, chimpanzee, gorilla, and simiang DNA (Manuelidis, 1978; Maio *et al.*, 1981a). *Eco*RI digestion of satellite III itself yields a 340-bp fragment (Mitchell *et al.*, 1979). Confirmation of the relation between the *Eco*RI dimer and sequences in the *Hae*III ladder by hybridization with cloned probes is required. Using uncloned total satellite III as a probe and rodent human somatic cell hybrids that contained a limited number of identified human chromosomes Beauchamp *et al.* (1979) found that certain size classes of *Hae*III generated fragments typical of total satellite are missing on specific chromosomes.

A second type of sequence included in satellite III is constructed of imperfect tandem repeats of the sequence 5'-TTCCA-3' (see sequence 2 and 3, Fig. 6) (Deininger *et al.*, 1981); one of these is localized to the Y-chromosome (see below). A third type contains interrupted imperfect repeats of 5'-TTCCA-3'; at least two quite different versions of this type have been characterized within cloned segments (see sequences 1 and 4, Fig. 6) (Cooke and Hindley, 1979; Deininger *et al.*, 1981). One of them is localized mainly to chromosome 1 (Cooke and Hindley, 1979). Since all these different cloned segments were chosen essentially at random, a very large number of different versions of the 5'-TTCCA-3' sequence may occur, each in long tandem arrays that constitute domains. Satellite II may contain related sequences (Manuelidis, 1978; Mitchell *et al.*, 1979). Inspection of the α-satellite sequences in Fig. 5 reveals a marked frequency of sequences like 5'-TTCC-3' or 5'-GGAA-3' (equivalent to 5'-TTCC-3' on the other strand). This may bespeak a relation between the satellite III sequences and α-satellite (Deininger *et al.*, 1981).

A fourth class of sequences included in satellite III occurs only in male DNA and thus resides on the Y-chromosome. Upon digestion with *Hae*III, *Eco*RI, or *Eco*RII, male DNA yields a 3.4-kbp fragment set that is not found in female DNA (Cooke and McKay, 1978; Cooke and Hindley, 1979; Bostock *et al.*, 1978; Kunkel *et al.*, 1979). This fragment set may account for 40% of the entire Y-chromosome. Experiments with both cloned and uncloned 3.4-kbp fragments show that at least 50% of the sequences within the 3.4 kbp set hybridize with satellite III segments that are also abundant in female DNA. The remainder are male specific. Both types of sequences may be linked together in the members of the set (Kunkel *et al.*, 1979). This suggests an interspersion of Y-specific and Y-nonspecific sequence elements in a domain with a 3.4-kbp repeat length. The Y-specific domain is located on the long arm of the Y-chromosome (Kunkel *et al.*, 1977; Bostock *et al.*, 1978). Other domains that include the Y-nonspecific portion of the Y-domains but not within a 3.4-kbp repeat unit are represented in different amounts and different arrangements (domains) in various autosomes (Cooke and McKay, 1978). The properties of two cloned segments picked at random from a population of repeated sequences confirm the earlier analyses (Deininger *et al.*, 1981). One of the two hybridizes only to male DNA (sequence 2, Fig. 6) while the other hybridizes both to the male 3.4-kbp fragment set and to

1. Located on chromosome 1

5' AATTCATTTGAAGACAATTCCATTCAATACCAATTGATGATGGTTATTTTTGATTCCATTTGATGATGATTACATTCCAT

TTCATCATAATTCCATTCGATTCCACTCGAGATTCCATTCGATTCCATTCAA..........................

.........CGAATGAATGAGTCCATCCATTTCAATTTCATGATAATTCCATTCGTTTCAATTCGATGGTGTTTCCATTC

GATT.................................TTCATTCGATTCATTTGATGATGATTCATGCGCGATTCA

TTAGATGATGACCCCTTTCATTTCCATTCAATGGAGGATTCCATTCGGTTCCAT...................3'

2. Located on Y chromosome

5' TTAATTCCATTCCATTCCATTCCATTCCATTCCATTCCATTCCATTCCATTCCGTTCCATTCCATTCGTGTTGATTCCAT

TCCATTCCATTCCACTCCATTCCAATCCATTACATTCCACTCGGGTTGATTCCATTCCTTTCCATTCCAATCCATTCCAT

TCCATTCC 3'

3. Located on Y chromosome and elsewhere

5' CATTCAAGAAAGTTCCATTCCAGTCCATTCCCTTCGATTCCATTCCATTCGATTCCATTCTACTCGATTCCAATCTTGTC

CATTCCGTTGCATTCCATTCTATTCCATTCCATTGCATTGCATTCCATTCCATTTGATTACATTCCATTATATTCCATTC

CATTCA 3'

4. Location unknown

5' TTATTCCATTAGATTCCATTCGATGATGATTCCATTCGATTCCATTTGATGATTGCATTCTATTTCATTTGATGATGATT

CCATTCGAGTCCACTCGATGATTCCATTCGAGTCCATTCATTGATTCCATCCGATTTCATTGGATGATGACTCCATTCGA

GTCCATTCGATGATTCCACTCGATTCCATTAGATGATTCCATTGGAGTCCATTTGATTGTTCCATTCGATTCCATTCGAT

TCCT 3'

FIG. 6.   Related but variant domains in human satellite III. (1) From Cooke and Hindley (1979); (2, 3, and 4) from Deininger *et al.* (1981).

female DNA (sequence 3, Fig. 6). Both are constructed of regular repeats of 5'-TTCCA-3' and variants thereof; the differing variations in the two cloned segments must explain the specificity of hybridization. Note that a total of four segments, all included in satellite III and all built from variations of 5'-TTCCA-3', have been described (Cooke and Hindley, 1979; Deininger *et al.*, 1981); their structures are summarized on Fig. 6. In the absence of primary sequence data, the different hybridization specificities of the fragments would have obscured their similarities.

Little is known about satellite structure in other primates except that α-type sequences are ubiquitous (see above) and that satellite sequences in anthropoid apes are similar to one another (Deininger and Schmid, 1976b; Marx et al., 1979; Mitchell et al., 1981).

E. Marsupials

Macropus rufogriseus, the red necked wallaby, contains a satellite that can be isolated on CsCl gradients containing actinomycin D. The bulk of the satellite is degraded by *Bam*HI to sets of fragments that are from 1 to 5 times 2.5 kilobase pairs in length (Dunsmuir, 1976; Dennis et al., 1980) and the repetition frequency is 5×10^5 in the genome. The 2500-bp repeat may be internally repetitive with a periodicity of about 300 bp. Secondary digests with other enzymes of both total satellite DNA and monomer units allow a division into distinct variant subsets of the basic structure. The variant subsets are clustered in eight domains, A through H, the characteristic structures of which are shown on Fig. 7. Within each domain, occasional repeat units have altered sequences that result in ladders of fragments with different enzymes. A model for the evolution of the satellite (see arrows on Fig. 7) was constructed on the principle that a restriction site common to the units in several domains was present in a common progenitor.

F. Carnivora

Centromeric C-banding is rare or nonexistent in members of the order Carnivora and early attempts to demonstrate a density satellite in *Felis catus* (Pathak and Wurster-Hill, 1977) failed. Matthews et al. (1980) have now reported that a cryptic satellite is revealed by density gradient centrifugation in the presence of netropsin. Analysis by reassociation kinetics after purification of the satellite by repeated centrifugation indicated a $C_0 t_{\frac{1}{2}}$ of 7×10^{-4} mole/seconds/liter.

G. Comparisons between Satellite Sequences

Brutlag (1980) has recorded some statistically significant similarities between short sequences in the complex satellites of various species including *Drosophila* and several mammals. Thus residues 31 through 53 in the AGM α-satellite consensus sequence (Fig. 5, #1) are homologous to a portion of the 1.706D bovine satellite and residues 73 through 98 of the human sequence (Fig. 5, #4a) are homologous to a portion of the rat consensus sequence (Fig. 2b). Mouse satellite can be added to the list. Residues between 7 and 17 of the AGM consensus

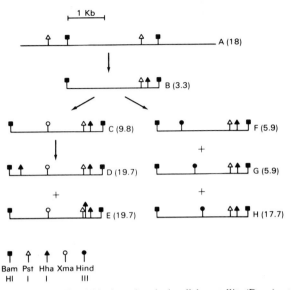

FIG. 7. Domains (A through H) within the red necked wallaby satellite (Dennis *et al.*, 1980). The numbers in parentheses represent the approximate percentage of total satellite in each domain.

(5'CTGAGAAACTG) are more than 87% homologous to multiple regions of the mouse consensus and are themselves repeated in part in 3 other positions of the AGM consensus. Further, the sequence is related to the progenitor nonanucleotides proposed for the mouse satellite (see above). Also, there is a significant similarity (93%) between residues 91 to 103 of the AGM sequence and 123 to 136 of the mouse. Unfortunately, statistical significance is neither necessary nor sufficient for biological importance. Questions about common and functionally significant satellite sequences in a range of species await new approaches.

H. Chromosomal Location

Are specific satellites or domains located on particular chromosomes? This important question remains largely unanswered except for the few instances already mentioned. Yet it is an important question for speculating about satellite evolution and is relevant to certain proposals about satellite function. Early experiments that used radioactive classical satellite preparations or RNA copies thereof as probes for *in situ* hybridization to chromosomes are now suspect not only because of possible contamination with other sequences but also because of the cross-hybridization of related but distinct satellites. In some instances, experiments with cloned and characterized probes should afford definitive data. But where different domains cross-hybridize (e.g., in the African green monkey and the wallaby) chromosome specificity will have to be determined with isolated chromosomes. Somatic cell hybrids are a stable source of some individual chromosomes and this approach is being used (as reviewed above). Furthermore, direct methods for the isolation of specific chromosomes are being improved (Davies *et al.*, 1981).

I. THE QUESTION OF SATELLITE FUNCTION

There is no experimental evidence regarding the function of mammalian satellite DNA. Speculation about function presently rests on the universal (but not necessarily exclusive) association of satellite DNA with heterochromatin (Brutlag, 1980; John and Miklos, 1979). However, heterochromatin function is itself not understood at the molecular level and most of the relevant information is limited to simpler eukaryotes that are amenable to either genetic or cytogenetic analysis or both. Reviews of earlier experiments and speculations are available (Bostock, 1980; Miklos and John, 1979; Hilliker *et al.*, 1980; Brutlag, 1980; John and Miklos, 1979; Orgel and Crick, 1980; Doolittle and Sapienza, 1980). The available data are most consistent with the idea that satellite functions in germ line processes (Bostock, 1980). The following two experimental systems may prove relevant to the question of satellite function.

1. *Transcription*

Transcription of satellite DNA occurs on specific loops in oocyte lampbrush chromosomes of newts (Varley *et al.*, 1980a,b; Diaz *et al.*, 1981; Gall *et al.*, 1981). Within these genomes, the satellite occurs both in centromeric regions and at the chromosomal regions corresponding to the loops. In *Notophthalmus viridescens*, hundreds of 9-kbp-long clusters of histone genes are each embedded within variable lengths of satellite sequence (estimated to be up to 100 kbp long) at the loops. Transcription may initiate at promotor sites preceding the histone genes and then proceed through a cluster continuing into the downstream satellite sequences (Varley *et al.*, 1980b; Diaz *et al.*, 1981). This is viewed as a failure of normal transcription termination but whether that failure is functionally significant remains to be learned.

There is one report that rat satellite sequences are transcribed in HTC rat tissue culture cells (Sealy *et al.*, 1981). This is of great interest, since earlier experiments with less sensitive techniques showed no transcription in a variety of species of organisms and cell lines.

2. *Centromeric Function*

The structure and function of yeast centromeres are now being investigated at the molecular level (Clarke and Carbon, 1980a,b; Hsiao and Carbon, 1981; Fitzgerald-Hayes *et al.*, 1982; Clarke *et al.*, 1981). The ability to couple precise information on DNA structure with detailed genetic analysis makes this an especially powerful system. Using recombinant DNA techniques small circular DNA molecules have been constructed that contain selectable yeast genetic markers (e.g., genes required for synthesis of a particular amino acid) and a yeast DNA replicator sequence. Such plasmids replicate autonomously in yeast cells but are mitotically unstable and are rapidly lost from the population in the absence of dependency on the marker gene. The addition of a DNA segment spanning a centromeric region imparts both meiotic and mitotic stability to the plasmid. It then behaves like a true minichromosome, and becomes a stable component of the cells even in the absence of the selective pressure. During crosses involving two minichromosomes bearing identical centromeres but distinguishable marker genes, the two can apparently pair, moving to opposite poles in the first meiotic division (Clarke et al., 1981). The centromeres of yeast chromosomes 3 (Clarke and Carbon, 1980a) and 11 (Fitzgerald-Hayes et al., 1982) have been characterized. Yeast does not contain satellite sequences and there are no markedly repetitious segments in the defined centromeric regions. The essential regions of centromeres 3 and 11 are not more than 627 and 900 base pairs in length, respectively, and do not cross-hybridize with one another. The centromere 3 sequence is not repeated elsewhere in the genome. Yet *both* contain long A·T-rich regions flanked by short (14 bp or less) regions of homology. One of the latter is reported to display limited sequence homology with the 1.688 gm/cm^3 satellite DNA of *Drosophila melanogaster* (Clarke et al., 1981; Hsieh and Brutlag, 1979a).

J. AMPLIFICATION OF SATELLITES

Most satellites are restricted to one or a closely related group of species and thus many appear to be of relatively recent origin. This in turn suggests that satellites change in evolution more than many other genomic regions. Compare, for example, the very close similarities between the β-globin regions (Barrie et al., 1981) and the *Alu* families (see below) of the primate genomes with their strikingly different though related satellite domains. Large scale deletions and amplifications were required to arrive at present day satellites. The similarities between satellites of related species led to the "library" hypothesis regarding satellite amplification and deletion (Salser et al., 1976; Fry and Salser, 1977); a common library (or set) of related sequences is available within related species and different members of the library are amplified to differing extents in the several organisms. At present there is no reason to assume a preexisting or common library. It is equally probable that newly divergent sequences were and are candidates for amplification (Gillespie et al., 1980).

Mammalian genomic sequences can be amplified. The organization of the multiple globin genes (Fritsch et al., 1980, 1981) demonstrates this phenomenon in evolutionary time. The amplification of certain eukaryotic genes also occurs in experimental time. For example, clones of murine tissue culture cells containing many copies of the gene for dihydrofolic acid reductase (Alt *et al.*, 1978; Dolnick *et al.*, 1979) or of the complex of genes responsible for synthesis of aspartyl transcarbamylase (Wahl *et al.*, 1979) are readily obtained under appropriate selective pressure. In at least one instance, amplification of the mouse dihydrofolate reductase gene was accompanied by amplification of mouse satellite (Bostock and Clark, 1980). Once amplified, extra copies of these genes are readily, but not necessarily, lost when selective pressure is removed. Several mechanisms for amplification of tandem repeats have been discussed (Tartof, 1975; Botchan *et al.*, 1978; Schimke *et al.*, 1980; Smith, 1976; Kurnit, 1979; Baltimore, 1981). Among these, unequal crossing-over (Smith, 1976) is a relatively simple mechanism that is consistent with available data and has been demonstrated experimentally with the tandemly repeated ribosomal RNA genes of yeast (Petes, 1980; Szostak and Wu, 1980). In brief, homologous crossing-over between nonallelic repeat units in a tandem array (on sister chromatids or homologous chromosomes or homologous regions on nonhomologous chromosomes) yields an unequal number of repeat units in the products of recombination. Besides resulting in reciprocal amplification and deletion, unequal crossing-over provides for the maintenance of sequence homogeneity in tandem arrays (Smith, 1976). It also predicts that repeat units near the ends of tandem arrays will be more divergent than those near the center (Smith, 1976; Brutlag, 1980), as found with the AGM satellite (McCutchan *et al.*, 1982). Another mechanism for assuring homogeneity, called gene conversion, is described in Section V,D. This mechanism may be relevant to the fact that satellites on different chromosomes are characteristic of the species, even though they may be distinct domains (compare for example the a-a-a-a type structure of African green monkeys with the a-b-a-b-a-b of baboons).

Amplified genes, besides being subject to deletion, can follow at least three additional courses. First, they may be used essentially as such, as are the two adult α-globin genes. Second, they may evolve independently into either related functional genes, as for example, the embryonic and fetal globin genes, or into two quite different genes. Finally, a duplicated gene may be nonfunctional, as for example, the pseudogenes in the globin gene clusters. Presumably, different selective conditions favor one or the other course.

In this context, the situation with satellites is very striking. Is there in the sequences an inherent tendency toward amplification greater than that of, say, the dihydrofolic acid reductase gene or is there positive or negative selective pressure that operates to maintain multiple copies? Is there a selective force that dictates a minimal number of copies? Is there a copy number at which additional amplification is rejected? Is the location at centromeric and telomeric positions influential in determining copy number? It is not sufficient to argue that these

DNA segments exist only for their own replication (Orgel and Crick, 1980; Doolittle and Sapienza, 1980), without asking how they get away with this profligate reproduction while other sequences do not. What distinguishes a sequence that is tolerable in millions of copies from all the other genomic segments that are not? Is it the inability to be transcribed? Do the satellites provide a particular environment for genes or special sequences buried within them? The minimum requirements for centromere function in mitosis and meiosis can be supplied without millions of tandem repeats, as is evident from the recent work on yeast centromeres described above. It is conceivable that only one or a few copies of the satellite sequences are essential to cell function and that, unlike many other genomic segments, extra copies are tolerable safeguards and collect by virtue of amplification and the lack of negative selective pressure. This hypothesis provides a middle ground between the bold suggestion that satellite has no function at all (Orgel and Crick, 1980; Doolittle and Sapienza, 1980) and the well grounded assumption that biological systems are efficient.

Another set of questions arises in relation to the heterochromatic location of satellite. Is late replication in the cell cycle a consequence of heterochromatin structure or is it dictated by the sequences themselves? Does heterochromatin structure supply a useful positive function or does it represent a defensive mechanism whose chief advantage is to protect the genome from the effects of satellite sequences while still permitting functioning of those sequences at appropriate times or places? The latter idea is consistent with other examples of selective heterochromatin formation, as in X-chromosome inactivation.

IV. Interspersed Repeated Sequences

A. Emerging Concepts

Interspersion analysis—the measurement of the percentage of denatured DNA that registers as double-stranded on hydroxyapatite at a fixed low C_0t value as a function of chain length (Davidson et al., 1973)—demonstrated that some highly repeated sequence families are dispersed among single-copy sequences in eukaryotic genomes. Two types of interspersion patterns were discerned. The Drosophila pattern (Manning et al., 1975) is characterized by families of repeat units several kilobase pairs in length separated by tens of kilobase pairs of single-copy sequence. This arrangement has been confirmed by molecular analysis and several repeated elements in Drosophila are known to be mobile units (for review see Calos and Miller, 1980; Spradling and Rubin, 1981). The Xenopus pattern (Davidson et al., 1973) is characterized by families of repeat units a few hundred base pairs in length separated by up to a few thousand base pairs of single-copy sequence. Very few organisms and not even all insects have the *Drosophila* type pattern but many, including most mammals, are of the *Xenopus* type. Several organisms including chickens (Musti *et al.*, 1981) show interspersion patterns between the two extremes. And within the *Drosophila* genome another kind of arrangement, scrambled clusters of different short, moderately repeated sequence elements, occurs (Wensink *et al.*, 1979).

Substantial analysis at the molecular level has now revealed additional complexity. Two different classes of mammalian interspersed and highly repeated sequences have thus far been distinguished on the basis of size and relative abundance. Dispersed families with unit lengths under 500 base pairs are found in as many as hundreds of thousands of copies; they are here termed SINES, for short interspersed repeated sequences. In some organisms (e.g., humans) the copy number of the major SINE family is higher than the copy number of some satellite sequences, thus emphasizing the fact that satellite is not always the most rapidly reannealing portion of a genome. Other dispersed mammalian families are several kilobase pairs in length and occur on the order of 10^4 times (or fewer); they are here termed LINES, for long interspersed repeated sequences.

B. Short Interspersed Repeated Sequences (SINES)

SINE families often contain more than 10^4 member sequences although smaller families also exist. The many members in a family are similar enough to hybridize to one another under stringent conditions, but are not identical (Rinehart *et al.*, 1981). It is quite possible that different families in a particular species are related to one another. Although a variety of functions have been suggested (Britten and Davidson, 1969; Davidson and Britten, 1979; Jelinek *et al.*, 1980) the physiological significance of these sequences remains unknown. The strongest available arguments for assuming a functional importance for at least some members of these families are their marked conservation and the fact that they are transcribed. The characterized families of rodent and primate SINES share common features but are, at the same time, quite different.

The repeat units occur in both possible directions within genomic segments; the significance of the direction, if any, is unclear. This arrangement accounts for many if not most of the abundant short inverted repeats previously described in mammalian genomes and is reflected in nuclear RNA molecules that also contain inverted copies of family members. Transcripts of family members are most abundant in heterogeneous nuclear RNA but also occur in cytoplasmic RNA.

1. *Human Alu Family*

The best characterized mammalian SINE family is the *Alu* family of the human genome. The presence of a large set of cross-hybridizing short DNA sequences dispersed throughout most of the human genome and occurring frequently in both possible orientations was demonstrated initially by renaturation kinetics, hyd-

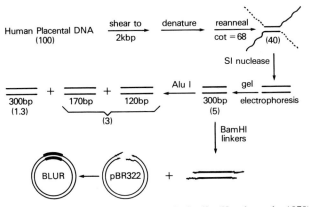

FIG. 8. The isolation and cloning of the human *Alu* family (Houck *et al.*, 1979). The cloning vector was the *E. coli* plasmid pBR322. The numbers in parentheses indicate the approximate percentage of total DNA in each fraction.

roxyapatite separation, interspersion analysis, and electron microscopy (Schmid and Deininger, 1975; Deininger and Schmid, 1976a). *Alu* family sequences were first isolated (Fig. 8) (Houck *et al.*, 1979) by renaturing sheared (2 kbp) denatured human placental DNA to a C_0t of 68 and removing single-stranded tails with S1 nuclease. Analysis of the size of the double-stranded products by gel electrophoresis revealed a broad range of sizes but about 5% of the products were 300 bp in length. More than 60% of the 300-bp-long set of duplexes was cleaved by *Alu*I into two fragments of about 170 and 120 bp in length, respectively (Houck *et al.*, 1979): thus the name, *Alu* family. More precise measurements now put the reiteration frequency of the family close to 3×10^5 copies or about 3% of the genome (Rinehart *et al.*, 1981). Other families of SINES may also be interspersed in the human genome (Fritsch *et al.*, 1981; Deininger *et al.*, 1981), but it seems unlikely that any approach the copy number of the *Alu* family (Rinehart *et al.*, 1981).

Direct evidence for the distribution of *Alu* sequences comes from studies with long cloned segments of the human genome. Recombinant libraries that comprise the bulk of the human genome divided into segments 15–20 kbp long, each inserted into a λ-bacteriophage vector, are available (Maniatis *et al.*, 1978; Lawn *et al.*, 1978). Screening of individual phage in a library by hybridization with a cloned *Alu* segment indicates that more than 90% of the human segments hybridize with *Alu* (Tashima *et al.*, 1981). Many phage that were selected from a library for other reasons also contain *Alu* sequences. There is, for example, an *Alu* sequence some 6 kbp downstream from the 3' end of the insulin gene (Bell *et al.*, 1980); it is the only *Alu* sequence within about 19 kbp. There are 8 *Alu* sequences within the 65 kbp region that contains the entire cluster of β-globin genes (Duncan *et al.*, 1979; Coggins *et al.*, 1980; Baralle *et al.*, 1980; Fritsch *et al.*, 1980). Eight cloned segments randomly selected from a human library have a minimum of 9 *Alu*s in a total of 88 kbp (Pan *et al.*, 1981). And 3 *Alu* segments are within a 15 kbp segment that contains DNA homologous to the transforming gene of simian sarcoma virus (Dalla-Favera *et al.*, 1981); all three are within putative intervening sequences (introns). Altogether this represents 21 *Alu* sequences in a total of 187 kbp of DNA or one *Alu* sequence every 9 kbp. If a similar distribution occurs in all the library segments then there are an average of 1–2 *Alu* family members in each of the 90% that hybridized, approximating the 3×10^5 copies estimated by other means.

The primary nucleotide sequence of 11 randomly selected and cloned members of the *Alu* family is known (Rubin *et al.*, 1980b; Deininger *et al.*, 1981; Pan *et al.*, 1981). Ten of the 11 are part of a group referred to as BLUR (Rubin *et al.*, 1980b; Deininger *et al.*, 1981) and were selected from the 300-bp-long fragments obtained from the total genome (Fig. 8). These 10 sequences were used to construct a consensus sequence (see Fig. 9,1). Although the individual BLUR sequences differ from each other (and from the consensus) both by insertions, deletions, transitions, and transversions, they are, on the average, 88% alike. The alterations are spread around the sequence thus accounting for the strong cross-hybridizations. In addition, the sequence of four *Alu* family members from known positions in the human genome have been determined: one near the insulin gene (Bell *et al.*, 1980) and one each preceding the ϵ-globin gene (Baralle *et al.*, 1980), the Gγ-globin, and the δ-globin genes (Duncan *et al.*, 1981). All are approximately 300 bp long and are about 80% homologous to the consensus.

Certain very striking features of *Alu* family members have emerged from the sequence analysis. First, the 300 bp unit is essentially a head-to-tail dimer of a sequence about 135 bp long. One-half of the internally repeated sequence is interrupted by about 30 extra bp (approximately residues 218 to 250, Fig. 9,1). Each half is terminated by a stretch of A·T base pairs though this is generally most striking toward the right end (as written, Fig. 9,1). Within each half is a 14-bp segment (residues 41 to 53 and 176 to 188, in Fig. 9,1) which is similar to a sequence around the origin of replication of papovaviruses (Jelinek *et al.*, 1980).

Short direct repeats flank the *Alu* sequence in each instance studied (Fig. 10); the BLUR clones do not include flanking sequences because of the way in which they were isolated. These flanking repeats are not part of *Alu* itself and are different in each case examined both in length and structure (Fig. 10). A similar direct repeat has been found in a sequenced *Alu* segment from the African green monkey genome (see below and Fig. 10).

2. *Transcription of Human Alu Sequences*

The percentage of cytoplasmic RNA that hybridizes to *Alu* is at least 10-fold less than that found in nuclear RNA. Heterogeneous nuclear RNA (hnRNA)

```
                 '               '             '              '              '              '
1   GCT GGGCGTG   G       TGGCTCACA   CCTGTAATCC   CAGCACTTTG   GGAGGCCGAG   GTGGGTGGAT
2   GCC GGGCGGG   A       TGGCTCATG   CCTGTAATCC   CAGCACTTTG   GGAGTTCGAG   GCGGGAGGAT
5       CGGGGC TG GAGAGATGGCTCAGC     GGTTAAGAGC   GCCC         GACTGCCTT
6      GAGGC TG   GAGAGATGGCTC GA     GGTTAAGAGC   ACCA         ACTGCTGT


                 '               '             '              '              '              '
1   CACCTGAGGT    CAGGAGTTCA    AGACC  AGCCT   GGCCAACATG   GTGAAACCCC   GTCTCTACTA
2   CACCTGAGGT    CGGGAGTTCG    AAACC  TGC T   GGCCAACATG   GCGAAACCCC   GTCTCTACTA
5    CCAGAGGT     CATGAGTTCA    ATTCCCAGC      AACCA CATG   GTGGCTCACA   ACCATCTGTA
6   TCCAGAGGT     CCTGAGTTCA    ATTCCCAGC      AACCA CATG   GTGGCTCATA   ACAATCTATA


                 '               '             '              '              '              '
5   AAGAGATCTG    ATGCCCTCTT    CTGGTGTATC    TGAAGACAGC   TACAGTGTAC   TTATATATAA
6   ATGAGATCTG    GTGCCCTCTT    CTGGTGTGCA    GATATATATG   GAAGCAGAAT   GTTGTATACAT


                 '
1   AAAATACAAA    AATTA
2   AAAATACAAA    AATTA
5   TAAATAAATA    AATCTTTAAA    AAAAACAAAA    CAAAAACAAA   AACAAAA
6    AATAAATA     AATAAAATCT    TAAAAAAA


         '               '             '              '              '              '
1   GCCGG         GCGT  GGTGGC    GCGCGCCTGT    AATCCCAGCT   ACTCGGGAGG   CTGAGGCAGG
2   GCCCG         GTGT  GGTGGC    GCATGCCTGT    AGTCCCGGCT   ACTCGGGAG    CTGAGGCAGG
3    CCGG         GCAT  GGTGGT    GCATGCCTTT    AATCCCAGC    ACTCGGGAGG   CAGAGGCAGG
4    CCAG         GCATT GGTGGT    ACACACCTTT    AGTCCCAGC    ACTCAGGAGG   CAGAGGCAGG
                              C


         '               '             '              '              '              '
1   AGAATCGCTT    GAACCCAGGA    GGTGGAGGTT    GCAGTGAGCC   GAGATCGCGC   CACTGCACTC
2   AGAGTTGCTT    GAACCT GGA    GGTGGAGGTT    TCAGTGAGAC   AAGATCACAT   CACTGCAC
3   CGGATTTCT     GAGTTCGAGG                                            C
4   AGGATCACTT    GAGTTCAAGA    G                                       C
                            C
                            T


         '                                                                      '
1   CAGCCTGGGC                                              AACAGAGCGA
2   CAGCCTGGGC                                              ACAGAGCAA
3   CAGCCTGGTC    TTC AGAGT     GAGTTCC      AGGACACCAG   GGCTA       CAGAGAAA
4   CAGCCTGGTC    TACCAGAGTT    CCTGAGTT C   AAGCCA       GGCTAT      ACAGAGAAA
                            T
                            A


         '                            '            '
1   GACTCCATCT    C             AAAAAAAAA    AAAAAAAAAA   AAA
2   CACTCTA       ACAGAGAG      AAAAAAAAA    ACCCACAAAA   AAA
3   CCCTGTCT   -  A rich
4   CCCTGTCT      [(A)nN]i
```

Fig. 9. Comparison of SINES. (1) is the human *Alu* consensus sequence (Deininger *et al.*, 1981); (2) is a cloned monkey *Alu* sequence (Grimaldi *et al.*, 1981); (3) is the mouse B1 consensus sequence (Krayev *et al.*, 1980); (4) is the Chinese hamster consensus sequence (Haynes *et al.*, 1981); (5) is the second half of the dimeric SINE found in the intervening sequence of the rat growth hormone gene (the sequence is reported in Page *et al.*, 1981 and the homologies were found using the computer program of Queen and Korn, 1980); (6) is a Chinese hamster sequence (Haynes *et al.*, 1981) which resembles the rat sequence shown in line 5. The marks above the human *Alu* sequence (1) indicate every tenth base pair in that sequence. Only one DNA strand is shown. Each base in each sequence is shown, in order, in the 5' (upper left) to 3' (lower right) direction. In sequence (4) frequent variations in the consensus sequence are noted under the line.

1. ∿∿AAGATTCACTTGTTTAG∿∿ Alu ∿∿AAGATTCACTTGTTTAG ∿∿

2. ∿∿AAAACAA GCAGGAG ∿∿ Alu ∿∿AAAACAA GCAGGAG ∿∿∿

3.   ∿∿GCTTTG ∿∿ Alu∿∿ GCTTTG ∿∿∿

4. ∿∿AAAAGAAACTTGGAAAGGA∿∿CH∿∿ AAAAGgAAACTTGGAAAGGA∿∿∿

5. ∿∿∿GAGACAACAAATCAgag∿∿Bl∿∿GAGACAACAAATCAaAt ∿∿∿

FIG. 10.   Direct repeated sequences surrounding members of the primate and rodent SINE families. Selected examples are given. (1) *Alu* 5′ to human Gγ-globin gene (Duncan *et al.*, 1981); (2) *Alu* 3′ to human insulin gene (Bell *et al.*, 1980); (3) a monkey *Alu* (Grimaldi *et al.*, 1981); (4) a Chinese hamster SINE (Haynes *et al.*, 1981); (5) a mouse B1 sequence (Krayev *et al.*, 1980). Imperfections in the repeats are indicated by lower case letters. Other examples may be found in the following references: Elder *et al.* (1981), Baralle *et al.* (1980), Haynes *et al.* (1981), Page *et al.* (1981), and Grimaldi and Singer (1982).

contains high levels of *Alu* transcripts in molecules ranging from 100 to more than 5000 nucleotides in length (Federoff *et al.*, 1977; Jelinek *et al.*, 1978, 1980; Elder *et al.*, 1981). At least some of the transcribed *Alu* sequences are found in intra- and intermolecular RNA·RNA duplexes, presumably because of inverted orientations and the transcription of both strands of *Alu* family members. *Alu* transcripts are found in both polyadenylated and nonpolyadenylated cytoplasmic RNA (Pan *et al.*, 1981; Elder *et al.*, 1981; Jelinek *et al.*, 1978; Tashima *et al.*, 1981). The presence of *Alu* transcripts in polyadenylated RNA cannot be taken as evidence that the RNAs are polyadenylated by posttranscriptional modification, as are most eukaryote messenger RNAs, because the long terminal A stretches in the *Alu* sequence may account for the binding of these molecules to the oligo(dT)-cellulose used in standard separations (Elder *et al.*, 1981). A discrete 7 S cytoplasmic RNA from HeLa cells (uninfected) hybridizes to *Alu* DNA and this is the only abundant HeLa cytoplasmic RNA that hybridizes with *Alu* (Weiner, 1980). Analysis of hybrids formed between 7 S RNA and cloned *Alu* DNA indicates imperfect duplexes suggesting (Weiner, 1980) that 7 S RNA may be encoded by an unusual subset of the *Alu* family.

Many but not all *Alu* sequences in cloned human DNA fragments are transcribed *in vitro* by RNA polymerase III (Duncan *et al.*, 1979; Elder *et al.*, 1981). The structure of the transcripts is consistent with initiation at the start of the *Alu* sequence and termination beyond the end of *Alu* and the RNA has the sequence of the DNA strand in Fig. 9,1. Transcriptional control regions are generally present internal to the coding sequences of genes transcribed by RNA polymerase III (reviewed by Duncan *et al.*, 1981). The *Alu* consensus sequence includes a segment 5′-GAGTTCPuAGACC-3′ (at about position 75) that is similar to a sequence noted by Fowlkes and Shenk (1980) (5′-GGGTTCGANNCC-3′) as important in control of transcription initiation by RNA polymerase III. This sequence does not appear clearly in the second half of the *Alu* consensus; the analogous position is interrupted by the extra base pairs. Similarly, termination of RNA polymerase III transcription of *Alu* occurs at a run of 4 or more T residues, as with other polymerase III templates (Duncan *et al.*, 1981). There is evidence suggesting that 7 S RNA is synthesized *in vivo* by RNA polymerase III (reviewed by Weiner, 1980; Zieve, 1981).

3. *Alu-like Sequences in Other Primates*

The bonnet monkey (Houck and Schmid, 1981) and African green monkey (Dhruva *et al.*, 1980; Grimaldi *et al.*, 1981) each has an abundant family of SINES closely similar to human *Alu* in copy number and sequence. The galago (*Galago crassicaudatus*) has a distantly related family of unknown copy number (Houck and Schmid, 1981). About 75% of recombinant phage representing a library of the African green monkey genome hybridize with a cloned human *Alu* sequence, corresponding to a minimum of 1.8×10^5 copies per haploid genome. Further, a group of 4 phage, chosen from the library essentially at random with reference to *Alu*, contain together at least 8 copies of *Alu* or an average of one every 8 kbp of DNA. The sequence of one of the monkey *Alu* segments was determined (Fig. 9,2); it is 84% homologous to the human *Alu* consensus sequence and is flanked by short direct repeated segments (Grimaldi *et al.*, 1981). Among the cloned monkey DNA segments were some in which the *Alu* sequence abutted or interrupted African green monkey α-satellite sequences (see Sections III,D and IV,B,6).

4. *SINES in Rodents*

One predominant family of SINES—the *Alu* family—have been described in primates while rodent genomes are known to contain several distinct SINE families of relative similar abundance. This is not to say that the *Alu* sequences are the only SINES in primate genomes; others that may be present on the order of 10^4 times have already been detected (Deininger *et al.*, 1981; Fritsch *et al.*, 1981).

Mouse heterogeneous nuclear RNA forms both inter- and intramolecular double-stranded regions (reviewed by Georgiev *et al.*, 1981). The double-stranded regions were isolated from intramolecular duplexes by digesting away single strands with ribonucleases T1 and A and the resulting RNA fragments were used to detect molecular clones of genomic DNA that hybridize with the RNA probes. Two distinct families of SINES termed B1 and B2 were isolated. Members of the two families each occur about 10^5 times scattered throughout the genome (Kramerov *et al.*, 1979; Georgiev *et al.*, 1981). These sequences, like *Alu*, occur in both orientations along the genome and the inverted configuration presumably accounts for much of the intramolecular double-stranded heterogeneous nuclear RNA. The sequences of 3 randomly chosen cloned members of the B1 family are known and were used to construct a consensus sequence

(Fig. 9,3) (Krayev *et al.*, 1980); the mismatch between any pair of the three is less than 8%. Two of the sequenced B1s represent separate units within a single 2.3-kbp-long cloned fragment of mouse genomic DNA.

The B1 consensus sequence is about 130 bp long followed by an A·T-rich region (A-rich at the 3'-end of the strand shown in Fig. 9,3). Beside the A-rich region there are other striking similarities between the mouse B1 sequence and the primate *Alu* sequences (Pan *et al.*, 1981). A 40-bp sequence that is repeated in each arm of *Alu* (residues 21 to 60 and 156 to 195, Fig. 9,1) is highly conserved in the B1 sequence (Fig. 9,3); this includes the 14-bp segment homologous to the origin of replication of papovaviruses. B1 sequences are also flanked by short direct repeats of DNA (see Fig. 10). Overall B1 can be seen as analogous to one half of the *Alu* dimer. However a human *Alu* sequence (clone BLUR 8) does not hybridize to mouse DNA under stringent conditions (Shih *et al.*, 1981).

The B2 family does not cross-hybridize with B1 (Kramerov *et al.*, 1979). Other evidence indicating multiple families of abundant mouse SINES is seen in the fact that three separate, noncross-hybridizing SINES were discovered interspersed among the 65-kbp-long mouse β-globin gene cluster; each of the three occurs twice or more within the segment (Haigwood *et al.*, 1981). The distribution of SINES in the β-globin gene cluster of two different *Mus musculus* strains (BALB/c and C57BL/10) appears to be essentially the same (Haigwood *et al.*, 1981). The three SINE families were labeled a, b, and c (Haigwood *et al.*, 1981) and it is not known if any of them are related to the previously described B1 and B2. Cloned members of the a and b families hybridized with essentially all the phage in a mouse genomic DNA library. Thus, there are at least 2×10^5 separate members in each of these families assuming that the inability to cross-hybridize extends to all members of each family. It is possible that mouse SINE family members are interspersed in mouse satellite (Haigwood *et al.*, 1981).

Five different SINE families occur at 20 positions within the 44-kbp segment containing the genes for the β-globins of rabbits (Shen and Maniatis, 1980; Fritsch *et al.*, 1981). Only one family, called C, hybridizes (weakly) to a human *Alu* sequence under nonstringent conditions. C sequences, like *Alu*, are transcribed by RNA polymerase III *in vitro*. Members of SINE families were found buried in long cloned segments of Chinese hamster DNA. The clones were selected because they hybridized with double-stranded heterogeneous nuclear RNA (Jelinek, 1978; Haynes *et al.*, 1981). Thirty-seven percent of the phage in a Chinese hamster genomic DNA library hybridized. A large proportion of the cloned segments had more than one hybridizing region and many of these occurred in opposite orientations. The sequences of 6 such regions were determined and a consensus derived from 5 of them (Haynes *et al.*, 1981) (Fig. 9, 4). The similarity with the mouse B1 consensus sequence is evident. One of the 6 (Fig. 9, 6) differs markedly from the others after the first 60 base pairs, but is, therefore, homologous to a rat sequence (see below). Like *Alu* and B1, the Chinese hamster SINES are surrounded by short direct repeats of DNA (Fig. 10).

Much less is known about the situation in rats. Houck and Schmid (1981) reported the presence of abundant 300-bp-long segments in reannealed rat DNA that was treated with S1 as outlined in Fig. 8 for the original isolation of human *Alu* sequences. One repeated DNA segment occurs within an intron of the rat growth hormone gene and again after the end of the last exon in the gene (Page *et al.*, 1981). The sequence of the segment in the intron (Fig. 9, 5) includes two direct repeats of a segment about 200 base pairs long. This unit may represent a rat SINE family. It is, as noted above, similar to one Chinese hamster sequence (see Fig. 9, 5 and 6). A repetitive sequence has also been identified downstream from the rat insulin I gene (Bell *et al.*, 1980); it does not hybridize with a human *Alu* probe.

5. Transcription of Rodent SINES

As already described, the mouse B1 and B2 sequences are abundantly transcribed; copies of both strands of the sequence occur in heterogeneous nuclear RNA (Kramerov *et al.*, 1979). A unique version occurs in mouse cytoplasm as a discrete 4.5 S RNA (Harada and Kato, 1980); this sequence differs from the consensus sequence derived from three cloned members of the B1 family (Krayev *et al.*, 1980). The 4.5 S RNA hybridizes to both cytoplasmic and nuclear polyadenylated RNA (Jelinek and Leinwand, 1978). It is striking that the 4.5 S mouse sequence is more similar (3 differences out of about 90 nucleotides) to a 4.5 S cytoplasmic RNA of the Chinese hamster (see below) than are the mouse and Chinese hamster consensus DNA sequences to one another (see Fig. 9). One, or a few members of these families may be the functional genes for 4.5 S RNA in each of the species, and if so, that gene has been markedly conserved (Haynes *et al.*, 1981). Both the mouse and Chinese hamster 4.5 S RNAs are copies of the same DNA strand (sequence equivalent to the DNA strands in Fig. 9).

6. Are SINES Functional?

As a background, it is interesting to recall proposals suggesting that highly repeated dispersed sequences may be without function (Orgel and Crick, 1980; Doolittle and Sapienza, 1980) and also disagreement concerning those proposals (Cavalier-Smith, 1980; Dover, 1980; T. F. Smith, 1980; Orgel *et al.*, 1980; Dover and Doolittle, 1980). Specific functions that have been suggested include the control of gene expression (Britten and Davidson, 1969; Davidson and Britten, 1979; Jelinek *et al.*, 1980), perhaps by involvement of transcripts of SINES in the maturation of messenger RNA (Georgiev *et al.*, 1981; Jelinek *et al.*, 1980; Zieve, 1981; Lerner and Steitz, 1981), and service as origins of DNA replication (Jelinek *et al.*, 1980; Georgiev *et al.*, 1981). The cytoplasmic 4.5 S RNA of rodent cells and the 7 S RNA of human cells appear to represent abundant transcripts of a particular member(s) of the respective SINE families. At least these SINE family members are likely to be functional genes. It is striking that the 4.5 S RNA of the hamster and mouse are identical in all but 3 of 90 nucleotides (Harada and Kato, 1980; Haynes *et al.*, 1981). They are both found in small nuclear ribonucleoprotein complexes, as well as cytoplasm and recent speculations regarding their function are reviewed by Zieve (1981) and Lerner and Steitz (1981). Although the 4.5 S RNAs have 5'-terminal triphosphates, consistent with *in vivo* transcription by RNA polymerase III, it is not clear to what extent RNA polymerase II transcripts that include SINES also contribute to homologous sequences in heterogeneous nuclear or cytoplasmic RNA.

Some clues regarding SINE function might be expected from a comparison of their distribution around corresponding genes in different organisms. Extensive data are available for the β-globin gene clusters of mouse (Haigwood *et al.*, 1981), rabbit (Shen and Maniatis, 1980; Hoeijmakers *et al.*, 1980; Fritsch *et al.*, 1981), and human (Fritsch *et al.*, 1981) and the α-globin cluster of humans (Fritsch *et al.*, 1981). Within each cluster are several coding regions, one or more pseudogenes, and various SINES. In the human β-cluster 5 genes are expressed in a timed sequence: ϵ in embryos, Aγ and Gγ during fetal development, and δ and β during adulthood. In the rabbit, two genes are expressed in embryos and one in adults, while in the mouse, two are expressed in adults and one in embryos (Jahn *et al.*, 1980). Both rat and human insulin genes (Bell *et al.*, 1980) have a SINE following the end of the last exon, although at different distances. There is no equivalent to the SINE within one intervening sequence of the rat growth hormone gene in a corresponding human gene (Page *et al.*, 1981). While some patterns can be discerned (e.g., in the rabbit the C-SINE family is clustered in the region of embryonic genes and in the human β-globin cluster no *Alu* occurs between simultaneously expressed genes) and may prove to be meaningful, no definitive statements can be made at this time.

Another interesting possibility is that some SINES are mobile or transposable sequence elements analogous to those described in yeast and *Drosophila* (reviewed by Calos and Miller, 1980; Spradling and Rubin, 1981). Mobile elements are flanked by short direct repeats of sequences that are not part of the element itself. In the case of characterized mobile elements the repeat lengths are characteristic of the element and constant, and the duplications reiterate the target site at which insertion occurred. Similarly, integrated DNA copies of RNA tumor virus genomes in birds and mammals are flanked by such short direct repeats (reviewed by Temin, 1980). These viral inserts as well as several well characterized mobile elements in bacteria, yeast, and *Drosophila* share other structural features. These include direct terminal repeats of several hundred base pairs at the two extremities of the movable element itself as well as a total length of several kilobase pairs. Also eukaryote mobile elements do not typically have the long terminal stretch of AT base pairs that is a common feature of mammalian SINES.

However, a recently discovered movable *Drosophila* element does have notable similarities to the mammalian SINES. This element, referred to as 101F (Dawid *et al.*, 1981), has no long terminal repeats, is flanked by short direct repeats of the target site, and contains a 3'-terminal poly(A) segment 18 nucleotides long. Unlike the mammalian SINES, 101F is about 4 kbp in length.

Several observations are consistent with the possibility that SINES are mobile in mammalian genomes, at least in evolutionary time. First, there is no counterpart in the human gene to the SINE within the intervening sequence of the rat growth hormone gene, although the genes themselves are likely to be homologous and the introns similarly placed. Thus, assuming a common ancestral sequence, the SINE was either lost during evolution of the human gene or acquired during evolution of the rat gene. Second, the very different distributions of SINES in the clusters of globin genes (comparing rodents with primates or the α- and β-clusters of humans) are consistent with mobility. Third, *Alu* sequences are found within African green monkey α-satellite segments (Grimaldi *et al.*, 1981). Because the *Alu* sequence is highly conserved and satellite was amplified after separation of the monkey and human evolutionary lines, the *Alu* sequence was probably acquired after amplification of the satellite. Finally, the direct repeats surrounding mammalian SINE segments (Fig. 10) are reminiscent of the target site duplications that flank mobile elements. Very recently the primary nucleotide sequence of an African green monkey α-satellite segment that is interrupted by an *Alu* family member was determined. A 13-bp-long repeat surrounds the *Alu* and this is a duplication of the α-satellite sequence at the point of interruption (Grimaldi and Singer, 1982).

Calos and Miller (1980) give an excellent summary of the possible significance of mobile elements in eukaryotes, emphasizing their potential for the generation of genetic diversity. The following additional point may be important, in view of the suggestions that highly repeated sequences have no function at all. A mobile element may generate diversity with a potential for selective advantage, but it can also generate disadvantage if it moves into an essential gene. Mutation by movable elements has been demonstrated in yeast (Roeder and Fink, 1980) and *Drosophila* (reviewed in Spradling and Rubin, 1981). The high frequency of mutation caused by the presence of large numbers of movable elements within a mammalian genome might have proven intolerable and been selected against, unless it was counterbalanced by some positive functional advantage.

Finally, the suggestion (Jelinek *et al.*, 1980; Georgiev *et al.*, 1981) that SINES may serve as origins for DNA replication should be considered. The basis for the suggestion is the presence in SINES of a short (14 bp) homology to a sequence associated with the origin of replication of murine and primate papovaviruses. Georgiev *et al.* (1981) describe some preliminary experiments that are consistent with this suggestion. However, in papovavirus genomes this region is part of a complex control region and may be involved in the control of transcription as well as replication. Only additional experiments will resolve these questions.

C. Long Interspersed Repeated Sequences (LINES)

Dispersed repeated sequences several kilobase pairs in length (LINES) have been described in both rodents and primates. Several LINE families may eventually be found in each genome. As yet none is completely characterized.

1. *Primates*

Several groups of investigators appear to have discovered the same family of LINES, called here the *Kpn*-LINE family (Adams *et al.*, 1980; Kaufman *et al.*, 1980; Schmeckpeper *et al.*, 1981; Maio *et al.*, 1981b; Manuelidis, 1981b; Rogers, 1981; J. C. Rogers and C. Milliman, unpublished; G. Grimaldi and M. F. Singer, unpublished). A segment about 6.4 kbp long and estimated to be repeated 3 to 5 thousand times in the human genome was found 3 kbp downstream from the 3′ end of the human β-globin gene (Adams *et al.*, 1980; Kaufman *et al.*, 1980). This and several other somewhat divergent randomly selected members of the family were cloned from a human library. Portions of this LINE hybridize under stringent conditions with a cloned 2.8 kbp *Kpn*I fragment of African green monkey DNA; this fragment was found interrupting and abutting α-satellite and was independently identified as a portion of a monkey LINE found in thousands of copies in both human and monkey DNA (G. Grimaldi and M. F. Singer, unpublished). Other members of the monkey family diverge in such a manner that the sequences in the 2.8-kbp *Kpn*I fragment are divided into two *Kpn*I fragments, 1.2 and 1.5 kbp in length, respectively. The 2.8-kbp fragment hybridizes to both monkey and human genomic *Kpn*I fragments 1.2, 1.5 (1.6), 2.8, 3.4, and 4.6 kbp long; it also hybridizes to monkey and human *Hin*dIII segments 1.9 and 2.6 kbp long. One member of a set of abundant 1.9-kbp human *Hin*dIII fragments was cloned separately and shown to hybridize to *Kpn*I fragments of the same size (Rogers, 1980; J. C. Rogers and C. Milliman, unpublished). Independent work by Schmeckpeper *et al.* (1981) identified a LINE family that also hybridized with the same size classes of *Hin*dIII and *Kpn*I fragments from the human genome. The LINE family was located on both X and autosomal chromosomes. The *Kpn*-LINE family appears to be common among primates since digestion of total DNA from several species yields the distinctive set of *Kpn*I bands (Maio *et al.*, 1981b).

The dispersed organization of the *Kpn*I-LINE family is inferred from (1) its presence on human X and autosomal chromosomes (Schmeckpeper *et al.*, 1981; Manuelidis, 1981b), (2) its distribution in the phage of the monkey (G. Grimaldi and M. F. Singer, unpublished) and human (Adams *et al.*, 1980) libraries, and

(3) the absence of any ''ladders'' upon restriction endonuclease cleavage (all papers already mentioned).

Five percent of the bacteriophage comprising an African green monkey genomic library (McCutchan et al., 1981) hybridized with the cloned 2.8 KpnI fragment, consistent with about 12,000 dispersed copies of the sequence. Some estimates for the human genome (Maio et al., 1981b; Rogers, 1981) are similar, although these numbers are markedly higher than the 3–5 thousand copies estimated by Adams et al. (1980). J. C. Rogers and C. Milliman (unpublished) have obtained data suggesting that the copy number of the Kpn-LINE family may vary from individual to individual.

It is already clear that there is extensive divergence among the members of the Kpn-LINE family although the various members cross-hybridize. Both the cloned human segments (Adams et al., 1981) and cloned monkey segments (G. Grimaldi and M. F. Singer, unpublished) show restriction endonuclease site variability. Also, it is evident from digests of total genomic DNA hybridized to cloned probes that the abundance of different variants is very different in monkeys and in humans (G. Grimaldi and M. F. Singer, unpublished). Data suggesting that Kpn-LINE family members cross-hybridize with α-satellite (Maio et al., 1981b) are puzzling since other investigations give no indication of such a relation (McCutchan et al., 1981; G. Grimaldi and M. F. Singer, unpublished; Manuelidis, 1981b). It is probable that the uncloned α-satellite probes were contaminated with Kpn-LINES.

2. Rodents

Several groups of workers have obtained data on what appears to be a single abundant LINE family in mice, called here the EcoRI-LINE family. Between 1 and 3% of the Mus musculus genome is converted to a set of fragments 1.3 kbp in length upon digestion with EcoRI; this is equivalent to as many as 50,000 copies per haploid genome (Hörz et al., 1974; Cheng and Schildkraut, 1980; Heller and Arnheim, 1980; Manuelidis, 1980; Brown and Dover, 1981; Meunier-Rotival et al., 1981). The 1.3-kbp segment is frequently if not always contained within repeat units that are at least 5.6 kbp in length (Meunier-Rotival et al., 1981; Brown and Dover, 1981). Although the sequence itself is conserved among the members of the family, subgroups of family members show marked divergence in restriction endonuclease sites (Brown and Dover, 1981; Meunier-Rotival et al., 1981). EcoRI-LINES probably occur on many mouse chromosomes. Homologous families that largely retain the internal EcoRI segment have been detected in other species of Mus, in Apodemus genomes, and in rats and Chinese hamsters (Heller and Arnheim, 1980; Brown and Dover, 1981). The relative abundance of different variants of the EcoRI-LINE family, however, differs in most of these species.

Preliminary data on other possible families of LINES have been published. One, estimated to be repeated about 4×10^4 times, comprises part of the nontranscribed spacer region in the array of mouse genes for ribosomal RNA and also occurs flanking genes for the constant region of mouse heavy chain immunoglobulin (Arnheim *et al.*, 1980). This family is not homologous to the 1.3-kbp *Eco*RI segment of the *Eco*RI-LINE family. Two relatively abundant nonsatellite fragment sets 1.5 and 1.7 kbp long, respectively, and estimated to comprise together about 0.2% of the genome are produced by *Eco*RII cleavage (Manuelidis, 1980); neither one hybridizes with the other. Preliminary evidence indicates the presence of an abundant LINE family in the Kangaroo rat (Liu and Lark, 1981).

3. *Are LINES functional?*

The discovery of LINE families in mammals is recent and there is very little information available regarding function. Adams *et al.* (1980) found no transcripts homologous to the human *Kpn*-LINE family in bone marrow cells and Manuelidis (1981b) also reports negative preliminary experiments. There is no information available regarding the possibility that LINES are mobile in mammalian genomes.

E. AMPLIFICATION AND DISPERSION

SINE family members are dispersed between genes, in introns, and within satellite DNA sequences. The SINES are also highly conserved within genomes and between relatively closely related mammalian groups such as members of the genus *Mus* or old world monkeys and man. This conservation is in marked contrast to the very different, noncross-hybridizing (under stringent conditions) satellites of, for example, various old world monkeys and man. On the other hand, when the SINE families of primates and rodents are compared, major differences are apparent even though the repeat units have some striking homologies. There are markedly different multiplicities of different SINE families in different genomes. Further, the B1-SINES of the mouse are, for example, about 130bp in length while the *Alu*-SINES of primates are more than twice that length. At some time after the rodent and primate evolutionary lines separated, one or the other or both lines acquired distinct families of SINES. The mechanism by which the acquisition occurred necessarily involved amplification into many dispersed positions. Providing experimental data relevant to an understanding of that mechanism(s) is a major challenge for future research. The conceptual problems have already been discussed (Scherer and Davis, 1980; Pan *et al.*, 1981; Dover, 1981; Baltimore, 1981; Grimaldi *et al.*, 1981). Unequal crossing-over, which can account for the amplification of satellite DNA (Section III,J), is less satisfactory with regard to amplification and dispersal of SINES.

The process called gene conversion may provide a better model. Gene conversion is a nonreciprocal recombination; a DNA sequence is duplicated at a (partly) homologous distinct genomic site without being lost from the original "donor" site. Both intrachromosomal (Scherer and Davis, 1980) and interchromosomal (Jackson and Fink, 1981; Klein and Petes, 1981) gene conversion have been demonstrated in yeast. Multiple gene conversions might account for the mass change from one SINE family to another partly homologous SINE family as well as the maintenance of homogeneity in an established SINE family. The different relative reiteration frequencies of SINE families might reflect differential rates of gene conversion in different species, but might also reflect more fundamental distinctions between genomes. However, by itself gene conversion does not easily explain the initial or any continuing dispersion of SINE family members into the genome. Dispersion could be explained if SINES prove to be mobile (or transposable) elements, but the mechanism of transposition would have to be such that the donor sequence remains in its original site in addition to being duplicated at a new one (as in gene conversion). The transposable elements of prokaryotes appear to act in just such a way (reviewed by Calos and Miller, 1980).

Similar considerations apply to the amplification and dispersion of LINES. Also, it will be important to understand the processes that maintain LINES in smaller numbers of copies than SINES. Recent data suggest that LINES may be associated with major components of mouse DNA (nonsatellite) that are separable by density gradient centrifugation (Soriano *et al.*, 1981; Meunier-Rotival *et al.*, 1981). Perhaps, unlike SINES, LINES are restricted to particular genomic regions. Second, the extensive polymorphism of LINES within a given species is interesting. Does it reflect a fluid, continually changing population or is it a fixed distribution? And what, if any, are the functional consequences of the polymorphism?

V. Concluding Remarks

The precise insights into the molecular structure and organization of highly repeated sequences that are afforded by modern techniques are rewarding but frustrating. This is especially so in mammals compared, for example, to *Drosophila* (Spradling and Rubin, 1981), because of the difficulties in wedding genetics or cytogenetics with molecular analysis. New experimental approaches to the question of function are essential. One such approach, the search for proteins that specifically bind repeated sequences, has been urged by Brutlag. A protein that interacts specifically with one satellite from *Drosophila melanogaster* has been described (Hsieh and Brutlag, 1979b), but unfortunately this approach has not yet been applied to other organisms. Recent progress in elucidating the structure of mammalian kinetochores provides interesting possibilities for studying proteins that might bind to satellites (Ris and Witt, 1981; Brenner *et al.*, 1981). Another avenue yet to be fully exploited involves analysis of the effects of repeated sequences on replication and transcription of small, constructed genomes. Genomes for use in mammalian cells have been designed (Hamer, 1980; Mulligan and Berg, 1980) and recombinant DNA techniques permit the introduction of repeated sequences. It is to be hoped that an emphasis on function will develop in the near future.

ACKNOWLEDGMENTS

I am grateful to the following investigators who provided me with preprints: Molly Fitzgerald-Hayes, John Carbon, Joe Maio, Giorgio Bernardi, Prescott Deininger, K. G. Lark, Carl Schmid, Laura Manuelidis, Hans Zachau, A. Plucienniczak, John Rogers, Rolf Streeck. My colleagues Antonella Maresca, Giovanna Grimaldi, Theresa N. H. Lee, and Ronald E. Thayer, kindly and critically reviewed the manuscript.

REFERENCES

Adams, J. W., Kaufman, R. E., Kretschmer, P. J., Harrison, M., and Nienhuis, A. W. (1980). *Nucleic Acids Res.* **8,** 6113–6128.

Alt, F. W., Kellems, R. D., Bertino, J. R., and Schimke, R. T. (1978). *J. Biol. Chem.* **253,** 1357–1370.

Alwine, J. C., Kemp, D. J., Parker, B. A., Reiser, J., Renart, J., Stark, G. R., and Wahl, G. M. (1979). *Methods Enzymol.* **68,** 220–242.

Appels, R., and Peacock, W. J. (1978). *Int. Rev. Cytol. Suppl.* **8,** 69–126.

Arnheim, N., Seperack, P., Banerji, J., Lang, R. B., Miesfeld, R., and Marcu, K. B. (1980). *Cell* **22,** 179–185.

Baltimore, D. (1981). *Cell* **24,** 592–594.

Baralle, F. E., Shoulders, C. C., Goodbourn, S., Jeffreys, A., and Proudfoot, N. (1980). *Nucleic Acids Res.* **8,** 4393–4404.

Barrie, P. A., Jeffreys, A. J., and Scott, A. F. (1981). *J. Mol. Biol.* **149,** 319–336.

Beauchamp, R. S., Mitchell, A. R., Buckland, R. A., and Bostock, C. J. (1979). *Chromosoma* **71,** 153–166.

Bedbrook, J. R., and Gerlach, W. L. (1980). *In* "Genetic Engineering" (J. K. Setlow and A. Hollaender, eds.), Vol. 2, pp. 1–19. Plenum, New York.

Bell, G. I., Pictet, R., and Rutter, W. J. (1980). *Nucleic Acids Res.* **8,** 4091–4109.

Bonner, J., Garrard, W. T., Gottesfeld, J., Holmes, D. S., Sevall, J. S., and Wilkes, M. (1973). *Cold Spring Harbor Symp. Quant. Biol.* **38,** 303–310.

Bostock, C. J. (1980). *Trends Biochem. Sci.* **5,** 117–119.

Bostock, C. J., and Clark, E. M. (1980). *Cell* **19,** 709–715.

Bostock, C. J., Gosden, J. R., and Mitchell, A. R. (1978). *Nature (London)* **272,** 324–328.

Botchan, M. R. (1974). *Nature (London)* **251,** 288–292.

Botchan, M., Topp, W., and Sambrook, J. (1978). *Cold Spring Harbor Symp. Quant. Biol.* **43,** 709–719.

Brenner, S., Pepper, D., Berns, M. W., Tan, E., and Brinkley, B. R. (1981). *J. Cell Biol.* **91**, 95–102.

Britten, R. J., and Davidson, E. H. (1969). *Science* **165**, 349.

Britten, R. J., and Kohne, D. E. (1968). *Science* **161**, 529–540.

Britten, R. J., Graham, D. E., and Neufeld, B. R. (1974). *Methods Enzymol.* **29**, 363–405.

Brown, S. D. M., and Dover, G. (1979). *Nucleic Acids Res.* **6**, 2423–2434.

Brown, S. D. M., and Dover, G. (1980a). *Nucleic Acids Res.* **8**, 781–792.

Brown, S. D. M., and Dover, G. (1980b). *Nature (London)* **285**, 47–49.

Brown, S. D. M., and Dover, G. (1981). *J. Mol. Biol.* **150**, 441–466.

Brutlag, D. L. (1980). *Annu. Rev. Genet.* **14**, 121–144.

Brutlag, D. L., Fry, K., Nelson, T., and Hung, P. (1977). *Cell* **10**, 509–519.

Calos, M. P., and Miller, J. H. (1980). *Cell* **20**, 579–595.

Carlson, M., and Brutlag, D. L. (1977). *Cell* **11**, 371–381.

Cavalier-Smith, T. (1980). *Nature (London)* **285**, 617–618.

Cheng, S.-M., and Schildkraut, C. L. (1980). *Nucleic Acids Res.* **8**, 4075–4090.

Christie, N. T., and Skinner, D. M. (1980). *Proc. Natl. Acad. Sci. U.S.A.* **77**, 2786–2790.

Clarke, L., and Carbon, J. (1980a). *Nature (London)* **287**, 504–509.

Clarke, L., and Carbon, J. (1980b). *Proc. Natl. Acad. Sci. U.S.A.* **77**, 2173–2177.

Clarke, L., Fitzgerald-Hayes, M., Buhler, J.-M., and Carbon, J. (1981). Stadler Symposium, Vol. 13. Univ. of Missouri, Columbia, Missouri, in press.

Coggins, L. W., Grindlay, G. J., Vass, J. K., Slater, A. A., Montague, P., Stinson, M. A., and Paul, J. (1980). *Nucleic Acids Res.* **8**, 3319–3333.

Cooke, H. J., and Hindley, J. (1979). *Nucleic Acids Res.* **6**, 3177–3197.

Cooke, H. J., and McKay, R. D. G. (1978). *Cell* **13**, 453–460.

Corneo, G., Ginelli, E., and Polli, E. (1968). *J. Mol. Biol.* **33**, 331–335.

Corneo, G., Ginelli, E., and Polli, E. (1970). *J. Mol. Biol.* **48**, 319–327.

Corneo, G., Ginelli, E., and Polli, E. (1971). *Biochim. Biophys. Acta* **247**, 528–534.

Corneo, G., Nelli, L. C., Meazza, D., and Ginelli, E. (1980). *Biochim. Biophys. Acta* **607**, 438–444.

Dalla-Favera, R., Gelmann, E. P., Gallo, R. C., and Wong-Staal, F. (1981). *Nature (London)* **292**, 31–35.

Davidson, E. H., and Britten, R. J. (1979). *Science* **204**, 1052–1059.

Davidson, E. H., Hough, B. R., Amenson, C. S., and Britten, R. J. (1973). *J. Mol. Biol.* **77**, 1–23.

Davies, K. E., Young, B. D., Elles, R. G., Hill, M. E., and Williamson, R. (1981). *Nature (London)* **293**, 374–376.

Dawid, I. B., Long, E. O., DiNocera, P. P., and Pardue, M. L. (1981). *Cell* **25**, 399–408.

Deininger, P., and Schmid, C. W. (1976a). *J. Mol. Biol.* **106**, 773–790.

Deininger, P. L., and Schmid, C. W. (1976b). *Science* **194**, 846–848.

Deininger, P. L., Jolly, D. J., Rubin, C. M., Friedmann, T., and Schmid, C. W. (1981). *J. Mol. Biol.* **151**, 17–33.

Dennis, E. S., Dunsmuir, P., and Peacock, W. J. (1980). *Chromosoma* **79**, 179–198.

Dhruva, B. R., Shenk, T., and Subramanian, K. N. (1980). *Proc. Natl. Acad. Sci. U.S.A.* **77**, 4514–4518.

Diaz, M. O., Barsacchi-Pilone, G., Mahon, K. A., and Gall, J. G. (1981). *Cell* **24**, 649–659.

Dolnick, B. J., Berenson, R. J., Bertino, J. R., Kaufman, R. J., Nunberg, J. K., and Schimke, R. T. (1979). *J. Cell Biol.* **83**, 395–402.

Donehower, L., and Gillespie, D. (1979). *J. Mol. Biol.* **134**, 805–834.

Donehower, L., Furlong, C., Gillespie, D., and Kurnit, D. (1980). *Proc. Natl. Acad. Sci. U.S.A.* **77**, 2129–2133.

Doolittle, W. F., and Sapienza, C. (1980). *Nature (London)* **284**, 601–603.

Dover, G. (1978). *Nature (London)* **272**, 123–124.

Dover, G. (1980). *Nature (London)* **285**, 618–620.

Dover, G. (1981). *In* "Mechanisms of Speciation" (C. Barigozzi, G. Montalenti, and M. J. D. White, eds.), Rome, in press.

Dover, G., and Doolittle, W. F. (1980). *Nature (London)* **288**, 646–647.

Duhamel-Maestracci, N., Simard, R., Harbers, K., and Spencer, J. H. (1979). *Chromosoma* **75**, 63–74.

Duncan, C., Biro, P. A., Choudary, P. U., Elder, J. T., Wang, R. R. C., Forget, B. G., de Riet, J. K., and Weissman, S. M. (1979). *Proc. Natl. Acad. Sci. U.S.A.* **76**, 5095–5099.

Duncan, C. M., Jagadeeswaran, P., Wang, R. R. C., and Weissman, S. (1981). *Gene* **13**, 185–196.

Dunsmuir, P. (1976). *Chromosoma* **56**, 111–125.

Elder, J. T., Pan, J., Duncan, C. H., and Weissman, S. M. (1981). *Nucleic Acids Res.* **9**, 1171–1189.

Federoff, N., Wellauer, P. K., and Wall, R. (1977). *Cell* **10**, 597–610.

Fittler, F. (1977). *Eur. J. Biochem.* **74**, 343–352.

Fitzgerald-Hayes, M., Buhler, J.-M., Cooper, T. G., and Carbon, J. (1982). *Mol. Cell. Biol.* **2**, 82–87.

Flavell, R. (1980). *Annu. Rev. Plant Physiol.* **31**, 569–596.

Fowlkes, D. M., and Shenk, T. (1980). *Cell* **22**, 405–413.

Fritsch, E. F., Lawn, R. M., and Maniatis, T. (1980). *Cell* **19**, 959–972.

Fritsch, E. F., Shen, C. K. J., Lawn, R. M., and Maniatis, T. (1981). *Cold Spring Harbor Symp. Quant. Biol.* **45**, 761–775.

Fry, K., and Salser, W. (1977). *Cell* **12**, 1069–1084.

Fuke, M., and Busch, H. (1979). *FEBS Lett.* **99**, 136–140.

Fuke, M., Davis, F. M., and Busch, H. (1979). *FEBS Lett.* **102**, 46–50.

Gall, J. G., Stephenson, E. C., Erba, H. P., Diaz, M. O., and Barsacchi-Pilone, G. (1981). *Chromosoma* **84**, 159–171.

Gaillard, G., Doly, J., Cortadas, J., and Bernardi, G. (1981). *Nucleic Acids Res.* **9**, 6069–6082.

Georgiev, G. P., Ilyin, Y. V., Chmeliauskaite, V. G., Ryskov, A. P., Kramerov, D. A., Skryabin, K. G., Krayev, A. S., Lukanidin, E. M., and Grigoryan, M. S. (1981). *Cold Spring Harbor Symp. Quant. Biol.* **45**, 641–654.

Gillespie, D. (1977). *Science* **196**, 889–891.

Gillespie, D., Pequignot, E., and Strayer, D. (1980). *Gene* **12**, 103–111.

Graf, H. (1979). *Hoppe-Seyler Z. Physiol. Chem.* **360**, 1029.

Graf, H., Fittler, F., and Zachau, H. G. (1979). *Gene* **5**, 93–110.

Grimaldi, G., and Singer, M. F. (1982). *Proc. Natl. Acad. Sci. U.S.A.,* in press.

Grimaldi, G., Queen, C., and Singer, M. F. (1981). *Nucleic Acids Res.* **9**, 5553–5568.

Haigwood, N. L., John, C. L., Hutchison, C. A., III, and Edgell, M. H. (1981). *Nucleic Acids Res.* **9**, 1133–1150.

Hamer, D. (1980). *In* "Genetic Engineering" (J. K. Setlow and A. Hollaender, eds.), Vol. 2. Plenum, New York.

Harada, F., and Kato, N. (1980). *Nucleic Acids Res.* **8**, 1273–1285.

Haynes, S. R., Toomey, T. P., Leinwand, L., and Jelinek, W. R. (1981). *Mol. Cell. Biol.* **1**, 573–583.

Heller, R., and Arnheim, N. (1980). *Nucleic Acids Res.* **8**, 5031–5042.

Hilliker, A. J., Appels, R., and Schalet, A. (1980). *Cell* **21**, 607–619.

Hoeijmakers-van Dommelen, H. A. M., Grosveld, G. C., de Boer, E., and Flavell, R. A. (1980). *J. Mol. Biol.* **140**, 531–547.

Hörz, W., and Altenburger, W. (1981). *Nucleic Acids Res.* **9**, 683–696.

Hörz, W., and Zachau, H. G. (1977). *Eur. J. Biochem.* **73**, 383–392.

Hörz, W., Hess, I., and Zachau, H. G. (1974). *Eur. J. Biochem.* **45,** 501–510.

Houck, C. M., and Schmid, C. W. (1981). *J. Mol. Evol.* **17,** 148–155.

Houck, C. M., Rinehart, F. P., and Schmid, C. W. (1979). *J. Mol. Biol.* **132,** 289–306.

Hsiao, C.-L., and Carbon, J. (1981). *Proc. Natl. Acad. Sci. U.S.A.* **78,** 3760–3764.

Hsieh, T., and Brutlag, D. (1979a). *J. Mol. Biol.* **135,** 465–481.

Hsieh, T., and Brutlag, D. (1979b). *Proc. Natl. Acad. Sci. U.S.A.* **76,** 726–730.

Hubbell, H. R., Robberson, D. L., and Hsu, T. C. (1979). *Nucleic Acids Res.* **7,** 2439–2456.

Igo-Kemenes, T., Hörz, W., and Zachau, H. G. (1982). *Annu. Rev. Biochem.* **51,** in press.

Jackson, J. A., and Fink, G. R. (1981). *Nature (London)* **292,** 306–311.

Jahn, C. L., Hutchison, C. A., III, Phillips, S. J., Weaver, S., Haigwood, N. L., Voliva, C. F., and Edgell, M. H. (1980). *Cell* **21,** 159–168.

Jelinek, W. R. (1978). *Proc. Natl. Acad. Sci. U.S.A.* **75,** 2679–2683.

Jelinek, W. R., and Leinwand, L. (1978). *Cell* **15,** 205–214.

Jelinek, W. R., Evans, R., Wilson, M., Salditt-Georgieff, M., and Darnell, J. E. (1978). *Biochemistry* **17,** 2776–2783.

Jelinek, W. R., Toomey, T. P., Leinwand, L., Duncan, C. H., Biro, P. A., Choudary, P. V., Weissman, S. M., Rubin, C. M., Houck, C. M., Deininger, P. L., and Schmid, C. W. (1980). *Proc. Natl. Acad. Sci. U.S.A.* **77,** 1398–1402.

John, B., and Miklos, G. L. G. (1979). *Int. Rev. Cytol.* **58,** 1–114.

Kaput, J., and Sneider, T. W. (1979). *Nucleic Acids Res.* **7,** 2303–2322.

Kaufman, R. E., Kretschmer, P. J., Adams, J. W., Coon, H. C., Anderson, W. F., and Nienhuis, A. W. (1980). *Proc. Natl. Acad. Sci. U.S.A.* **77,** 4229–4233.

Kit, S. (1961). *J. Mol. Biol.* **3,** 711–716.

Klein, H. L., and Petes, T. D. (1981). *Nature (London)* **289,** 144–148.

Kopecka, H., Macaya, G., Cortadas, J., Thiery, J. P., and Bernardi, G. (1978). *Eur. J. Biochem.* **84,** 189–195.

Kornberg, R. (1981). *Nature (London)* **292,** 579–580.

Kramerov, D. A. Grigoryan, A. A., Ryskov, A. P., and Georgiev, G. P. (1979). *Nucleic Acids Res.* **6,** 679–713.

Krayev, A. S., Kremerov, D. A., Skryabin, K. G., Ryskov, A. P., Bayev, A. A., and Georgiev, G. P. (1980). *Nucleic Acids Res.* **8,** 1201–1215.

Kunkel, L. M., Smith, K. D., Boyer, S. H., Borgoankar, D. S., Wachtel, S. S., Miller, O. J., Breg, W. R., Jones, H. W., and Rary, J. M. (1977). *Proc. Natl. Acad. Sci. U.S.A.* **74,** 1245–1249.

Kunkel, L. M., Smith, K. D., and Boyer, S. H. (1979). *Biochemistry* **18,** 3343–3353.

Kurnit, D. M. (1979). *Hum. Genet.* **47,** 169–186.

Kurnit, D. M., and Maio, J. J. (1973). *Chromosoma* **42,** 23–36.

Kurnit, D. M., and Maio, J. J. (1974). *Chromosoma* **45,** 387–400.

Lapeyre, J. N., and Becker, F. F. (1980). *Biochim. Biophys. Acta* **607,** 23–35.

Lapeyre, J.-N., Beattie, W. G., Dugaiczyk, A., Vizard, D., and Becker, F. F. (1980). *Gene* **10,** 339–346.

Lauer, J., Shen, C-K. J., and Maniatis, T. (1980). *Cell* **20,** 119–130.

Lawn, R. M., Fritsch, E. F., Parker, R. C., Blake, G., and Maniatis, T. (1978). *Cell* **15,** 1157–1174.

Lerner, M. R., and Steitz, J. A. (1981). *Cell* **25,** 298–300.

Lipchitz, L., and Axel, R. (1976). *Cell* **9,** 355–364.

Liu, L.-S., and Lark, K. G. (1981). *Fed. Proc. Fed. Am. Soc. Exp. Biol.* **40,** 1648.

Long, E. O., and Dawid, I. B. (1980). *Annu. Rev. Biochem.* **49,** 727–764.

Macaya, G., Thiery, J.-P., and Bernardi, G. (1977). *In* "Molecular Structure of Human Chromosomes" (J. J. Yunis, ed.), pp. 35–58. Academic Press, New York.

Macaya, G., Cortadas, J., and Bernardi, G. (1978). *Eur. J. Biochem.* **84,** 179–188.

McCutchan, T., Hsu, H., Thayer, R. E., and Singer, M. F. (1982). *J. Mol. Biol.,* in press.
Maio, J. J. (1971). *J. Mol. Biol.* **56,** 579–595.
Maio, J. J., Brown, F. L., and Musich, P. R. (1977). *J. Mol. Biol.* **117,** 637–655.
Maio, J. J., Brown, F. L., and Musich, P. R. (1981a). *Chromosoma* **83,** 103–125.
Maio, J. J., Brown, F. L., McKenna, W. G., and Musich, P. R. (1981b). *Chromosoma* **83,** 127–144.
Maniatis, T., Hardison, R. C., Lacy, E., Lauer, J., O'Connell, C., Quon, D., Sim, G. K., and Efstratiadis, A. (1978). *Cell* **15,** 687–701.
Manning, J., Schmid, C., and Davidson, N. (1975). *Cell* **4,** 141–155.
Manuelidis, L. (1976). *Nucleic Acids Res.* **3,** 3063–3075.
Manuelidis, L. (1978) *Chromosoma* **66,** 1–21.
Manuelidis, L. (1980). *Nucleic Acids Res.* **8,** 3247–3258.
Manuelidis, L. (1981a). *FEBS Lett.* **129,** 25–28.
Manuelidis, L. (1981b). *In* ''Genome Evolution and Phenotypic Variation'' (G. A. Dover and R. B. Flavell, eds.), Academic Press, in press.
Manuelidis, L., and Wu, J. C. (1978). *Nature (London)* **276,** 92–94.
Marx, K. A., Allen, J. R., and Hearst, J. E. (1976). *Biochim. Biophys. Acta* **425,** 129–147.
Marx, K. A., Purdom, I. F., and Jones, K. W. (1979). *Chromosoma* **73,** 153–161.
Matthews, H. R., Pearson, M. D., and MacLean, N. (1980). *Biochim. Biophys. Acta* **606,** 228–235.
Maxam, A., and Gilbert, W. (1980). *Methods Enzymol.* **65,** 499–560.
Mazrimas, J. A., and Hatch, F. T. (1977). *Nucleic Acids Res.* **4,** 3215–3227.
Meunier-Rotival, M., Soriano, P., Cuny, G., Strauss, F., and Bernardi, G. (1982). *Proc. Natl. Acad. Sci. U.S.A.* **79,** 355–359.
Miklos, G. L. G., and John, B. (1979). *Am. J. Hum. Genet.* **31,** 264–280.
Miklos, G. L. G., Willcocks, D. A., and Baverstock, P. R. (1980). *Chromosoma (Berlin)* **76,** 339–363.
Mitchell, A. R., Beauchamp, R. S., and Bostock, C. J. (1979). *J. Mol. Biol.* **135,** 127–149.
Mitchell, A. R., Gosden, J. R., and Ryder, O. A. (1981). *Nucleic Acids Res.* **9,** 3235–3249.
Moore, G. P., Constantini, F. D., Posakony, J. W., Davidson, E. H., and Britten, R. J. (1980). *Science* **208,** 1046–1048.
Morrow, J. F. (1979). *Methods Enzymol.* **68,** 3–24.
Mulligan, R., and Berg, P. (1980). *Science* **209,** 1423–1427.
Musich, P. R., Brown, F. L., and Maio, J. J. (1980). *Chromosoma* **80,** 331–348.
Musti, A. M., Sobieski, D., Chen, B. B., and Eden, F. (1981). *Biochemistry* **20,** 2989–2999.
Orgel, L. E., and Crick, F. H. C. (1980). *Nature (London)* **284,** 604–607.
Orgel, L. E., Crick, F. H. C., and Sapienza, C. (1980). *Nature (London)* **288,** 645–646.
Page, G. S., Smith, S., and Goodman, H. M. (1981). *Nucleic Acids Res.* **9,** 2087–2104.
Pan, J., Elder, J. T., Duncan, C., and Weissman, S. M. (1981). *Nucleic Acids Res.* **9,** 1151–1170.
Pardue, M. L., and Gall, J. G. (1970). *Science* **168,** 1356–1358.
Pathak, S., and Würster-Hill, D. H. (1977). *Cytogenet. Cell Genet.* **18,** 245–254.
Pech, M., Streeck, R. E., and Zachau, H. G. (1979a). *Cell* **18,** 883–893.
Pech, M., Igo-Kemenes, T., and Zachau, H. G. (1979b). *Nucleic Acids Res.* **7,** 417–432.
Petes, T. D. (1980). *Cell* **19,** 765–774.
Philippsen, P., Streeck, R. E., and Zachau, H. G. (1974). *Eur. J. Biochem.* **45,** 479–488.
Posakony, J. W., Scheller, R. H., Anderson, D. M., Britten, R. J., and Davidson, E. H. (1981). *J. Mol. Biol.* **149,** 41–67.
Pöschl, E., and Streeck, R. E. (1980). *J. Mol. Biol.* **143,** 147–153.
Queen, C., and Korn, L. (1980). *Methods Enzymol.* **65,** 595–609.
Rinehart, F. P., Ritch, T. G., Deininger, P. L., and Schmid, C. W. (1981). *Biochemistry* **20,** 3003–3010.
Ris, H., and Witt, P. L. (1981). *Chromosoma* **82,** 153–170.

Roberts, R. J. (1980). *Methods Enzymol.* **65**, 1–15.
Roeder, G. S., and Fink, G. R. (1980). *Cell* **21**, 239–249.
Rogers, J. C. (1981). *Fed. Proc. Fed. Am. Soc. Exp. Biol.* **40**, 1649.
Roizes, G. P. (1974). *Nucleic Acids Res.* **1**, 1099–1120.
Roizes, G. P. (1976). *Nucleic Acids Res.* **3**, 2677–2696.
Roizes, G. P., Pages, M., and Lecou, C. (1980). *Nucleic Acids Res.* **8**, 3779–3792.
Rosenberg, H., Singer, M. F., and Rosenberg, M. (1978). *Science* **200**, 394–402.
Rubin, C. M., Deininger, P. L., Houck, C. M., and Schmid, C. W. (1980a). *J. Mol. Biol.* **136**, 151–167.
Rubin, C. M., Houck, C. M., Deininger, P. L., Friedmann, T., and Schmid, C. W. (1980b). *Nature (London)* **284**, 372–374.
Sadler, J. R., Tecklenburg, M., and Betz, J. L. (1980). *Gene* **8**, 279–300.
Sakano, H., Maki, R., Kurosawa, Y., Roeder, W., and Tonegawa, S. (1980). *Nature (London)* **286**, 676–683.
Salser, W., Bowen, S., Browne, D., El Adli, F., Federoff, N., Fry, K., Heindell, H., Paddock, G., Poon, R., Wallace, B., and Whitcomb, P. (1976). *Fed. Proc. Fed. Am. Soc. Exp. Biol.* **35**, 23–35.
Scherer, S., and Davis, R. W. (1980). *Science* **209**, 1380–1384.
Schimke, R. T., Brown, P. C., Kaufman, R. J., McGrogan, M., and Slate, D. L. (1980). *Cold Spring Harbor Symp. Quant. Biol.* **45**, 785–797.
Schmeckpeper, B. J., Willard, H. F., and Smith, K. D. (1981). *Nucleic Acids Res.* **9**, 1853–1872.
Schmid, C. W., and Deininger, P. L. (1975). *Cell* **6**, 345–358.
Sealy, L., Hartley, J., Donelson, J., Chalkley, R., Hutchison, N., and Hamkalo, B. (1981). *J. Mol. Biol.* **145**, 291–318.
Segal, S., Garner, M., Singer, M. F., and Rosenberg, M. (1976). *Cell* **9**, 247–257.
Sen, S., and Sharma, J. (1980). *Chromosoma* **81**, 393–402.
Shen, C.-K. J., and Maniatis, T. (1980). *Cell* **19**, 379–391.
Shih, C., Padhy, L. C., Murray, M. J., and Weinberg, R. A. (1981). *Nature (London)* **290**, 261–264.
Singer, D. S. (1979). *J. Biol. Chem.* **254**, 5506–5514.
Singer, D. S., and Donehower, L. (1979). *J. Mol. Biol.* **134**, 835–842.
Smith, A. J. H. (1980). *Methods Enzymol.* **65**, 560–580.
Smith, G. P. (1976). *Science* **191**, 528–535.
Smith, T. F. (1980). *Nature (London)* **285**, 620.
Soriano, P., Macaya, G., and Bernardi, G. (1981). *Eur. J. Biochem.* **115**, 235–239.
Southern, E. M. (1970). *Nature (London)* **227**, 794–798.
Southern, E. M. (1975a). *J. Mol. Biol.* **94**, 51–69.
Southern, E. M. (1975b). *J. Mol. Biol.* **98**, 503–517.
Southern, E. M. (1979). *Methods Enzymol.* **68**, 152–176.
Spradling, A. C., and Rubin, G. M. (1981). *Annu. Rev. Genet.* **15**, 219–264.
Stambrook, P. J. (1981). *Biochemistry* **20**, 4393–4398.
Streeck, R. E. (1981). *Science* **213**, 443–445.
Streeck, R. E., and Zachau, H. G. (1978). *Eur. J. Biochem.* **89**, 267–279.
Streeck, R. E., Pech, M., and Zachau, H. G. (1979). *In* "FEBS Special Meeting on Enzymes" (P. Mildner, ed.). Pergamon, Oxford.
Sturm, K. S., and Taylor, J. H. (1981). *Nucleic Acids Res.* **9**, 4537–4546.
Szostak, J. W., and Wu, R. (1980). *Nature (London)* **284**, 426–430.
Szybalski, W. (1968). *Methods Enzymol.* **12B**, 330–360.
Tartof, K. D. (1975). *Annu. Rev. Genet.* **9**, 355–385.
Tashima, M., Calabretta, B., Tovelli, G., Scofield, M., Maizel, A., and Saunders, G. F. (1981). *Proc. Natl. Acad. Sci. U.S.A.* **78**, 1508–1512.
Temin, H. M. (1980). *Cell* **21**, 599–600.

Thayer, R. E., McCutchan, T., and Singer, M. F. (1981). *Nucleic Acids Res.* **9,** 169–181.
Varley, J. M., MacGregor, H. C., and Erba, H. P. (1980a). *Nature (London)* **283,** 686–688.
Varley, J. M., MacGregor, H. C., Nardi, I., Andrews, C., and Erba, H. P. (1980b). *Chromosoma* **80,** 289–307.
Wahl, G. M., Padgett, R. A., and Stark, G. R. (1979). *J. Biol. Chem.* **254,** 8679–8689.
Weiner, A. M. (1980). *Cell* **22,** 209–218.
Wensink, P. C., Tabata, S., and Pachl, C. (1979). *Cell* **18,** 1231–1246.
Wu, J. C., and Manuelidis, L. (1980). *J. Mol. Biol.* **142,** 363–386.
Zachau, H. G., and Igo-Kemenes, T. (1981). *Cell* **24,** 597–598.
Zieve, G. W. (1981). *Cell* **25,** 296–297.

NOTE ADDED IN PROOF

The mouse interspersed repeated family described by Arnheim *et al.* (1980) is probably a SINE family, not a LINE family [Miesfeld, R., Krystal, M., and Arnheim, N. (1981). *Nucleic Acids Res.* **9,** 5931–5947].

INTERNATIONAL REVIEW OF CYTOLOGY, VOL. 76

Moderately Repetitive DNA in Evolution

ROBERT A. BOUCHARD

Department of Biology, B-022, University of California, San Diego, La Jolla, California

I. Introduction

It has long been known that the genomes of eukaryotes are characterized by a tremendous range of nuclear DNA content even among closely related members of phyletic lineages (cf. Rees and Jones, 1972; Hinegardner, 1976; Bennett and Smith, 1976). A substantial amount of this variation is attributable to various types of sequences which can be shown to be present in the genome in a number of copies of more or less similar sequence (Britten and Kohne, 1968; Lewin, 1974). Such repeated sequence components include, usually as a minor component, those genes (rRNA, 5 S RNA, tRNAs, histones, etc.) known to be present in multiple copies, but are generally dominated by repeated segments without such readily assignable functions. These are the simple-sequence or satellite DNAs, and the moderately repetitive DNAs which form the subject of this article.

The definition of the "moderately repetitive" component of genomic DNA is not entirely straightforward. The range of repetition frequencies in repeated classes which have been termed "moderate"- or "middle"-repetitive in different studies of different genomes is quite impressive, ranging from 10^1 to over 10^5 copies per haploid genome (Davidson and Britten, 1973); for extremes, compare the human *Alu* family at $2-3 \times 10^5$ copies (Houck *et al.*, 1979) with *Caenorhabditis* repeats averaging less than 10 (Emmons *et al.*, 1980). Even in a single genome components with distinctly different repeat abundances or even a continuum of abundances may be assigned to this class. Rather than any particular level of repetition frequency, it is certain features of organization within the genome, distinct from those shown by simple-sequence repeated DNAs, which are generally employed to assign a particular class of repeats to the moderately repetitive DNA component. The more important and universal of these characteristics may be summarized as follows (Lewin, 1974):

1. Moderately repetitive sequences are generally found as dispersed segments among unrelated, often single-copy DNA sequences throughout the genome, while simple-sequence DNAs occur in long tandemly repeated clusters, frequently at the centromeric or telomeric regions of chromosomes.

2. Moderately repetitive sequences do not generally consist of highly similar derivatives of a very short cardinal base sequence, as do simple-sequence DNAs.

3. As a class, the moderately repeated component of a genome has about the same GC content as the mass of the genome, whereas simple-sequence components frequently deviate widely.

4. Sequences of the moderately repetitive class frequently exhibit a spectrum of relatedness to one another, while simple-sequence components are often highly uniform.

5. The predominant simple-sequence components of closely related genomes can be entirely different in sequence and amount, whereas the moderate repeats are generally more conserved.

6. Simple-sequence components are generally transcriptionally silent, while sequence elements of the moderately repetitive DNA are found to be transcribed in cell nuclei.

Taken together, these attributes convey the impression that the moderately repetitive sequences are an integrated and established component of eukaryote genomic DNA, especially when compared with the labile simple-sequence class. In a series of intellectually stimulating reviews, Britten and Davidson in particular have drawn on this sort of circumstantial evidence, together with their own group's detailed observations on the moderate-repeat components of particular organisms, to argue that the moderately repeated elements have a central role in the functioning of the eukaryote genome. Proposed roles have included the regulation of transcription (Britten and Davidson, 1969; Davidson and Britten, 1973), production of evolutionary novelty (Britten and Davidson, 1971), the encoding of regions within large RNAs which allow them to function as regulatory activators (Davidson *et al.*, 1977), or the regulation of specific processing of large RNA transcripts (Davidson and Britten, 1979). Other proposals as to possible function have related the moderate repeat segments to the integration of chromosome structure or recognition (Zuckerkandl, 1976; Fedoroff, 1979).

Contrasting points of view have however been put forward and have recently provoked spirited discussions. Doolittle and Sapienza (1980) and Orgel and Crick (1980), in simultaneous essays, argued that much of the repeated DNA of eukaryotes might consist of what they termed "selfish" or "parasitic" sequences, segments analogous (or even homologous) to prokaryotic transposons, whose only function is to maintain and increase their numbers in the genome. Cavalier-Smith (1980) responded by advancing arguments based on his earlier discussions (see Cavalier-Smith, 1978) of non-sequence-specific selection for increases or decreases in DNA content as a modulator of nuclear and cellular size. Dover (1980) pointed out that evolutionary change in the number of repeated DNAs often appears to occur in a fashion "ignorant" of the sequence content of the segments involved, and suggested that such events may be an unavoidable byproduct of the complexities of replication and recombination in eukaryotic chromatin. Several of these disputants have since spoken in the same forum (Orgel *et al.*, 1980; Dover and Doolittle, 1980) to point out that these various postulates are not mutually exclusive, and that the main difference of opinion may lie in the importance each is believed to have in evolution. In the same discussion, Jain (1980) put forward as another postulate the idea that the continual generation of repeated sequence families derived from randomly chosen segments could be an incidental but not accidental product of a particular sort of mutational mechanism preserved in eukaryote evolution because it periodically generates selectively beneficial variability.

The reasoning of all of these authors is of course both less distinct and extreme and more elaborate and sophisticated than it is possible to convey in the space available here. A point which can be made concerning these recent debates in particular, however, is that the discussions of the various models do not always clearly differentiate between observations made on simple-sequence components and those for moderate repeats. In addition, little distinction is made between the moderate repeat elements which show clearcut similarities to prokaryotic transposons found in some eukaryote genomes, and the predominant repeat classes of other eukaryotes, the repeat segments of which do not seem at all similar to known transposons.

In this respect, both functionalists and nonfunctionalists tend to lump observations, treating all eukaryote genomes, and sometimes even all repeats, as being in some sense equatable. This emphasis on unity and on grouping phenomena under unifying explanations is eminently scientific and has long proven its utility in molecular biology. Nevertheless, in dealing with objects as complex as some eukaryote genomes, or entities as diverse as those grouped together in the eukaryote kingdom, there is the danger of inadvertently stressing such facts as permit one to infer more unity than may actually exist.

In light of this, the present article is intended to survey the varieties of available information bearing on the evolution of moderately repetitive DNA in the genomes of representatives of a broad spectrum of eukaryote groups. In surveying these data, I have tried to avoid fitting data to any model, attempting rather to establish an ongoing dialectic between the accumulating observations on various genomes and some of the models or postulates of repeat origin, organization, and function which have been put forward to account for observations on particular genomes. The first section examines data on the organization of the moderately repetitive component in representatives of a spectrum of eukaryote lineages, paying particular attention to the differences seen in groups of relatively established phyletic affinity. This is followed by a treatment of information derived from studies involving more direct experimental comparisons within taxa and genomes as this bears on the evolution of moderately repetitive DNA components in different groups. In this way, group-specific or convergent peculiarities may be more readily identified, and more universal aspects should stand out. In the closing sections, key aspects of the various postulates concerning moderate-repeat evolution are assessed against the background provided by this information. Distinctions are drawn with respect to the application of different proposals to the question of evolutionary *origin* of moderate repeats versus evolutionary *accumulation* of these elements in genomes, and a tentative unified schema is developed. Brief consideration is also given to aspects of population genetics and speciation processes which may influence the evolution of moderately repetitive components.

II. Technical Background and Exemplars

A. COMMONLY USED TECHNIQUES AND TERMINOLOGY

In the majority of cases, mass investigations of the complexity, segment length, and interspersion of whole moderately repetitive classes have relied on optical monitoring of reassociation, on the ability of hydroxyapatite (HAP) columns to bind DNA fragments bearing a portion of reassociated duplex without regard to the size of any covalently linked single-stranded regions, and on the ability of S_1 nuclease, under properly controlled conditions, to digest away single-stranded regions while leaving reassociated stretches intact. Some studies have also been performed using electron microscopy to visualize reassociated regions. In the last several years, repeat segments have also been identified by the more generally familiar mocular cloning, Southern blotting, and DNA sequencing techniques. This section describes the approaches which can be made using the reassociation and sizing techniques and the types of information they have revealed in some prototype studies. Readers desiring further information are directed to the detailed methods paper of Britten *et al.* (1974) and the excellent brief treatment of reassociation reactions and kinetics found in Ham and Veomett (1980).

Estimates of moderate repeat abundance are generally derived by monitoring the rate of reassociation of an initially single-stranded preparation over time, measured either on HAP columns, by S_1 digestion, or by optical monitoring of the change in UV absorbance as reassociation proceeds. A brief nonmathematical discussion of the theory in such experiments will serve as a useful introduction to commonly encountered concepts and terms. At a given starting concentration (C_0) of single-stranded DNA, the rate of reassociation of any given sequence component over time (t) will occur with a rate dependent on the effective concentration of that component in the mixture. Since this reassociation is a function of encounters between two complementary strands moving freely in solution, it follows second order kinetics. Therefore, it is most convenient to plot the course of reassociation as the percentage or proportion of strands remaining single stranded on an arithmetic axis versus time normalized for starting concentration on a logarithmic scale. This parameter is C_0 times t or $C_0 t$. In cases where all sequence components are represented once in every genomic unit in the mixture, all will have the same effective concentration, and first reassociation events for fragments in the mixture will follow a single second-order curve which has a symmetrical S shape, according to the formula, $C/C_0 = 1/1 + kC_0 t$. The overall rate of this reassociation decreases in proportion to the total number of different sequence components, and can therefore be used as a measure of the number of such components, or *sequence complexity,* usually given in either nucleotide pairs (ntp) or daltons. This sequence complexity is inversely proportional to the $C_0 t_{\frac{1}{2}}$, or time required to reach the half point of reassociation at a given DNA concentration, and can be derived from the experimental $C_0 t_{\frac{1}{2}}$ by comparison to the $C_0 t_{\frac{1}{2}}$ of a DNA of known sequence complexity.

For the case of a DNA mixture derived from a genome containing some sequences which are repeated, that is present in multiple copies in each genomic unit, the repeated sequence components will have a higher effective concentration and therefore reassociate faster than those components present only once per genome (the *single copy*). From the total weight of DNA in a given genome (usually given in picograms, pg, 1 pg $= 10^{-12}$ gm) and the observed reassociation of genomes of known weight not containing repeats, it is possible to calculate the expected $C_0t_{\frac{1}{2}}$ for reassociation of the single copy; detection of significant component showing reassociation at a much lower $C_0t_{\frac{1}{2}}$ is diagnostic for the presence of repeated sequence components.

Consider the hypothetical case of a genome in which 30% of the weight of DNA present consists of sequences repeated 10^3 times per genome, while the remaining 70% is sequences present once per genome. Upon reassociation, 30% of the DNA would be seen to reanneal 1000 times faster than predicted from the weight of the genome, while the remaining 70% would follow the predicted single-copy kinetics. This genome would then be said to contain a *moderately repetitive* component comprising 30% of its weight, with an average sequence repetition frequency of 1000. The complexity of this genomic class could also be calculated from its rate of reassociation relative to known standards and would be a measure of the total number of distinct sequence entities which together comprise the component (see Fig. 1).

Such a measurement, however, only gives the frequency and number of sequences making up the moderately repetitive component; it provides no information as to the length of the repeated segments or their arrangement among other sequences in the DNA. The most precise mass measurement of this latter aspect of moderate-repeat organization is obtained by construction of a so-called "R versus L" curve (Davidson *et al.*, 1973). This approach requires a good prior knowledge of the range of C_0t values over which reassociation of the moderately repetitive component takes place in the genome of interest. The experimenter employs a series of radioactively labeled whole DNA probes of increasing lengths (L) which are reannealed through the completion of reassociation for the moderately repetitive class in the presence of a large excess of uniformly sheared relatively short cold driver DNA. The proportion of the probe scored by HAP as reassociated (R) at a C_0t where reannealing of the moderately repetitive component is complete is then scored for each length (L) of probe used, and the two are plotted against one another. To correct for the effects of foldback and very simple sequence DNA binding, a second set of determinations is done at a C_0t where the moderate repeats have not yet begun to reassociate, and the R values are normalized.

The rationale of this technique is reasonably self-evident. At short lengths, with random shearing, some probe fragments derived from the portion of the genome characterized by interspersed repeats will overlap the junction between a repeated segment and a nonrepeated one, and will thus be bound to HAP. Others will be derived entirely from the nonrepeated intervals and will not be bound, since no portion of their length reassociates with the repetitive portion of the driver. With increasing probe length, a larger and larger proportion of randomly

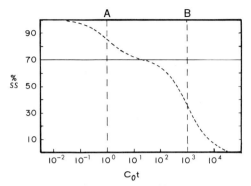

FIG. 1. Idealized C_0t curve for a genome containing a 30% moderate-repeat component comprised of sequence families repeated 1000 times relative to the 70% single-copy class. A and B mark the observed $C_0t_{\frac{1}{2}}$ points for these components in the whole DNA reassociation, which are $C_0t_{\frac{1}{2} A} = 1.0$ and $C_0t_{\frac{1}{2} B} = 1000$, respectively, for this example. The observed kinetics follow the general two-component compound equation

$$C/C_0 = P_A (1/1 + k_A C_0t)) + P_B (1/1 + k_B C_0t)$$

where P_A and P_B are the proportions of each of the components, and k_A and k_B are the observed rate constants for each in the mixture, equal to $1/C_0t_{\frac{1}{2}A}$ and $a/C_0t_{\frac{1}{2}B}$. For this example, then

$$C/C_0 = 0.3 (1/1 + 1/1.0 \ C_0t) + 0.7 (1/1 + 1/10^3 \ C_0t)$$

The actual kinetic complexities of the components are then proportional to the true $C_0t_{\frac{1}{2}}$s they would exhibit in isolation, given by:

$$C_0t_{\frac{1}{2}A'} = C_0t_{\frac{1}{2}A} \ P_A \text{ and } C_0t_{\frac{1}{2}B'} = C_0t_{\frac{1}{2}B}P_B$$

Here these would be 0.3 and 700. If a bacterial standard of known complexity 2×10^6 ntp were found to exhibit a $C_0t_{\frac{1}{2}}$ of 3.0 under the same conditions, this would indicate complexities of 2×10^5 ntp for the moderate-repetitive component and 4.67×10^8 ntp for the single copy component. If S_1 or other measurements for length of paired duplex show the average size of reassociated moderate-repeat segments is 400 ntp, then the approximate number of families is given by:

$$\text{number of families} = \text{complexity of class/repeat length}$$

$$= 2 \times 10^5 \text{ ntp}/4 \times 10^2 \text{ ntp} = 500 \text{ families.}$$

sheared molecules will contain part or all of a repeated segment. Eventually, a fragment size is reached which exceeds the length of the interspersed non-repeated segments, at which point all fragments from the interspersed portion of the genome are bound, and further length increases will result in no further increase in binding. The point of inflection of the graph of R (proportion bound) as a function of L (probe length) therefore represents the length of nonrepeated segments between adjacent interspersed repeats. Extrapolation to the Y axis gives the proportion of the genome which is truly composed of moderate repeats. In a more complex case, where a proportion of the genome is interspersed with a longer interval separating the repeated segments, the proportion of R will continue to rise after the inflection but with a lesser slope, until the average spacing of repeats in this second proportion is attained (see Fig. 2).

A less precise method can be employed when the objective is simply to identify genomes with short-term interspersion of short repeats (Goldberg *et al.*, 1975). In this case, C_0t curves are run using short (250–300 ntp) versus long (2000–3000 ntp) DNA fragments, and the proportion bound to HAP over the moderately repetitive kinetic range is determined for each shear size. With short-term or "*Xenopus*-type" interspersion, the proportion bound at the long fragment size will be much larger than that seen with short fragments, and the two proportions can be used to derive a reasonable estimate of the proportion of the genome characterized by such short-term interspersion.

The actual length of the interspersed repeats can be estimated in several ways. In aqueous solution, double-stranded DNA at a given concentration absorbs a fixed amount of ultraviolet light at 260 nm, while the same weight of DNA shows a 36% higher absorption after being rendered single-stranded at high temperature. This change in UV absorption is called the hyperchromicity and will be proportionally less for a solution initially containing a mixture of double- and single-stranded regions. Therefore, the hyperchromicity of an HAP-bound fraction, when compared with that observed for native versus entirely single-

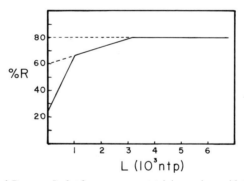

FIG. 2.   Hypothetical R versus L plot for a genome containing regions with two different spacings of repeat segments, and a third component of single copy sequence without interspersed repeats. The Y intercept of the plot indicates the proportion of the genome which would be bound due to moderate repeats alone, or the true proportion of moderate repeats, in this case 25%. (In an actual determination, this is of course arrived at by extrapolation to the R axis using the slope observed for the portion of the curve preceding the first inflection, since reassociation of fragments below length L of about 50–100 is experimentally impossible.) Position of the first inflection at $L = 1000$ indicates a genomic component comprised of single-copy regions of length 1000 ntp, flanked by moderate repeats, while extrapolation to the R axis shows that this component is 60% of the genome. The second inflection at $L = 3200$ ntp implies that another 20% of the genome consists of single-copy segments of this length flanked by moderate repeats. The remaining 20% of the genome is very long single-copy stretches uninterrupted by moderate repeats.

stranded DNA fragments of the same length, gives a measurement of the proportion actually present as reannealed duplex. A more direct measurement may be obtained by digesting a mixture of long fragments which has been allowed to reanneal through the moderately repetitive range with S_1 nuclease and sizing the resistant duplexes. These will represent the spectrum of full length repeat segments when the original fragments are much longer than the average length of interspersed repeat (Flavell and Smith, 1976) and may be sized in sedimentation gradients or on agarose bead columns (Davidson et al., 1974). Such measurements produce an estimate of the weight average repeat length. Electron microscopy has been employed in a few instances to visualize the proportion of duplexed versus unpaired region in formamide spreads (see below), producing a number average estimate of repeat length.

B. Exemplars of Genome Organization: "Short-Term" and "Long-Term" Interspersion

The earliest detailed measurements employing the above approaches were performed on the genomes of Xenopus (Davidson et al., 1973) and sea urchin (Graham et al., 1974). These revealed a rather similar pattern of moderate repeat organization for these animal genomes. C_0t curve analyses for each organism indicated the presence of substantial moderately repetitive DNA components, whose complexities indicated the presence of many distinct sequence families (800 for Xenopus, 5000 for sea urchin), with an average repetition frequency in the thousands of copies. Extrapolation of the R versus L curves to the Y axis implied that 25–31% of these genomes consisted of moderate repeats, while the first inflection of the plots indicated that a large proportion of their single-copy DNA existed in 700–1050 base pair segments among the interspersed repeats. Hyperchromicity and S_1 measurements confirmed by EM for Xenopus (Chamberlin et al., 1975) implied that the bulk of these repeats were 300–400 nucleotide pairs long. Thus a large proportion of each of these genomes is organized in alternating moderately repetitive and single-copy segments of short length, and this type of organization has been termed "Short-term" or "Xenopus-like" interspersion, a pattern which was quickly found to prevail in a number of evolutionarily diverse animals (Goldberg et al., 1975; Davidson et al., 1975a; Bonner et al., 1974). The widespread occurrence of a similar mode of sequence organization, together with the demonstration that in sea urchin a large proportion of the single copy sequences coding for embryo RNA are contiguous to interspersed repeats (Davidson et al., 1975b), were taken as strong circumstantial evidence of a central role for this organization in gene regulation.

The first organism convincingly shown to possess a different type of moderate-repeat organization was Drosophila, which was studied by Manning et al. (1975). A first indication of the different nature of Drosophila moderate repeats, previously noted by Laird and McCarthy (1969), is their much lower repetition frequency, averaging less than 100 copies per genome. When Manning *et al.* examined the proportion of the *Drosophila* genome bound to purified short (0.4 kb) moderate-repeat driver with DNA fragment populations of different lengths, ranging from 4.0 to 12.5 kb, they found only a 15% increase in binding due to moderate-repeat reassociation. Reassociation experiments were also performed with unfractionated DNA of 400 and 3500 nucleotide pairs. In contrast to the pattern observed in genomes with short-period interspersion, only a modest (from 12 to 15%) rise in the proportion bound due to reassociation of moderate repeats was observed, despite the order of magnitude range in fragment sizes employed. When the length of paired duplex was scored by electron microscopy of very long (4.2, 11.6, or 17.4 kb) fragments reassociated through the moderately repetitive range, the number average length of interspersed repeats was found to be 4.1 and 5.6 ntp in two different experiments. This measurement, in conjunction with the proportion binding to HAP at different fragment lengths, allowed Manning *et al.* to calculate that these long repeat segments were separated by single-copy regions in excess of 13,000 nucleotide pairs in length. Thus *Drosophila* possesses, as its predominant mode of interspersed sequence organization, an arrangement in which both repetitive and single-copy regions are more than an order of magnitude longer than seen in the *Xenopus*-like genomes. The existence of this apparently atypical pattern in *Drosophila* was subsequently confirmed by Davidson and Britten's group (Crain *et al.,* 1976a), using the same methods which had previously been employed for *Xenopus* and sea urchin. This mode of organization was soon christened ''long-term'' or ''*Drosophila*-like'' interspersion.

The existence of this different mode of organization for moderately repeated and single-copy segments in an organism widely employed in genetic and developmental studies posed a severe challenge to the models which had assumed the universality of the short-term interspersion mode (cf. Davidson and Britten, 1973; Laird *et al.,* 1974) and refocused interest on comparative studies of genome organization. In the following section are summaries of results which have since been obtained in a number of studies covering a great variety of eukaryote genomes and the implications these have for general models of genome organization.

III. Patterns of Moderate-Repeat Organization in the Evolutionary Spectrum

The early discovery of two apparently distinct modes of moderately repetitive sequence organization has had a substantial influence on the attitude of subsequent workers analyzing the organization of various eukaryote genomes. Results on particular genomes have generally been referred to this apparent dichotomy, and their organization classified as being characterized overall by either short- or long-term interspersion. In the following subsections, we will examine representative studies now available for a spectrum of eukaryote genomes in the context of this convenient paradigm and assess its adequacy as a system for classifying the actual diversity of situations encountered in nature. Where enough information exists for representatives of lineages whose phylogenetic affinities are considered well understood, these observations are compared. This approach allows some assessment to be made as to the ease and frequency of evolutionary transitions between one mode of genomic organization and another.

A. CHORDATA

1. *Mammalia*

With the possible exception of the Syrian hamster (Moyzis *et al.*, 1977), the mammalian genomes which have been studied appear to have short-term interspersion. Close examination of the nature of the interspersed repeats, however, has recently revealed some novel features for several mammals. An apparently typical short-term S_1 nuclease definable 300 ntp repeat class was shown by Schmid and Deiniger (1975) to be interspersed in the human genome, as determined by the usual mass techniques. This "class," however, has more recently been demonstrated to consist largely of a single repeated family of some 250,000–300,000 members. This repeated family has been intensively studied using current techniques of sequencing analysis and molecular cloning. It has been dubbed the *Alu* family because of the characteristic internal *Alu*I restriction site (Houck *et al.*, 1979) seen in the major portion of bulk S_1 defined 300 ntp repeat DNA. The predominance of this family is further attested by the fact that member sequences comprise the large majority of random clones from the 300 ntp S_1 class which are not attributable to satellite sequence (Deininger *et al.*, 1981). The member sequences are 15–20% diverged from one another (Houck *et al.*, 1978; Deininger and Schmid, 1979; Rubin *et al.*, 1980), while the consensus sequence itself is a diverged dimer (Deininger *et al.*, 1980). Member repeats have been detected in the β-like globin gene group (Fritsch *et al.*, 1981) and are represented in the double-stranded regions in hnRNA as well as the small RNA class found associated with it by hydrogen bonding (Jelinek *et al.*, 1980). Within the simian primates, the bonnet monkey (*Macaque radiata*) has a predominant 300 ntp family closely similar to the human *Alu* repeats, while the prosimian *Galago* has a more diverged prominent 300 ntp family which hybridizes to *Alu* family DNA on Southern blots under reduced stringency (Houck and Schmid, 1981). The *Alu* family's origins are probably of great antiquity, since it also shows a high degree of sequence similarity to regions within an abundant 130 ntp monomeric repeat found in the mouse genome (Krayev *et al.,* 1980).

The observations of the *Alu* family just described raise the possibility that the ''predominantly short-term organization'' of interspersed repeats in human and primate genomes may be largely due to the predominant role of this single short-sequence family. In this context, it is interesting to note the recent discovery of a long interspersed moderately repetitive family in human DNA by Adams *et al.* (1980). This family was initially identified by virtue of the presence of one member sequence to the 3′ side of the human β-globin gene in a genome clone. Other clones containing similar repeats were found by screening a human genomic library and found to be interspersed among unrelated DNA sequences thousands of ntp long, most of which are much less repeated or unique. Further analysis indicates that this family appears to consist of some 3000–4800 copies of 6400 ntp repeat, comprising 1–2% of the human genome, and characterized by moderate (about 3%) member sequence divergence. These repeats are generally interspersed among other sequences and do not appear to contain any major internal repetitions. As information is rapidly accumulating from studies of specific cloned long genomic fragments, other regions characterized by long interspersed repeats may well be identified.

While *Alu* family-type segments appear to account for much of the interspersed 300 ntp repeat component in primate DNAs, other mammalian genomes may be slightly more complicated. The bovine genome has been recently analyzed by Mayfield *et al.* (1980), employing HAP binding and EM to analyze repeat segment length and interspersion, and $Ag^+ - CS_2 SO_4$ to identify the noninterspersed satellite repeats. Their results show that the interspersed moderately repetitive DNA component of cow comprises 6–10% of the genome, with the repeat segments averaging 350 ntp and separated by single-copy regions with a mean length of 4000 ntp. The data on repeat length, together with the repetition frequency of the class (60,000), and the proportion of the genome it constitutes, as determined by reassociation, indicate that the interspersed repeat segments in the bovine genome belong to only 8 to 14 families. Five interspersed repetitive families from the rabbit genome have been identified through member segments occurring among the β-like globin genes of this animal in clones of genomic DNA (Shen and Maniatis, 1980); four of these are short repeats, while the fifth is 1400 ntp. In mouse, a sizable fraction of the interspersed repeats are between 1000 and 1500 ntp (Cech and Hearst, 1976). Nevertheless, investigations by Georgiev and co-workers (Georgiev *et al.,* 1981) have detected a small class of ubiquitous short repeats in mouse with many properties analogous to the primate *Alu* repeats. For example, these short segments are interspersed throughout the genome, exist in both isolated and palandromic configurations, and are transcriptionally expressed at high levels in mouse hn RNA. At least one family also exhibits sequence homology to the replication origin of SV40.

Longer repeat segments also exist in mouse, however. Georgiev *et al.* have detected a number of relatively low copy number repeats which appear to code for a class of longer double-stranded nuclear RNA. A prominent long repeat family which does not appear to be expressed also exists. This family has been identified through the presence, in restriction digests of whole mouse genome DNA, of an *Eco*RI excisable 1350 ntp segment containing an internal *Bgl*II site and characterized by member sequence heterogeneity (Heller and Arnheim, 1980). This segment has been shown to represent an interior portion of a 3000 ntp interspersed family accounting for 1–2% of the genome. Member sequences of this repeated family are generally interspersed in the genome of *Mus musculus,* and a closely related family exists in *Mus spretus,* while somewhat more diverged relatives are found in *Apodemus sylvaticus* and *A. mystacimus* (Brown and Dover, 1981). This repeated family also shares homology with repeated sequences in rat and Chinese hamster. In rat, Wu *et al.* (1977) have produced evidence suggesting that long and short repeats contain homologous regions.

2. *Sauropsidia*

a. *Reptilia.* Epplen *et al.* (1979) examined three reptilian genomes representing major subdivisions: *Python reticularis* (Lepidosauria), *Caiman crocodilus* (Archosauria), and *Terrapene carolina triungius* (Anapsida). The methods employed in their analysis were HAP C_0t curves of short versus long fragments, hyperchromicity determinations of HAP-isolated moderate-repeat containing fragments, and agarose column sizing of S_1 digested reassociated repeat-containing fragments. For all three genomes, the results indicated a substantial short-term interspersed component.

b. *Aves.* Epplen *et al.* (1978) studied the moderate-repeat content and interspersion in the duck *Cairina domestica* (Anseriformes), the chicken *Gallus domesticus* (Galliformes), and the pigeon *Columba livia* (Columbiformes), using the same methods as in their study on reptile DNAs, but found that the birds were characterized by a long-term mode of interspersion, in which repeats averaging more than 1500 ntp alternated with single-copy segments over 2300 ntp in length. Arthur and Straus (1978) independently examined the chicken genome and also concluded that it is characterized by long-term interspersion. These three birds represent diverse avian lineages, suggesting that bird genomes in general may well be characterized by a long-term mode of interspersion. An additional peculiarity of avian genomes noted by Epplen *et al.* (1979) is that the proportion of interspersed single-copy DNA actually declines with increasing genomic DNA content.

Eden and Hendrick (1978) have performed a somewhat more detailed study on chicken DNA. These workers employed a range of probe sizes to construct an *L* versus *R* curve and examined the length of duplexed repeat segments by EM and hyperchromicity as well as the S_1 nuclease/Agarose chromatography technique. Their results suggested that about 40% of the chicken genome consists of moderately repeated segments greater than 2000 ntp in length interspersed with single-copy segments of about 4500 ntp. About half of the genome is certainly comprised of much longer ($\geq$ 17,500 ntp) single-copy segments. While this pattern is more complex than the uniform mode inferred for the *Drosophila* long-term exemplar, it would appear to be far more closely allied to this genomic mode than the one seen in *Xenopus*. In this context, it should be noted that Murray *et al.* (1979) have suggested that some of the apparent interspersion detected in Eden and Hendrick's experiments may be due simply to the accelerated binding of long single-copy fragments to their short driver at longer fragment sizes. Certainly Eden and Hendrick's S_1 experiments show little or no repeat component of short length, a result which seems quite reliable, since these experimenters performed parallel experiments with sea urchin DNA and found the expected predominantly 300 ntp pattern.

3. *Amphibia*

The detailed studies on the *Xenopus* genome described above have of course provided one of the definitive examples of short-term interspersion. Many other Amphibia, however, have larger genomes characterized by a much higher proportion of moderately repetitive sequences ranging up to 80% of the total genome even at short shear size (Straus, 1971; Mizuno and Macgregor, 1974). Such genomes logically cannot have short-term interspersion of alternating 300–400 ntp repeats and 800–1000 ntp single-copy sequences as their predominant mode of organization, and experimental studies have subsequently indicated that they do not. Baldari and Amaldi (1976) and Bozzoni and Beccari (1978) have studied the large genomes of *Bufo bufo, Triturus cristatus,* and *Necturus maculosus* and compared them to the results obtained with *X. leavis* DNA, using their experimental methods. Their results suggest that these larger genomes are characterized by extensive interspersion of repeats with one another, as well as with fast binding-foldback sequences. Surprisingly, a large proportion of the single-copy DNA in these massive genomes appears not to be intimately interspersed with any other kinetic class. Results obtained for the genome *Bufo boreas* using HAP and EM techniques (Bouchard and Swift, in preparation) suggest that much of the interspersion for this species is of a complex repeat–repeat variety also.

4. *Cephalochordata*

The genome of *Amphioxus* has been examined by Schmidtke *et al.* (1979a), employing HAP reassociation of short (250 ntp) versus long (1200 ntp) fragments with results which indicate that this organism's genome is characterized by short-term interspersion of the moderate repeats.

5. *Summary—Chordata*

For the Chordata as a whole, the short-term mode of interspersion would appear to be the rule, with the Class Aves a notable exception. The comparative organization of avian and reptilian genomes if of particular evolutionary interest, since these two groups are sufficiently similar that they are often combined in one superclass. Indeed, phylogenetically the several avians found to have long-term interspersion are all more closely related to the Crocodylian *Caiman* (short-term interspersion) than the latter is to the other two reptiles, *Python* and *Terrapene*, also found to have short-term interspersion (Epplen *et al.*, 1979). From an evolutionary point of view, the avian transition to a long-term mode of interspersion might be considered a relatively recent event in one Sauropsidian lineage, while other lineages still classified as reptilian (including presumably the therapsid ancestors of the Mammalia) may be viewed as having the predominant short-term Chordate mode.

Close examination of the evidence, however, suggests in addition that it is inaccurate to consider the short-term organization found in many of the rest of the vertebrates and *Amphioxus* as in any sense uniform. The more complex organization of repeats in larger amphibian genomes has already been noted. The detailed knowledge now available concerning the interspersed repeats in human and other mammalian genomes makes it clear that these also differ notably from the apparently similar organization found in the prototypical short-term *Xenopus* genome, when their relative information content is taken into account. In the case of *Xenopus*, the proportion of its 4.9×10^8 ntp genome which consists of interspersed 300 ntp moderate repeats (18%), together with the average repetition frequency for members of this class (2000), may be used to calculate the approximate number of distinct moderately repetitive families which together comprise this class; this number is 800 (Galau *et al.*, 1976). By comparison, as described above, a *single* sequence family repeated some 300,000 times accounts for much of the 300 ntp interspersed repetitive component of man and primates, while a relatively small number of repeated families may account for this component in several other orders. A subsequent section deals with evidence indicating that the informationally complex moderately repetitive components of genomes in which many individual sequence families are each repeated a relatively small number of times undergo frequent augmentation during evolution by saltational addition of "new" repeated families. For the present, the simple comparisons made above suggests that the short-term interspersion in various vertebrate classes, while superficially similar, may be qualitatively quite different (Fig. 3).

B. ECHINODERMATA

The one member of this phylum, and indeed the only nonchordate deuterostome, whose genome has been extensively studied, is the sea urchin *Stroglyocentrotus purpuratus*. As noted above, the genome of this organism has been considered the other prototype of short-term interspersion. The nature of this interspersion and its role in transcription have been further explored in several

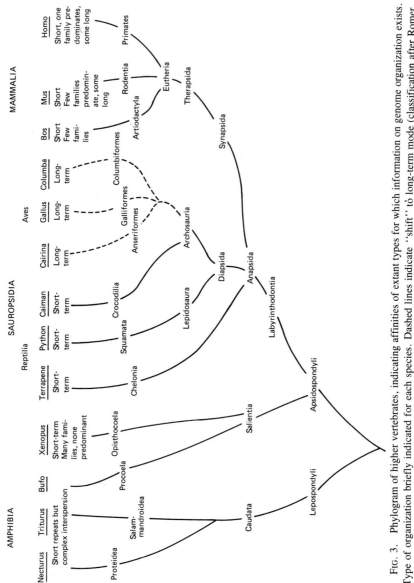

Fig. 3. Phylogram of higher vertebrates, indicating affinities of extant types for which information on genome organization exists. Type of organization briefly indicated for each species. Dashed lines indicate "shift" to long-term mode (classification after Romer, 1959; and Storer and Usinger, 1965).

ways. Davidson *et al.* (1975b) hybridized labeled embryo mRNA with the fraction of the single copy copurifying with reannealed moderate repeats on HAP and found that the majority of these expressed gene sequences appear to be enriched in this fraction and are thus repeat-contiguous. Studies on cloned lengths of genomic DNA containing several alternating segments of repeat and single-copy sequence have shown that the neighboring repeats are unrelated (Lee *et al.*, 1978). Transcripts of 10–20% of the sea urchin's 5000 diverse moderate-repeat families are found in the egg RNA (Constantini *et al.*, 1978), and these transcripts are found to be covalently associated with most of the poly(A)(+) maternal RNAs of the egg, which comprises more than 10^4 single-copy transcripts (Constantini *et al.*, 1980). Therefore, in sea urchin, a large number (500–1000) of moderately repetitive sequence families include members which lie adjacent to some 10^4 expressed single-copy sequences and are cotranscribed with them during the production of the maternal RNAs stored in the egg. Interestingly, RNA driven reactions with probes from six cloned repeats show that their quantitative level of expression is similar in the egg RNAs of *S. purpuratus* and *S. fransiscana*, although the number of copies of these families differ substantially in the two genomes (Moore *et al.*, 1980). Thus the pattern of transcription appears to be conserved. A given repetitive family must be represented on 10 or more distinct transcripts on the average, but no information is available on the disposition of the rest of the family members, which comprise the vast majority. It appears that the various egg RNAs on which a given repeated family is represented include transcripts of both strands, as shown by DNA/RNA reassociation with single-strand DNA probes (Constantini *et al.*, 1978; Moore *et al.*, 1980) and EM of reannealed egg RNA (Constantini *et al.*, 1980).

It should be noted that the presence of repeat transcripts covalently linked to poly(A)(+) RNA is not general in the sea urchin life cycle. The newly synthesized polysomal RNA of the gastrula shows no significant portion bearing covalently linked repetitive and single-copy segments (Davidson *et al.*, 1975b), while the only abundant transcripts of interspersed repeats are found in the nuclear RNA (Scheller *et al.*, 1978). It may therefore be suggested that the prevalence in the egg of transcripts spanning single-copy regions and contiguous repeats may be due to these stored maternal RNAs being unprocessed or partially transcripts, analogous to the hnRNAs of synthetically active cells.

The similarity between these findings and the demonstration of repeat region transcripts in the hnRNAs of higher vertebrates (Jelinek *et al.*, 1974, 1978) is of course striking. The same apparent contrast in the complexity of the expressed information exists as is seen in the total informational content of the moderately repetitive components, however, in that the expressed covalently linked repeats in sea urchin represent many hundreds of distinct repeated families, while the vertebrate repeats give a consensus fingerprint (Jelinek *et al.*, 1978) and are probably mostly referrable to members of the single *Alu* family in man (Jelinek *et al.*, 1980) and a small number of families in mouse (Georgiev *et al.*, 1981).

C. Protostomes and Lower Invertebrate Metazoa

1. *Arthropoda*

a. *Insecta.* Most investigations of arthropod genome organization have dealt with insects. *Drosophila* has already been discussed as the type genome for long-term interspersion. Subsequent investigations, particularly employing cloned repeats, have demonstrated many more peculiarities in moderate repeat components of this genome (see Rubin, 1978; and Spradling and Rubin, 1981, for review). It now appears that many of the large interspersed repeats of *Drosophila* exhibit many of the characteristics of eukaryotic transposible elements. Many appear to code for abundant discrete poly(A)(+) RNAs of their own, as opposed to segments of hnRNA also bearing covalently linked single-copy transcripts. A large body of evidence, reviewed in the next section, indicates that much of the *Drosophila* repeat component does in fact consist of mobile elements which transpose frequently during evolution.

The pattern of repeat organization in several relatives of *Drosophila* has been investigated. Crain *et al.* (1976b) examined the interspersion pattern of *Musca* (housefly), using their standard techniques, and found that this dipteran genome is characterized by short-term interspersion. They therefore suggested that the appearance of the long-term mode of organization may well have been a recent event in the evolution of the branch of the Diptera which includes *Drosophila.* Against this, however, stand observations on *Sarcophaga* (flesh fly—Samols and Swift, 1979; French and Manning, 1980) and *Chironomus* (bloodworm midge—Wells *et al.,* 1976), both of which exhibit a long-term organization very similar to that seen in *Drosophila,* at least when viewed at the mass level of repeat versus single-copy length and interspersion.

This pattern of occurrence of long-term versus short-term interspersion in these dipteran species is most intriguing. From the standpoint of systematics, the Sarcophagidae and the Muscadae are placed among the calyptrate Diptera, and are.considered far more closely allied to one another than either is to the acalyptrate Drosophilidae, though all three families are grouped in the advanced dipteran suborder Cyclorrhapha. The Chironomidae, on the other hand, are placed in the more primitive suborder Orthorrhapha (Borror *et al.,* 1976). If the short-term pattern of organization is considered ancestral and the long-term derived, it is necessary to postulate three independent shifts to the long-term pattern in the lineages leading to the modern *Chironomus, Drosophila,* and *Sarcophaga* to account for these observations. The last such shift would postdate the separation of the sarcophagid and muscid lineages, since *Musca* retains the primitive type of interspersion. Alternatively, it may be supposed that the shift to long-term interspersion occurred once in the remote ancestor of all these modern Diptera, but this requires that the *Musca* lineage have secondarily reverted to the short-term mode. Either choice implies considerable plasticity at the level of genome organization over the course of dipteran evolution (Fig. 4). One important question which awaits an answer is whether the long-term repeats of other Diptera will prove to be primarily transposon-like elements, as they are in *Drosophila*. It is to be hoped that future investigations will elucidate more fully the patterns of evolution at the level of genome organization in this insect group.

Data for the rest of the class Insecta are sparse. *Antheraea pernyi*, the only Lepidopteran subjected to detailed examination, exhibits a quite typical short-term repeat pattern (Efstratiadis *et al.*, 1976). The genome of the Hymenopteran *Apis mellifera* (honey bee), on the other hand, is characterized by a long-term intersper-sion (Crain *et al.*, 1976b). This would appear to indicate a completely independent appearance of this pattern of organization in the Insecta, since the Hymenoptera are generally held to have separated very early from the rest of the holometabolic higher insects (Borror *et al.*, 1976; though see Malyshev, 1968, for a different view). A recent study by French and Manning (1980) of the DNA of the Thysanuran *Thermobia domestica* reveals a typical short-term organization for the genome of this primitive insect. These authors suggest, based on a considera-tion of known insect genomes, that a large genome characterized by short-term interspersion has arisen several times in more advanced insect groups charac-terized by relatively low DNA content.

b. *Other Arthropoda.*   The genome of the land crab *Gecarcinus lateralis*, studied by Holland and Skinner (1977), has several interesting features. With respect to repeat segment length, at least two-thirds of the weight of duplexed repeats are found to be 1500 ntp in S_1 nuclease experiments, while the remainder shows a 300 ntp modal size. Nevertheless, about two-thirds of the single-copy DNA segments are found to be interspersed among repeats with an average length of 1000 ntp. The genome may therefore be regarded as having predomi-nantly short-term interspersion, albeit with certain peculiarities. The genome of the primitive Chelicerate *Limulus* (horseshoe crab) was examined by Goldberg *et al.* (1975), as part of a survey study on the DNAs of five Protostome inverte-brates, and found to be predominantly short-term in its interspersion, with the bulk of repeats in the 300 ntp modal size class.

2. *Mollusca*

Two bivalve mollusk genomes were also investigated as part of the survey cited immediately above, while the DNA of the gastropod *Aplysia* (sea hare) has been examined in more detail in a separate study (Angerer *et al.*, 1975). *Aplysia* and the Eulamellibranch Pelecypod *Spisula* (surf clam) both exhibit predomi-nantly short-term interspersion. The case of the more primitive Filobranch *Cras-sostrea* (oyster) is more unusual and rather resembles the situation seen in the land crab genome. At least 65% of the mass of moderately repetitive sequence in this small (0.69 pg) genome consists of repeats longer than 1500 ntp. A compo-nent with a modal length of 300 ntp is present, however, and at the long shear

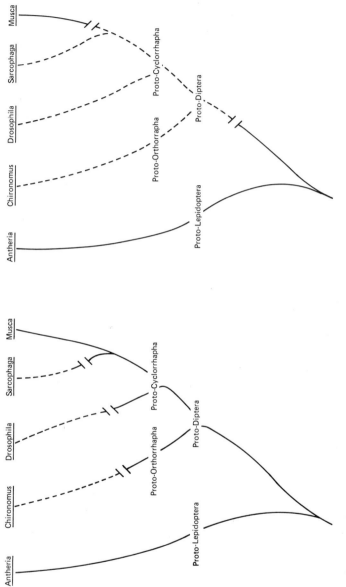

FIG. 4. Parallel phylograms for higher insects, indicating alternative hypotheses concerning shifts between short- and long-term modes of interspersion in lineages. Solid lines indicate short-term, dashed lines long-term interspersion (categories according to Borror et al., 1976).

size of 3000 ntp virtually all the single-copy DNA can be bound to HAP due to reannealing of adjacent repeated segments. This result suggests that the single-copy regions are finely interspersed with components of the repetitive, so that overall the oyster genome bears closer affinity to the short-term interspersion mode, despite the predominance of long repeated segments.

3. Nemertina (Cerebratulus) and Coelenterata (Aurelia)

These unrelated lower invertebrates were also looked at by Goldberg *et al.* and found to possess a high degree of short-term interspersion. For both organisms, most of the single-copy sequence detectable in HAP reassociation of short fragments if found to be linked to moderate repeats at a shear size of 2000–2800 ntp.

4. Aschelminthes/Nematodes

Because of the interest in these animals for genetic studies, and also due to the phenomenon of chromosome diminution which exists in some groups, several nematode genomes have been studied. Beauchamp *et al.* (1979) examined the HAP binding and hyperchromicity of the repeat kinetic component of *Panagrellus silusiae* DNA at short and long fragment sizes and also performed R vs L studies in which a range of tracer fragment sizes were annealed with short length driver. They concluded that the repetitive segments were greater than 2000 ntp and could extend to 10,000 ntp, while the single-copy segments are over two times longer. Thus *Panagrellus* appears to possess long-term interspersion. *Ascaris suum* and *Parascaris equorum* were examined by Roth (1979) in the course of a study on diminution. The soma DNA of *Ascaris,* which has been purged of highly repetitive sequences during diminution, shows a long-term interspersion pattern for the moderate repeats; the upper limit on the length of repeat segments was not evident even at a shear size of 7200 ntp.

In contrast to these two examples stands the repeat arrangement found in *Caenorhabditis* itself. The authors of an early brief report on the DNA of this organism (Schachat *et al.*, 1978) looked at HAP binding due to repeat reassociation at two different fragment lengths (380 and 3850 ntp) and concluded that *Caenorhabditis* did not show short period interspersion. However, when Emmons *et al.* (1979) performed random cloning of *Caenorhabditis Bam*HI genomic fragments ranging from 400 to several thousand ntp in size, they found that virtually every cloned fragment contained a portion of sequence repeated elsewhere in the genome. This unexpected observation indicated fine scale interspersion of some type of repeated element. In consequence, Emmons *et al.* (1980) performed a detailed study employing electron microscopy and careful kinetic analysis to examine the length and interspersion of *Caenorhabditis* single copy and repeat. Their EM analyses revealed extensively interspersed short moderate repetitive segments. These appear to have been missed in the earlier analysis because their average copy number is quite low (in the 10-fold range).

As a result, at the longer fragment size employed by Schachat *et al.*, reassociation of the interspersed repeats was retarded to near the single-copy rate due to the excluded volume effect (Wetmur and Davidson, 1968) produced by the long covalently linked single-copy segments. Accepting the results of Emmons *et al.* as the more accurate, we can aver that results on the genomic DNA of only three nematode species imply that this Class is also characterized by the occurrence of both short-term and long-term interspersion.

5. *Summary*

With the exception of the Diptera, detailed studies of genome organization on protostomes and lower invertebrate Metazoa are sparse, with many major taxa unrepresented. It is nevertheless clear that multiple instances of both short- and long-term interspersion have been detected. The known phylogenetic distribution of these modes of organization, while sparse, is sufficient to suggest that the evolutionary transition from one to the other is surprisingly easy and has occurred polyphyletically in several lineages.

In the case of the Insecta, French and Manning (1980) have pointed out that occurrence of longer term interspersion appears to be correlated with lower nuclear DNA content, a point made for vertebrates by Epplen *et al.* (1979) in their comparisons of bird and reptile genomes. The meaning of this correlation is not entirely clear, for the number of insect genomes investigated is small (seven). The break point is sharp, with *Sarcophaga* (genome 0.60 pg) showing long-term interspersion, while *Musca* (0.80 pg) has a short-term mode. In nematodes, the correlation appears to break down. *Caenorhabditis,* with short-term interspersion, possesses the smallest genomic DNA content of the species examined: 8×10^7 ntp, or 0.06 pg (Emmons *et al.*, 1980). The long-term genomes of *Panagrellus* and *Ascaris suum* are considerably larger, the estimate for *Panagrellus* being 0.09 pg (Beauchamp *et al.*, 1979), and for the soma DNA of *Ascaris,* 2.45×10^8 ntp, or 0.25 pg (Roth, 1979).

D. HIGHER PLANTS (ANTHOPHYTA)

Data on moderate-repeat interspersion in higher plants are still modest. Walbot and Goldberg (1979), reviewing the information then available, suggested that the patterns for the Dicotyldonae and Monocotyledonae which had been studied could all be characterized as short-term with respect to the spacing of repeats flanking single-copy segments. Some interesting differences in organization emerge, however, when attention is turned to the average length of repeat segments and their organization with respect to one another in some plant genomes. In *Gossypium hirsutum* (cotton), for example, Walbot and Dure (1976) found that much of the single copy was interspersed at spacings of 1800 ntp (60% of genome), or 4000 ntp (20% of genome). Nevertheless, 20% of the cotton genome consists of single-copy segments longer than 4000 ntp, and the modal size of repeated segments is 1250 ntp rather than 300 ntp. The genome of *Glycine max* (soybean) (Goldberg, 1978; also Walbot and Goldberg, 1979) shows interspersion of the single-copy regions, but only with a fraction of very low frequency (19 copies/genome) repeats; the major repeat components are long and not interspersed with single-copy segments. Within the genome of the monocot *Triticum aestivum* (wheat), the modal size of S_1 resistant repeat segments is 600 ntp, with the bulk falling between 400 and 800 ntp (Flavell and Smith, 1976). In contrast to the simpler type of short-term interspersion seen in the *Xenopus* exemplar, the majority of repeats are not interspersed individually among single-copy regions, but are instead interspersed with one another in a complex fashion. At least four-fifths of the small (25%) single-copy class is however interspersed among the repeats at intervals of a few thousand ntp, two-thirds being found in segments of about 1000 ntp, so in this respect the wheat pattern may be regarded as short-term. Surprisingly, in the related *Secale cereale* (rye), the range of repeat lengths (500–5000 ntp) and their modal size (2000 ntp) are much larger; however, most of the single-copy regions still appear to be interspersed at spacings between 400 and 3500 ntp (Smith and Flavell, 1977).

Current evidence suggests that different and complex modes of sequence organization can emerge in closely allied phyletic lineages in plants. In a study of the DNA of the dicot *Pisum sativum* (pea) Murray *et al.* (1978) found that 90–100% of the single-copy fraction of this large genome occurs as interspersed segments of 1000 ntp or less (modal size only 300 ntp). The modal repeat segment size in pea was also found to be 300–400 ntp. Since 85% of 300 ntp fragments reanneal in the repetitive range on HAP, this implies that many repeats must be interspersed with each other, and HAP-C_0t analyses with longer fragments verify this. Nevertheless, with regard to adjacency of single-copy and repeated segments, the *Pisum* genome can be considered highly short-term. In marked contrast, the far smaller genome of *Vigna radiata* (mung bean, a member of the same subfamily) is characterized by a completely different organization (Murry *et al.,* 1979). Here repeat lengths exhibit a broad size range with a mode at 1260 ntp and a substantial proportion ranging from 200 to above 8000 ntp. Thirty-five percent of single-copy sequences are 300–1200 ntp, and 18% are in the 1200–6700 ntp range, but 46% are not detectably interspersed at a length of 6700 ntp. Thus the mung bean genome includes fractions ranging over the spectrum from short- to long-term interspersion, as seen in cotton.

Both mung bean and cotton are dicots, but a situation of comparable complexity has been described for the relatively large (6.3 pg) genome of the monocot *Zea mays* (corn) by Hake and Walbot (1980). In this plant, the mass average repeat length is found to be 970 ntp by S_1 digestion, with EM and hyperchromicity measurements suggest most repeats fall in the range of 500–1000 ntp. Two levels of sequence repetition are found in HAP reassociation of short fragments, a class comprising 20% of the genome with average repetition of 800,000 copies and a class comprising 40% with average repetition 1000. HAP reassociation followed with long fragments and an R vs L experiment showing that one-third of the genome consists of interspersed repeats of the two classes, while another one-third appears to be comprised of interspersed moderate repeat and single-copy segments, the latter having an average length of 2100 ntp. The remaining third contains single-copy segments not found to be interspersed with repeats at a shear size of 5000 ntp, as well as long repeats and palindromic elements; at least one-quarter of the single-copy DNA is found in this long segment class.

A model repeat length of 1100 ntp has been estimated by S_1 digestion and neutral gradient sedimentation for the very large ($\sim$ 36 pg) genome of *Lilium* $\times$ Enchantment (Bouchard and Stern, 1980), indicating that longer repeats can occur even in plant genomes of large size. Preliminary HAP reassociation with DNA fragments ranging from 400 to 5000 ntp indicate that half or more of the single-copy component of this genome is not interspersed even at the largest fragment size used (Bouchard, unpublished observations). It therefore appears that most of the repeats in *Lilium* are interspersed with other repeats, and much of the single copy exists in long segments.

Even the few examples discussed here suggest that attempts to fit the organization of plant genomes into the stereotyped ''short-term'' or ''long-term'' modes derived from studies of Metazoan genomes may often be misdirected. A particular genome, such as wheat, may for example be regarded as ''short-term'' with respect to the length of most single-copy segments lying among repeats. In other respects, however, it differs from the *Xenopus* archtype both quantitatively (greater length of repeat segments and much larger proportion of repeats) and qualitatively (interspersion of most repeats among other repeats rather than single-copy segments). Other genomes, such as that of mung bean, appear to exhibit some characteristics which might be regarded as transitional between modes, and others peculiar to themselves. Plant genomes not only contain moderately repetitive DNA, they are frequently dominated by it (Flavell *et al.,* 1974), and it appears that under these conditions the organization of this component often exhibits a higher level of complexity than is seen in the traditional animal exemplars.

E. Eumycota—True Fungi

If the complexity and diversity of repeated DNA interspersion encountered in higher plants pose a challenge to the dichotomous classification of genome organization, the situation found in fungi appears to overthrow it entirely. Information now available on filimentous habit representatives from all three of the perfect fungal Classes suggests that repeat interspersion, where it is detectable at all, at best involves a small number of repeat segments tens or hundreds of thousands of ntp in length, interspersed among single-copy segments ranging up to 10^6 ntp.

The most detailed investigation of interspersion and expression of moderately repetitive DNA has been performed on the water mold *Achyla* (Class Pycomyces, subclass Oomycetidae). The genome of this fungus was analyzed by HAP-monitored self-reassociation of different sized fragments and reassociation of such fragments with excess short moderately repetitive driver, as well as kinetic determination of the single-copy content of long fragments bound to HAP by reassociation of repetitive regions and S_1 measurements of reassociated repeat segments (Hudspeth *et al.*, 1977). Within the experimental limits, neither of the interspersion tests with different fragment size could detect any linkage of repeat and single-copy segments. S_1 measurements also indicated that reassociated repeat duplexes are very long. Only the experiment in which long fragments were bound at repetitive C_0t, then sheared and examined for single-copy content, implied any linkage, with 3.6% of the total single-copy appearing in this fraction at a starting fragment size of 5000 ntp. Even if interspersion is presumed to be uniform, the authors' interpretation of these results shows that repeat segments averaging 2.7×10^4 ntp would be interspersed with single-copy segments of 1.35×10^5 ntp, and there is room for only 250 such units in the genome.[1] In a complementary study of the nuclear and polysomal poly(A)(+) transcripts of *Achyla*, Timberlake *et al.* (1977) were unable to detect any linkage of repetitive and single-copy transcripts in either population of RNAs, nor did they find any evidence of RNA processing in the transition from nuclear poly(A)(+) hnRNA to polysomal mRNA. The total complexity of the expressed mRNA was found to be equivalent to 2000 genes, an order of magnitude more than the maximum number of interspersed repeat/single-copy units. A separate study of the genome of *Phycomyces blakesleeanus* (Subclass Zygomycetidae) by Harshey *et al.* (1979) has shown that this Phycomycete is also characterized by a similar mode of very long-term interspersion. It therefore appears that these primitive fungi are characterized by a mode of moderate repeat interspersion and expression different from anything seen elsewhere.

Within the Class Ascomycetes the genomes of two Euascomycetidae, both extensively employed in genetic analyses of gene organization and control, have been examined with even more striking results. In *Aspergillus nidulans*, Timberlake (1978) was unable to demonstrate the presence of any moderately repetitive

[1]An interesting calculation can be made using these figures together with the computations given by Hudspeth *et al.* (1977) for the sequence complexity (7.6×10^4 ntp) and repeat frequency (80) of the major component of the moderately repetitive DNA. Taking the sequence complexity of 7.6×10^4 ntp and the estimate of 2.7×10^4 ntp for the length of an average repeat segment, we may calculate that only three 80-copy interspersed families of this segment length can account for the entire component and essentially all of the 250 repeat/single-copy units in the genome.

DNA in the HAP-monitored kinetic pattern of whole nuclear DNA. When DNA reassociating at low C_0t was purified, he was able to demonstrate the existence of a small repetitive component comprising 2–3% of the genome and repeated about 60 times. The compexity of this entity, however, was only 11,000 ntp, which can largely be accounted for by the repeated sequences coding for ribosomal RNA and their spacers. A similar situation was demonstrated for *Neurospora crassa* (Krumlauf and Marzluf, 1979), where the 8% moderately repetitive component has a complexity of 15,300 ntp, and little evidence of interspersion could be found in careful experiments employing a large range of fragment sizes. The small increase in HAP binding due to repeat reassociation which was seen with increasing fragment size may well be due to the 5 S gene sequences, which have been shown to be interspersed among unrelated segments in *Neurospora* (Free *et al.*, 1979), rather than being clustered. A somewhat different state of affairs prevails in the Hemiascomycedid *Saccharomyces cerviseae* (yeast). The genome of this organism contains interspersed moderately repeated segments, a few thousand base pairs in length, and a class of 250 ntp segments found both flanking these elements and in isolation, but these have been shown to be associated with transposable elements rather than fixed components of the DNA (Cameron *et al.*, 1979; Eibel *et al.*, 1981). Their lack of fixed location and the fact that they can be virtually lost from the genomes of isolated strains (Fink *et al.*, 1981) suggests it may be best not to consider these repeats to be firm components of the yeast genome, despite the fact that they are found within the yeast genome.

Length and interspersion of repeat segments has also been examined in the DNA of the Basidiomycete *Schizophyllum commune*, using HAP binding at different fragment lengths of repetitive C_0t and EM visualization of reassociated repeated duplex (Ullrich *et al.*, 1980). The results of this work suggest that if interspersion exists at all, there are a probable maximum of 16 repeat-containing segments in the genome, with an average length of 225,000 ntp, an order of magnitude longer term organization than in the enormous units calculated for the two Phycomycetes.

The results for the different representatives of filimentous fungal groups are summarized in Table I. Viewed together, they show little resemblance to any of the modes of sequence organization seen in groups previously dealt with. The organization of the two Pycomycete genomes might possibly be described as "ultra-long term," but even this degree of moderate-repeat interspersion is apparently absent in the higher fungi. Fungal systems are well-known for exhibiting, at least in simple form, the typical "complex eukaryote" phenomena of gene regulation involving coordinate activity of unlinked genes or induction of differentiative pathways by discrete environmental or hormonal signals, and have been widely exploited in genetic and developmental studies of such events. It appears, however, that the regulation of such events in the fungi cannot be accommodated

TABLE I

GENOME ORGANIZATION IN FILIMENTOUS FUNGI

Class and subclass/order[a]	Species (estimated genome size[b]) (ntp)	Estimated repeat segment length (ntp)	Proportion genome (%)	Estimated single copy segment length (ntp)	Maximum estimated number of repeat units	Estimated repeat class complexity (ntp)
Pycomycetes						
Oomycetidae	$Achlya$ (4.19×10^7)	2.7×10^4	2 (two 14 kinetic classes)	1.35×10^5	250	380 / 7.6×10^4
Zygomycetidae	$Pycomyces$ (6.6×10^7)	1.6×10^4	35	2.5×10^4	1440	$(5.5 \times 10^5)^c$
Ascomycetes						
Eurotiales	$Aspergillus$ (2.6×10^7)	—[d]	2–3	—	—	1.1×10^4
Xylariales	$Neurospora$ (2.7×10^7)	7.3×10^{4e}	8	—	30	1.53×10^4
Basidiomycetes	$Schizophyllum$ (3.6×10^7)	2.25×10^{6e}	12	—	16	—

[a] System of Scagel et al. (1965).

[b] All estimates are those of the authors of the papers except where noted.

[c] My estimate based on the author's data.

[d] (—) indicates estimate not given or impractical.

[e] A very approximate estimate.

to models involving the intimate interspersion of sizable repetitive segments with expressed single-copy DNA regions.

F. PROTISTA

1. Myxomycota—Slime Mold (Dictyostelium)

The precise relationship between the slime molds and the true fungi, and even among the various myxmycetes themselves, is a matter of ongoing discussion (Scagel et al., 1965). In light of this, it is interesting that the interspersion and expression of moderate repeats in Dictyostelium is entirely different from that found in the Eumycota, resembling far more closely the pattern seen in short-term Metazoa, though with some novel features of its own. The genome organization, as demonstrated by Firtel and Kindle (1975), is characterized by interspersion of 75% of the single-copy DNA with repeats at a fragment length of 3000 ntp. The average length of the interspersed single-copy segments is 1500 ntp, and S_1 analysis shows that half the repeats are 250–450 ntp, while the rest are longer than 2000 ntp. A distinct property of the transcriptional expression of repeats in Dictyostelium, however, is the persistence of covalently linked repeat and single-copy transcripts, not only on hnRNA, but on cytoplasmic mRNA, as has been convincingly demonstrated through the use of cloned probes (Kimmel and Firtel, 1979; Kindle and Firtel, 1979).

2. Ciliophora—Ciliated Protists

While the organization of repetitive and single-copy segments in the transmission genomes of ciliates is not well understood, a fascinating phenomenon which pertains to moderate repeats occurs in the Hypotrichidan ciliates Oxytricha and Stylonichia. These two organisms have been studied respectively by Prescott and co-workers (Prescott and Murti, 1974; Lauth et al., 1976; Lawn et al., 1978) and by Ammerman et al. (1974) with essentially identical results. In these ciliates, the diploid transmission genome, which will ultimately engage in meiosis and provide the haploid nuclei which fuse with those of the conjugant mate to produce a new diploid genotype, is carried in the transcriptionally inert micronucleus. All transcriptional activities necessary for the life and maintenance of the active cell appear to reside in the highly amplified DNA of the macronucleus. After mating, the old macronucleus degenerates, and a mitotic product of the new diploid micronucleus undergoes extensive endoreduplication to form a new one. Kinetic, isopycnic, and thermal stability studies on the DNA of isolated micronuclei show that the transmission genome is characterized by a complex moderately repetitive component. However, during the two cycles of endoduplication which occur to form the new macronucleus, all of the moderately repetitive segments (and much of the apparent single copy) are destroyed, so that the mature macronucleus contains sequences representing only a few percent of the micronuclear complexity, each repeated over 1000 times. These sequences occur in discrete size classes believed to correspond to individual genes and appear to be free of repeated regions, except for a terminal 26 ntp inverted repeat which is found on all of them and also occurs in the interior of longer segments (Lawn *et al.*, 1978). Aside from this single short sequence, all of the moderate repeats in the *Stylonichia* and *Oxytricha* genomes appear to be completely dispensible for the genetic activities involved in the physiological maintenance, activities, and asexual reproduction of these complex cells.

This drastic jettisoning of DNA sequences is not found in all ciliate genomes. Production of the macronucleus of *Tetrahymena pyriformes* is accompanied by loss of no more than 10–20% of all micronuclear sequences, though this loss may also occur preferentially in the moderately repetitive component (Yao and Gorovsky, 1974). The precise mode of formation is unknown for the macronucleus of *Paramecium,* but its moderately repetitive component is very small (<2%) and of low complexity (McTavish and Sommerville, 1980).

3. *Dinophyta—Dinoflagellates*

Analysis of the genomic DNA of the dinoflagellate *Crypthecodinium cohnii* has been performed by Allen *et al.* (1975). Sequence organization in the genome of a dinoflagellate is of particular interest, because a number of aspects of the nuclear ultrastructure, DNA packaging, and division cycle of these organisms suggest that the group may be a surviving lineage representing a transitional stage between prokaryote and eukaryote cellular organization. Despite the many "bacterial" aspects in the packaging of this organism's DNA, it is found to possess a eukaryote-like moderately repetitive component showing kinetic heterogeneity and comprising more than half the *C. cohnii* genome. These repeated elements show a modal S_1 resistant size of 400–450 ntp and appear to be interspersed among single-copy segments. Therefore, if the Dinophyta are accepted as a surviving example from the period of prokaryote-to-eukaryote transition, the sequence organization of this genomic Coelocanth suggests that the phenomenon of generation of relatively short interpolated moderately repetitive DNA segments may have been among the early particularizing characteristics of basic protoeukaryote types.

G. Summary and Discussion

The data for the array of genomes surveyed in this section are quite variable with respect to the detail and precision of the analyses performed and represents the merest sampling of the evolutionary spectrum. Nevertheless, certain points suggest themselves on the basis of the available information.

To begin, the practice of classifying genomes as short-term or long-term may tend to obscure important types of variability. Mammals on the one hand and *Xenopus* and several reptiles on the other have "short-term" interspersion, but the complexity and character of the interspersed repeats appear to be drastically different. In the large moderate-repeat dominated genomes of many higher plants and probably some amphibia, while a considerable proportion of the single-copy sequence is interspersed among repeats, a large fraction of the moderately repeated segments are found to be interspersed with one another in often complex fashion. The modal average length of the repeat segments also tends in these cases to be two to four times larger than the "typical" 300 ntp short repeat. Genomes may be characterized by aspects of both types of organization, as in mung bean or possibly chicken, or contain cryptic short-period interspersion of repeats of low abundance while the more abundant classes show a different organization, as in *Caenorhabditis* or soybean. In the filimentous fungi, when moderate repeats not accounted for by known genetic function are found at all, they appear to be "interspersed" only on a very coarse scale which dwarfs the pattern seen in the *Drosophila* exemplar and will leave most genes far from any repeat.

A second observation, which may be made by reference to the situations seen in the higher vertebrates, the Diptera and other insects, the nematodes, and the two legumes *Vigna* and *Pisum,* is that the evolutionary transition from one mode of organization to another appears to be a repeated event arising independently in different evolutionary lineages. There does appear to be some correlation between relative genome size and such shifts. Even this apparent unity may however be specious, for while the long-term repeats of *Drosophila* itself are now largely attributed to transposon elements, it has not yet been determined what those found in the other organisms may be.

It would therefore appear that the genomic DNAs of a wide variety of eukaryotes are characterized by a phenomenological similarity, in that they contain relatively short moderately repetitive segments interpolated among unrelated segments within the genome. Closer examination, however, reveals particularizing differences with respect to the average length of segments, proportion of long versus short segments, interspersion of segments, interspersion of segments among each other versus among repeats, the number and abundance of distinct repeated families, and the extent and character of their transcriptional expression. (The examples of *Drosophila* and yeast suggest in addition that genomes which are relatively devoid of shorter repeats may contain, or come to contain, classes of autonomous repeated elements which may be a rather distinctive type of sequence entity.) These considerations suggest that a tendency to generate and interpolate moderately repetitive segments is a characteristic of genomes in many lineages. But genomes (and perhaps lineages) differ with respect to the number of distinct sequence families, the number of repeats per family, and the average length of family member segment which accumulates in the genome, as well as the extent to which interpolation occurs among single-copy regions as compared to other repeats. The rate and consistency with which these phenomena occur in particular lineages cannot readily be inferred from the simple comparison of genome organizations determined for phyletically distinct organisms. In the next section, we therefore turn our attention to experimental studies which examine aspects of the evolutionary accumulation of moderate repeats within groups of related organisms in more direct fashion.

IV. Evolution of Moderately Repetitive DNA Components within Groups

Several approaches have been taken in addressing the question of the origin and accumulation of moderately repetitive DNA elements in related organisms. One type of analysis focuses on the size and complexity of the moderate repeat components of related genomes characterized by rather different DNA content, in order to assess the contribution of alterations in the moderate-repeat class to the change. More direct comparisons employ reassociation of radiolabeled probes from one species with its own and related DNAs to assess the proportion and degree of similarity of shared moderate repeats. The advent of molecular cloning has permitted analysis at another level of precision, since cloned repeat segments representing individual families can now be used as evolutionary probes.

The following pages review the results of research work by several groups who have employed one or more of these approaches to assess evolutionary differences in the amount of moderate-repeat component in related genomes and the inter- and intragenomic affinities of added repeats. In the first subsection are grouped the results of a number of studies which suggest that closely related genomes can exhibit large differences in their proportion of moderately repetitive DNA component, often accounting for most of their total differences in DNA content. These observations imply that large saltational changes in moderate-repeat amounts can occur over short evolutionary time spans, often unaccompanied by any obvious qualitative changes in phenotype or karyotype, but generally say little about the origins of the added repeats. The second subsection therefore surveys the results of detailed studies on several groups with large moderately repetitive components. These analyses employ sequence hybridization of moderate-repeat probes with homologous and heterologous genome DNAs to examine the proportion of shared repeat sequence families with different extents of evolutionary divergence and suggest that the added repeats in these genomes arise largely from sequences not previously found at detectable levels of repetition. The third section treats results on genomes containing smaller moderate-repeat components, particularly sea urchin, where the families which dominate the moderately repetitive component in a particular species appear to be derived by resaltation from a subset of families already characterized by some level of repetition in the ancestral genome of the extant lineages. Results for the rather distinct moderate-repeat elements of *Drosophila* are dealt with in their own subsection. A concluding subsection examines data on the *intra*genomic affinities of repeated segments in the large moderate-repeat components found in representatives of several plant groups, results which also suggest that different modes of moderate-repeat evolution may characterize different genomes.

A. QUANTITATIVE DIFFERENCES IN MODERATELY REPETITIVE COMPONENTS IN GENOMES OF RELATED ORGANISMS

The phenomenon of substantial genome size changes in closely related species with the same chromosome number is met with in a number of higher plant and animal groups (Stebbins, 1971). Several changes of this type have been investigated, with results that suggest that many of these changes are due to massive increases in the amount of moderately repetitive DNA.

An early comparative study employing filter hybridization methods was performed in the genus *Vicia* by Chooi (1971), who examined the relatedness of repetitive DNAs among six species having the same chromosome number but spanning a 6-fold range of DNA nuclear content. She concluded that many of the differences among species could be accounted for by differential increases in sequences which were present in both. In some cases, however, particularly that of *V. faba* vs *V. sativa,* a large proportion of the repetitive sequences present in either species did not seem to be represented in the moderately repetitive DNA of the other. Straus (1972) compared this same species pair (which differ in nuclear content by the ratio of 5:1) with respect to the C_0t curve shown by each DNA over the moderately repetitive region and found that both genomes are largely comprised of moderately repetitive sequences spanning a broad kinetic spectrum, but that *V. faba* has more moderately repetitive DNA and that this reanneals about three times faster than that of *V. sativa,* indicating a higher average reiteration frequency for repeats in the larger genome. When labeled repetitive DNA from each species was hybridized with cold driver from the other, only about 20% was found to cross-hybridize in either direction, and the thermal stabilities of these heterologous hybrids were lower than the modal value for reassociated homologous moderately repetitive sequence when the DNA was melted on HAP columns. Taken together, these data suggest that the majority (75–80%) of the moderately repetitive sequences in each genome have been added since the divergence of their respective lineage, and that the average number of repeats added by each event in the *V. faba* lineage has perhaps been 10- to 15-fold more than the average number added in *V. sativa.*

Coniferous species exhibit a uniform chromosome number, yet vary by a factor of 3.4 in nuclear DNA content. When Miksche and Hotta (1973) examined the HAP reassociation of DNA from seven species ranging in DNA content from 39.1 pg (*Thuja occidentalis*) to 138.6 pg (*Pinus resinosa*), the genomes of all their species consisted largely of a kinetically complex moderately repetitive DNA component. As in *Vicia*, this component appeared to reanneal faster in DNA from the larger genomes, implying a larger average number of repeats.

In the wild pea genus *Lathyrus*, Narayan and Rees (1976) employed optical techniques to examine the reassociation of nuclear DNA from species spanning a 3-fold difference in DNA content. They found that the changes could largely be accounted for by increases in the amount of moderately repetitive DNA, though a lesser increase in low-copy sequences was also observed. In a subsequent study (Narayan and Rees, 1977), repetitive and nonrepetitive DNA from *L. hirsutus*, the species with the largest DNA value, was prepared and labeled. These probes were then reannealed with homologous DNA and the heterologous DNAs of congeners, and the relative proportions of reannealed probe (index of homology) and the T_ms of the hybrids formed were determined for the various species by hydroxyapatite techniques. The index of homology was found to decrease as species of progressively lower DNA content were compared with *L. hirsutus*. The difference in T_m of the heterologous hybrids which did form was found to decrease in approximately parallel fashion. Taken together, these data indicated that species of lower DNA value share fewer sequences with *L. hirsutus* and show greater divergence in those they do share. These results therefore suggest that the change in DNA values within the genus *Lathyrus* has been accomplished by the progressive addition of new sequence families to the moderately repetitive DNA component, with what appears to be a far more modest but highly correlated rate of addition to the low-copy class as well.

Observations indicating genome size increase largely through augmentation of the moderately repetitive class are not confined to plant groups. The work of Mizuno and Macgregor (1974), which is discussed in detail below, suggests that precisely this mode of evolution has been operative in the salamander genus *Plethodon*. In another amphibian study, Baldari and Amaldi (1976) compared two Anuran and two Urodelan genomes, each pair differing by about a factor of two in DNA content, and found that the amount of single-copy DNA was about the same for the two Anura or the two Urodela, with increases in the repetitive component accounting for most of the additional DNA in the larger genome of each pair.

Schmidtke *et al.* (1979b) have observed an interesting situation in the teleostan family Cyprinidae, where genome sizes appear to have undergone evolution both by augmentation and the more traditionally understood mode of polyploidization. These workers compared the HAP reassociation curves of six Cyprinid species: four diploids spanning a 2-fold difference in DNA value and two very ancient tetraploids. For the diploids, there data indicate that most of the supplementary DNA found in the larger genomes is additional moderately repetitive component

DNA. The two ancient tetraploid genomes did not exhibit these proportions; rather, they had about twice the total single-copy DNA content as the smaller diploid genome and showed some additional moderately repetitive DNA, suggesting a probable derivation from small genome diploid ancestors.

An extremely interesting experiment has been performed with plants of the genus *Lolium* (rye-grass) by Hutchinson *et al.* (1979), in an attempt to detect genetic effects of supplementary DNA on various parameters of the phenotype. This genus can be divided into species characterized by a high level of inbreeding and species which preferentially outcross. The inbreeding species have about 40% more nuclear DNA on the average, though all species are diploids with $2n = 14$, and most of the difference is due to an increase in the moderately repetitive DNA class. Several of the high-DNA inbreeding and low-DNA outbreeding species were crossed, and the F_1 plants were then either backcrossed to one of the parent species or bred to produce an F_2. Backcross or F_2 progeny were scored with respect to 19 phenotypic characters which differed significantly in the parentals, including such presumed fitness components as rate of growth and flowering time, and the DNA values were also determined for the same plants. The data were treated statistically, to determine if there were any correlation between any of the characters and segregation with respect to DNA value. Rather surprisingly, no effect on any of the nineteen phenotypic characters was detected which could be attributed to the supplementary DNA. The only significant deviation from expectation was in the DNA value histograms of the F_2 and backcross progenies, which were slightly lower than the predicted midparent values. Aside from this modest bias in transmission, chromosomes from genomes which had experienced a 40% addition of moderately repetitive DNA appeared fully able to interact in conjunction with those of an unsupplemented genome to carry out a successful and undisrupted developmental program producing viable and fertile plants.

The pattern of repeat class increase occurring without change in chromosome number, while widespread, is probably not universal. Cullis and Schweizer (1974) followed the reassociation of the DNA of six *Anemone* species with genome sizes ranging from 19.9 to 32.0 pg, using optical techniques, and found no such simple correlation. Nevertheless, an examination of the available data for a number of plant and animal taxa by Hutchinson *et al.* (1980) leads to the conclusion that increases in the repetitive class very frequently account for the bulk of genome size changes within groups.

Claims for a contrasting mode of evolution have been made by Bachmann and Price (1977), who examined the reannealing properties of the DNA of eleven Cichoriea, including two species of *Crepis,* four species of *Microseris* and the related *Agroseris grandiflora,* and two *Pyrrhopappus* species, dividing the genomes of each into fast, intermediate, and slow reannealing components. The proportions of the genome in each fraction were compared with respect to the total genomic DNA content. These parameters were then considered from the perspective of relative evolutionary positions, based on ancestral versus derived characteristics of the various species as assessed by classical methods. The authors concluded that evolution in this group has involved considerable changes in the relative proportions of these fractions, including massive saltational *losses* from both the intermediate and slow classes in some derived types. They argue that these losses are regular, involving about half of the particular class in each case. A criticism of this study which must be made, however, pertains to the method used to determine the proportions in the various fractions. In the study, reassociation of each DNA was monitored optically until an apparent plateau was reached in the intermediate fraction; the remaining DNA was then assigned to the slow fraction. Most of this data is given in tabular form, but the reassociation of *Agroseris* DNA is shown as an example of the observations used in constructing the table. Examination of this figure reveals that the entire transition assigned to the intermediate fraction covers only a single decade of C_0t from the end of the plateau following the reassociation of the fast fraction to the termination of the experiment. Since the reannealing of even a kinetically uniform DNA component requires two decades of C_0t to complete the middle 80% of its transition (Britten and Kohne, 1968; see also Bachmann and Price's own curve for T4 DNA on the same figure), it appears that at best about half of the intermediate class of *Agroseris* has been scored in this experiment. (It may be suggested that the apparent plateau reached at termination may be due to thermal degradation and depurination, which can occur during long incubation times; Britten *et al.*, 1974.) The quality of the data for the other species cannot be assessed from the table given. In any event, the case for large saltational *decreases* in the moderately repetitive class, which would be most exciting if confirmed, requires further substantiation.

B. Augmentation of Large Moderate-Repeat Components by Saltational Generation of "New" Sequence Families

The studies described above were largely focused on groups characterized by large genomes containing large proportions of complex moderate-repeat components, where large-scale evolutionary changes were presumed and generally found to have occurred. In a few cases, more detailed direct sequence hybridization studies have been performed to examine the evolutionary origin of the added repeats. For the examples dealt with below, the results suggest that the newer repeated families in a given lineage arise from sequences not detectably repeated in other related lineages.

1. *Salamanders*

The genomes of Urodele amphibians are generally characterized by high DNA content (Hinegardner, 1976) and large moderately repetitive components (Straus, 1971; Moreschali and Serra, 1972; Baldari and Amaldi, 1976). Important studies of DNA evolution in this vertebrate group have been performed by Macgregor and co-workers as part of a program examining many aspects of cytomolecular organization in the salamander genus *Plethodon*. This North American genus is comprised of 21 species arranged in three major species groups, the Eastern small, Eastern large, and Western plethodons, as well as the geographically and evolutionarily isolated southwestern species *P. neomexicanus*. All species have 14 chromosomes and a similar karyotype, but they range in C-value from 18 to 69 pg. They thus provide excellent material for the comparative study of C-value, karyotype, and DNA sequence homology. In the first of two massive and detailed papers, Mizuno and Macgregor (1974) examined the evolution of the moderate-repeat component in the context of these other parameters, with results summarized in the following paragraphs.

With respect to C-values determined for 15 species, the four Eastern small plethodons examined all fell close to 20 pg. The five Eastern large plethodons included three species with similar genome sizes and two more with 70–80% more nuclear DNA. Five representatives of the Western plethodon group were also studied, of which three showed values of 33.8, 36.8, and 38.8 pg (*P. dunni*), while the other two were substantially larger. *P. neomexicanus* was intermediate at 30.7 pg.

Karyotypically, the plethodons show a much greater uniformity. Lampbrush chromosomes were examined in five species representing the three major groups: *P. c. cinereus* (Eastern small, 20.0 pg), *P. glutinosus* (Eastern large, 22.5 pg), and *P. vehiculum, P. dunii,* and *P. elongatus* (Western, 36.8, 38.8, and 33.8 pg, respectively). Significant differences in relative length (P 20.05) were found on only 3 chromosomes: chromosome 2 differing between *cinereus* and *dunii* and chromosome 5 and 11 differing between *cinereus* and both *dunni* and *vehiculum*. Thus the large changes in DNA content appear to have been distributed rather uniformly over the genome.

On the basis of this background information, C_0t analyses were performed on the DNAs of *P. cinereus, P. vehiculum,* and *P. dunni,* using HAP separation to monitor reassociation over time. A large and complex moderately repetitive class was found to exist in all three species' DNAs. For *P. c. cinereus* (20 pg), about 40% of the genome is in the slow reassociating fraction, while for the two Western species (*P. vehiculum,* 36.8 pg, and *P. dunni,* 38.8 pg) only about 20% of the genome is in this low-copy fraction. It therefore appears that much of the difference in DNA content has been due to changes in the moderately repetitive component.

The proportion of shared repeats in these components was examined by direct DNA comparisons, employing radioactive probe for the reiterated portions of the moderately repetitive classes of *P. c. cinereus* and *P. vehiculum* in hybridization and remelting studies with cold DNAs from species representing all groups. Additional studies were performed with *cinereus* probe employing low stringency reassociation and HAP binding conditions. In the stringent studies, species of the same group were found to have 60–90% of the repetitive sequences scored in common, while Eastern small and Eastern large species share 40–60%, and Eastern groups have less than 10% of their repeats in common with Western groups (Table II). The remelting studies indicate that these shared sequences differed by 10–15% in base sequence. The comparable values in reciprocal comparisons using Eastern small (*cinereus*) versus Western (*vehiculum*) probe were equivalent. Under low stringency conditions, detectable homology between *P. c. cinereus* repeat probe and Eastern large species' DNA rose from about 50 to nearly 80%, while cross-hybridization with Western group DNAs rose from the approximate 10% value to around 35%. Overall, these results imply that much of the moderately repetitive sequence component scored for a *Plethodon* species is

TABLE II
MODERATELY REPETITIVE PROBE DNA, CONDITIONS AND LABEL[a]

Whole driver DNA	P. cinereus high stringency		P. vehiculum high stringency ^{125}I *in vitro*	P. cinereus low stringency ^{125}I *in vitro*
	^{3}H *in vivo*	^{125}I *in vitro*		
Eastern small group				
P. c. cinereus	100	100	10.3	100
P. c. polycentratus	102.8	—	—	—
P. shenandoah	97.0	—	—	—
P. hoffmani	95.8	92	12.3	—
P. richmondi	97.4	—	—	—
P. n. hubrichti	79.1	90.8	9.6	104.2
Eastern large group				
P. wehrlei	48.7	54.7	11.4	81.8
P. glutinosus	45.0	52.4	9.6	76.3
P. punctatus	44.3	58.9	9.0	—
P. ouachitae	54.5	—	—	—
P. yonahlossee	54.8	57.4	6.9	—
P. jordani	59.0	55.1	4.7	—
Western group				
P. elongatus	6.4	13.8	57.5	39.3
P. vehiculum	8.6	12.1	100	34.9
P. dunni	7.5	12.1	81.6	33.3
New Mexican				
P. neomexicanus	8.2	9.3	20.4	—

[a] Relative percentage of moderately repetitive *Plethodon* probes reassociating with moderately repetitive kinetics to driver DNAs of variously related species. Homologous reaction set as 100% (data from Mizuno and Macgregor, 1974).

either not present in the equivalent repeat class of more distantly related congeners or else is highly diverged.

The more recent study of Mizuno *et al.* (1976) focused attention on the 10% component of the *P. c. cinereus* moderately repetitive class which is shared with the diverged genome of Western group species even under reasonably stringent criteria. In hybridizations to fractions of $CS_2SO_4-Ag^{2+}$ gradients of *P. dunni* DNA, the *cinereus* common component is found to track the main band peak, while probes for rDNA, 5 S DNA, and 4 S DNA sequences show skewed or satellite distribution, indicating that these known conserved repeat sequences do not constitute a significant portion of the homologous moderate repeats. The fidelity of homologous pairing for the shared component was found to be high by two different criteria. When *cinereus* repetitive probe was reassociated with cold *cinereus* or *dunni* DNA and digested with S_1 nuclease under fairly rigorous conditions, little difference was found in the susceptibility of the reassociation products. Complementary results from remelts showed that the T_m of reassociation products with the two drivers differed by only 3.5°C before S_1 digestion or 2°C afterward.

In order to examine this conserved component more closely, it was partially purified by hybridizing labeled *cinereus* repetitive DNA to high-molecular-weight *P. dunni* DNA, isolating that portion of the label which cosedimented with the longer driver in neutral sucrose gradients, and then refractionating the longer driver from the shorter probe in alkaline sucrose. This enriched probe was then reannealed with DNAs from *P. cinereus, P. dunni,* several other plethodons, and more distantly related species. The level of reassociation and HAP-monitored thermal stabilities were compared with those seen for unfractionated *cinereus* probe (see Table III).

With respect to *P. dunni* and other Western species, the *cinereus–dunni* probe exhibited a 3- to 5-fold higher cross-reactivity than an unfractionated *cinereus* probe. In cross-reaction with the more closely related Eastern *plethodon* species, where total repetitive cross-homology had previously been shown to be high, a more modest (1.3 to 1.4)-fold increase was found, while no detectable difference was seen with *cinereus'* close relative, *P. n. hubrichti*. Cross-homology of the *cinereus–dunni* probe was also readily detectable in reassociation to the DNAs of *Aneides* and *Ensatina,* other genera of the tribe Plethodontini, which exhibited about half the level seen with Western plethodons, while about 10% of this level was seen with *Pseudoeurycea,* a member of a different Plethodontid tribe, and with more distantly related salamanders. Thus little conservation of this repeated component is found beyond the tribal level, a result which was confirmed by filter hybridization/competition experiments.

The arrangement of the conserved repeats in the genome was not studied directly. However, when *cinereus* moderately repetitive DNA was nick-translated and annealed to *cinereus* or *dunni* meiotic chromosomes, the pattern

TABLE III

REASSOCIATION OF UNFRACTIONATED *P. cinereus* MODERATELY REPETITIVE PROBE VERSUS PROBE ENRICHED FOR SEQUENCES SHARED WITH *P. dunni;* RELATIVE PERCENTAGE TO THAT SEEN WITH HOMOLOGOUS (*P. cinereus*) DRIVER

| | [125]I Moderate-repetitive probe | | | |
| | Unfractionated *cinereus* | | *cinereus–dunni* "common" fraction | |
	C_0t 20	C_0t 145	C_0t 20	C_0t 145
Western group				
P. dunni	6.35[a]	10.3	26.5[a]	32.3
P. vehiculum	8.2	9.2	33.5[a]	29.5
P. vandykeyi	5.65[a]	—	19.7	—
P. larselli	6.05[a]	—	22.0[a]	—
P. elongatus	—	11.1	—	34.6
Eastern large				
P. glutinosus	39.6	43.9	55.0	58.7
Eastern small				
P. n. hubrichti	—	91.9	—	92.3
P. c. cinereus	100	100	100	100

[a] Mean of two experiments.

of *in situ* hybridization was distributed broadly over all chromosomes of both species, suggesting that the shared repeats are widely though perhaps not randomly dispersed within the *Plethodon* chromosomes.

2. Cereals

Impressively detailed information has been obtained on evolution and organization of the extremely large moderately repetitive components of wheat and its relatives by Flavell, Smith, and Rimpau, employing clever modifications of mass reassociation techniques, in particular the *L* vs *R* approach, to analyze the intergenomic as well as the intragenomic distribution of repetitive sequence segments. Their methodology is outlined in Smith *et al.* (1976).

Briefly, labeled tracer DNA of a variety of lengths from one species (in this case wheat) is reannealed with a large excess of homologous or heterologous driver DNA through the C_0t range for reannealing of the moderately repetitive component. The resulting curve with homologous driver can be used to infer the interspersion pattern of all repeats in the parent genome in the usual way. With the heterologous driver, however, only those segments of probe containing repeats also found in the other genome will be bound to HAP. The extrapolated *Y* intercept of the heterologous curve thus provides an estimate of the proportion of the *second* species' genome comprised of shared repeats, while its inflection points and eventual plateau with increasing size of probe give information as to the interspersion of these shared repeats among other heterologous DNA segments. For the case of wheat vs oats examined in the Smith *et al.* paper, the homologous (wheat/wheat) curve shows nearly 80% binding at the shortest probe fragment size of 250 nucleotides, reaches an inflection point at 90% binding and 900 nucleotides, and exhibits 95% binding at 4000 nucleotides. With oat DNA driver, however, initial binding is 30%, an inflection is seen at 50% and 700 nucleotides, and a plateau is reached at 67% and about 2300 nucleotides. When these curves are extrapolated to give Y intercepts, the homologous curve indicates that about 75% of the wheat genome consists of moderately repetitive segments, while the heterologous curve implies that about 22% of the wheat genome is comprised of repeats also found in oats DNA. Since 67% of the wheat probe can be bound to HAP when wheat fragments longer than 3200 nucleotides are annealed to oats driver, these shared repeats must be extensively interspersed through the wheat genome at intervals of less than 3500 nucleotide pairs. In conjunction with the homologous curve pattern, this indicates that the wheat repeats shared with oats must be interspersed among repeats not shared with oats and probably single-copy segments as well. The inflection at 750 nucleotides extrapolates to indicate that 46% of the wheat genome consists of such shared repeats separated by 750 nucleotides of unshared DNA, while the 67% binding plateau attained by 3500 nucleotides implies that the remaining shared repeats occupy 21% (67–46%) of the wheat genome and are separated by about 3500 nucleotides of unshared DNA.

In subsequent studies, this approach has been employed to dissect the genomes of wheat, barley, rye, and oats. Description of the complex details of these experiments is beyond the scope of this article, but the results may be summarized with respect to two facets of the evolution of the moderately repetitive DNA class in these species, sequence relationships and sequence interspersion.

In regard to sequence relationships, Flavell *et al.* (1977) performed homologous and heterologous hybridizations with probe and driver DNAs of all four species over a range of smaller fragment sizes sufficient to extrapolate a Y intercept and thus determine the true proportion of shared repeat sequence for each genome compared to the other three. This allowed them to classify the repeats within these genomes into seven groups, and these groupings are consistent with the evolutionary relationships of the four species. In group I are placed all repeats common to the most distantly related species, oats, and the other three. Group II comprises repeats shared by barley, rye, and wheat, but not oats, while group III consists of repeats held in common only by the more closely related rye and wheat. Groups IV through VII are those repeats peculiar to wheat, rye, barley, and oats, respectively. The proportion of the several genomes falling within these groups varies, as expected from the considerable difference in their C-values even when these are corrected for level of polyploidization (see Table IV). It appears however that all group I repeats are found in all four species, all group II repeats are found in barley, rye, and wheat, while all group III sequences are by definition present in both rye and wheat. It seems that the preponderance (over 70%) of moderate repeats in these genomes has entered this class since the separation of the oats lineage, while more than 40% have been added since the divergence of barley, and perhaps 30% since the separation of wheat and rye.

The interspersion patterns of these shared and unshared repeats have subsequently been examined using a full range of radiolabeled probe sizes for barley and oats (Rimpau et al., 1980) and for wheat and rye (Rimpau et al., 1978). The DNA bound to HAP by reassociation to given heterologous drivers at large fragment size was also sheared and rereannealed with the same or less heterologous drivers to determine experimentally the interspersion of the various groups. Separate experiments were performed in these studies to examine the nature of fragments not bearing a segment able to cross-react with probes from other species. Taken together, these experiments allow the construction of schematic maps illustrating the average interspersion structure of different classes of DNAs, which indicate that moderate-repeat sequences of recent evolutionary origin are predominantly finely interspersed among both single-copy regions and other repeats of more remote evolutionary origin.

To summarize these results, more than half of the total weight of each of these four plant genomes consists of moderately repetitive sequences which have been added since the divergence of ancestral oats from the mutual ancestor of the other three species. The process of augmentation would appear to have been continual over the course of evolution, for substantial subclasses can be identified as having been added during the intervals between each bifurcation in the lineages. It also appears to have been rather regular in each lineage when the total percentage of new DNA is summed for each species: oats—58% (group VII), barley—

TABLE IV

PERCENTAGE OF THE TOTAL MODERATE REPEAT COMPONENT IN EACH CEREAL GENOME FALLING INTO THE VARIOUS CROSS-REACTING GROUPS, AND TOTAL PERCENTAGE OF EACH GENOME IN EACH GROUP[a]

Species	I Common to all	II All but oats	III Wheat and rye only	IV Wheat only	V Rye only	VI Barley only	VII Oats only
Wheat	29.7 (22)	13.5 (10)	35.1 (26)	21.6 (16)	—	—	—
Rye	25.7 (19)	25.7 (19)	18.9 (14)	—	29.7 (22)	—	—
Barley	28.2 (20)	32.4 (23)	—	—	—	39.4 (28)	—
Oats	22.7 (17)	—	—	—	—	—	77.3 (58)

[a] In parentheses.

51% (groups II and VI), rye—55% (groups II, III, and V), and wheat—52% (groups II, III, and IV) (see Table V).

More specific information pertaining to these observations has recently been obtained by examination of a 5200 ntp random cloned fragment of wheat DNA plasmid, pTA8 (Flavell *et al.*, 1981). This segment was found to consist entirely of repeated components, but hybridizations of four subfragments produced by restriction of the insert to the DNA of *T. monococcum* and *Aegilops squarrosa* (modern representatives of two of the three wheat genomes) showed that the different fragments represent unrelated repeat families. The frequencies of these families vary greatly between the genomes, including one family virtually absent from *A. squarrosa* but repeated 800 times in *T. monococcum*. The particular ordering of these repeats found in the plasmid is not common in the wheat genome, rather Southern hybridizations with the subfragment showed hybridization to several large genomic *Hin*dIII bands, implying the presence of homologous repeats in longer repeated families.

Flavell *et al.* account for the results seen in their bulk cross-hybridizations and cloned repeat by postulating recurrent saltatory amplifications in which a segment representing an already repeated family is amplified together with the regions which happen at that time to lie adjacent to the particular repeat (see also Bedbrook *et al.*, 1980). Repeats of a group peculiar to a lineage would thus represent a low or single-copy segment adjacent to a repeat found in the ancestral genome, this pair having undergone amplification in the one lineage but not

TABLE V

PERCENTAGE OF WHOLE RADIOACTIVE CEREAL DNAS REASSOCIATING AS MODERATELY REPETITIVE WITH HOMOLOGOUS AND HETEROLOGOUS DNAS, AND CALCULATED PERCENTAGE OF GENOME COMPRISED OF MODERATE REPEAT FAMILIES ADDED TO EACH SINCE THE SEPARATION OF THE OATS LINEAGE

Unlabeled DNA	Labeled DNA			
	Wheat	Rye	Barley	Oats
Wheat	74	52	42	14
Rye	58	74	44	17
Barley	32	38	71	20
Oats	22	19	20	75
Percentage of genome which consists of families added since oats' divergence	52	55	51	58[a]

[a] Proportion of moderate-repeat component in oats minus average proportion shared with the other three genomes.

others. Rounds of recombination and deletion are then presumed to result in the variety of linear permutations which exist for most repeats.

The extent of applicability of this or any other model to cereal genome repeat organization will require much additional data to determine. One potential point of difficulty is that under the postulated system the repeat segments characterized by the lowest sequence extents divergence, presumably the most recent products of saltation, ought to be substantially longer than the average of all repeats, as this average would be dominated by the older families which have had more time to undergo rearrangement. Earlier data for wheat (Flavell and Smith, 1976) and rye (Smith and Flavell, 1977), however, suggest that these undiverged segments are as small or smaller than the average repeats. Further investigation or elaboration of the model may dispose of this quibble, or it may be that saltation and interpolation are in fact virtually coincident in cereal genomes. In any event, over evolutionary time the process by which new repeats have been added appears to be both dispersive and interpolative, so that by the time they are observed they are found to be intimately interspersed among other genomic sequences of different evolutionary origin.

3. *Lilium*

Members of the genus *Lilium* possess a large moderate-repetitive component with one of the highest informational complexities known: over five times the sequence information content of the entire genome of *E. coli* (Hotta and Stern, 1975). Particular interest has focused on this *Lilium* genomic DNA class since the demonstration (Smyth and Stern, 1973) that much of the DNA repair activity associated with recombination in pachytene meiocytes occurs preferentially in sequences of the moderate-repeat component. Recently, investigations of the repeat segments labeled at pachytene as compared with the total moderate-repeat component have been performed which have evolutionary implications (Bouchard and Stern, 1980).

In these studies, the reassociation and thermal stability of reannealed pachytene versus S-phase labeled moderate repeats were followed on HAP. The results indicate that the bulk genomic moderate repeats labeled at S-phase consist of sequence families covering the full spectrum of member sequence divergence which could have been detected at the criterion used. Virtually all of the pachytene labeling, however, was confined to repeat families showing little or no sequence divergence among their members. Both types of repeats were shown to be generally interspersed among unrelated flanking sequences by S_1 gradient experiments and were of similar size, with a mode between 1000 and 2000 ntp. These results imply that the isolated pachytene repeats were held under strong constraint with respect to sequence divergence, whereas the bulk of the genomic moderate repeats appear to be free to diverge through time.

Friedman and Bouchard have now performed comparative studies which confirm and extend these inferences. The details of these experiments are being submitted for publication, but I shall summarize the general observations here. *In vivo* labeled P-DNA probe prepared from the *Lilium* cultivar Enchantment was reassociated with the DNAs of *L. speciosum, L. longiflorum,* and *L. henryi,* representing different intrageneric lineages, and also with DNA from the more distantly related *Asparagus, Iris, Triticum* (wheat), *Zea* (corn), and *Vicia faba* (broad bean). Though the abundance of P-DNA-complementary repeats was found to differ in the more distantly related genomes, all these plant DNAs were found to hybridize with the pachytene repeat probe. When a probe prepared from all high thermal stability Enchantment repeats was employed, however, half failed to react even with DNA from other *Lilium* lineages. Virtually no cross-reaction took place with the more distantly related *Asparagus,* and none with wheat and corn, members of another monocot order, nor with the dicot *Vicia.* Thus the pachytene repeats have been highly conserved in seed plant evolution, whereas the bulk of undiverged repeats in Enchantment are families which have recently become members of the repeat component of the genome and have simply not had time to accumulate substantial member sequence divergence. Thermal stability studies indicate that the evolutionary conservation of pachytene repeats is probably quite high, with less than 15% divergence in the *Vicia* genome.

It therefore appears that the large and complex moderately repetitive sequence component of the *Lilium* genome, like the less complex components of the cereal genomes, is largely composed of families which have been saltationally added during the course of evolution, some of them quite recently. However, it also includes families implicated in the process of meiotic recombination which have been highly conserved through evolutionary time.

C. Origin of Segments Undergoing Saltational Increase in the Smaller Moderately Repetitive Components of *Xenopus,* Chicken, and Sea Urchin

The weight of evidence discussed thus far indicates that, for many groups, moderately repetitive DNA components are augmented continually in evolution by the addition of new sequence families. These observations naturally raise the question of the origin of those sequences which supply the new families produced by saltational events. For the large moderate-repeat component genomes dealt with in the preceding pages, these families appear to arise largely from segments not dectectably repetitive in related lineages. For the smaller moderately repetitive components of several intensely studied smaller genomes, however, a different state of affairs appears to prevail.

Data with direct bearing on this question for the *Xenopus* genome have been obtained by Galau *et al.* (1976), in a comparison of the moderately repetitive sequence components of *X. laevis* and *X. borealis* (*X. mulleri*). In their experiments, purified moderately repetitive DNA from each species was reassociated with homologous and heterologous driver over a broad range of C_0t and at an excess where cross-reaction could be detected even if the homologous sequences were present in far lower copy number in the heterologous DNA. The results indicated that the bulk of the repeat families present in one genome are complementary to sequences which exist in the other, but that about half of these shared sequences are found at a 10- to 100-fold lower copy number in the heterologous DNA. Therefore, half of the families now comprising the major moderately repetitive component in each of the *Xenopus* species are different from families found in the major component of the other, and have been added to that component by saltational increase since the separation of the two species' ancestors. The substrate sequences for the saltational increase in each lineage are, however, detectably repetitive in the other lineage. No direct information was obtained concerning the interspersion of the added repeats, but since at least three-fourths of the major class in *X. laevis* is known to be interspersed (Davidson *et al.*, 1973), a substantial number are presumably interpolated.

As noted in the section surveying repeat organization, the rather small moderate-repeat component of the chicken and certain other birds is characterized by a "long-term" mode of interspersion as compared with other vertebrates and deuterostome groups. Eden *et al.* (1978) have examined the reannealing of the chick moderate-repeat components to chick, Japanese quail, duck, and ostrich driver DNA and found that 100, 68, 66, and 23% of the whole repetitive probe reassociated with these respective drivers. To examine more closely the nature of the homology between chick repeats and the complementary sequences in the very distantly related ostrich genome, they then added homopolymer oligo(dA) tails to chick moderate-repeat DNA, purified the repeat portion cross-annealing to ostrich driver on HAP, and recovered the homologous repeats on oligo(dT)-cellulose. When these shared repeats were driven by whole chick DNA they were found to comprise 27% of the moderate-repeat component (or 3.5% of the genome) with kinetic abundance of 3000 repeats per haploid genome, about twice the average frequency of abundant moderate repeats in this genome. With ostrich driver, however, the portion which cross-reacted reannealed to sequences repeated only 30 times in the ostrich genome. Thus the shared repeat component was more abundant than the total abundant moderate-repeat component of chick but was complementary to sequences present at only a very low level of repetition in ostrich, implying that these particular shared families predate the evolutionary separation of these avian lineages but have reached high abundance only in the line leading to the chicken genome.

The most detailed information now available concerning evolution of short moderate repeats is for the genome of sea urchin. Harpold and Craig (1977) examined the homologous and heterologous reassociation of labeled *Strongylocentrotus purpuratus* repetitive probe and cold *S. centrotus, S. droebachiensis, S. franciscanus,* and *Lytochinus pictus* driver DNAs, and determined that the proportion of shared families was 90, 83, and 40%, respectively. This indicates a considerable degree of evolutionary stability for this component in sea urchin genomes, since these species are probably separated by divergence times of 20, 25–40, and 120–200 million years, respectively. HAP thermal stability studies indicate that the shared sequences are more diverged between than within genomes, the relative degree of divergence roughly paralleling the elapsed time since lineages diverged. Remelts of reactions reassociated to differing extents further suggested that the more repeated families tend to be less conserved.

These findings were largely confirmed and extended by Moore *et al.* (1978), in a study on *S. purpuratus, S. franciscanus,* and *L. pictus.* These workers employed purified moderately repetitive DNA fractions from more than one species and also used cloned moderate-repeat segments from *S. purpuratus* previously prepared by Scheller *et al.* (1977). In one set of experiments with repeat fractions, tracers were prepared from all three genomes representing the more repetitive families of the middle-repetitive class, sequences present in 1000–2000 copies per genome. In the reciprocal comparisons between *S. purpuratus* and *S. franciscanus,* reaction of each probe with its homologous DNA indicated a 2- to 3-fold higher abundance than seen with the heterologous DNA. The reciprocal reactions involving the more distantly related *S. purpuratus* and *L. pictus* showed heterogeneous and incomplete reassociation in each heterologous reaction. For the sequences which do cross-react, the average representation in the heterologous genome was more than 100-fold less than in the homologous DNA. Despite the fact that the reassociations were done at a driver excess which should have detected complementary sequences even if these resided in the single-copy component of the heterologous DNA, at least a third of the probes failed to react to the heterologous drivers, indicating very high divergence or absence. When similar comparisons were done with *S. purpuratus* probe representing the broader spectrum of repetition, the general pattern of slower annealing with the more distant species was still observed, though the heterologous reaction is more complex, particularly in the case of *S. purpuratus* vs *S. franciscanus.* Nevertheless, it may be concluded that in the sea urchins the dominant repetitive DNA families in one genome have higher frequencies than in related genomes, and that different sets of repetitive families have come to comprise this dominant class in the different genomes. When the two highly diverged genomes of *S. purpuratus* and *L. pictus* are compared, it becomes evident that a majority of the moderate repeats in each represent sequences which are repeated only one or a few times in the other.

When individual families were followed using strand separated probe prepared from cloned repeats, the results were compatible with those seen in the mass repeat fraction comparisons. The clones used here have been shown to represent families covering a representative spectrum of repeat frequencies and internal sequence divergence (Klein *et al.*, 1978). Of nine *S. purpuratus* clones whose relative degree of repetition was measured for *S. purpuratus* vs *S. franciscanus*, the ratios of kinetically determined repetition frequencies ranged from 1 to over 20, with a mean of 6.6 (Table VI). Corrections of these ratios for the effects of heterologous sequence divergence were made in six cases, but the qualitative pattern of lower sequence abundance in the *franciscanus* genome remains intact. Interestingly, despite the difference in genomic copy number, the level of transcription of several of the repeats has been found to be the same in egg RNA (Moore *et al.*, 1980). For the four clones compared to *L. pictus*, both the uncorrected and (for two) corrected ratios show that these are present in very low copy number in this distantly related genome. These results were further substantiated for three of the clones by a solution based filtration hybridization method, the results of which agree within experimental error.

An interesting conclusion follows from the observation that predominant repeat families in the sea urchin *S. purpuratus* genome appear to have arrived at their present high copy number through saltational events confined to the *S. purpuratus* line and postdating the separation of the *S. purpuratus* and *S. franciscanus* lineages. If a particular family has undergone a recent substantial increase in membership in the ancestral lineage of a particular genome, the bulk of its members should be fairly similar to one another in that genome. In a second lineage which branches off before the saltation, however, only the lower copy number, internally diverged ancestral family will be found.[2]

On the basis of these interspecific comparisons, it appears that the complex moderately repetitive DNA components of the "type" short-term genomes of *Xenopus* and sea urchin arise by a complex process of progressive saltation and interpolation. Certain repeated families become more repetitious with time, but the particular families chosen to form the predominant repetitive class appear to be peculiar to a given evolutionary lineage. The process thus appears to combine aspects of both specificity (families which are already repetitive are the source of the sequences which become more repetitive) and randomness (the repeat families chosen for an increase in membership are a different subgroup in different descendants of an ancestral genome). Some evidence for at least partial operation of a similar system also exists for the longer repeats of birds. The process by which this occurs is apt to be quite complex, as attested by three recent studies employing cloned repeats to examine the nature of related repeats within a genome at a level of precision previously impossible.

In the first of these investigations, Anderson *et al.* (1981) employed three

---

[2]As this article was being completed, results of further evolutionary comparisons involving a number of these same clones were published which accord well with this view (Moore *et al.*, 1981).

TABLE VI

Comparison of Frequencies for Repeated Families Identified by Nine Cloned Moderate Repeats from *S. purpuratus* DNA in the Genomes of *S. purpuratus* (Sp), *S. franciscanus* (Sf), and *L. pictus* (Lp) Derived from Reassociation Kinetics with Each DNA[a]

Clone	Frequency Sp	Frequency Sf	Ratio Sp/Sf	Corrected ratio[b]	Frequency Lp	Ratio Sp/Lp	Corrected ratio[b]
CS2007	400	45	9	7	Low	—	—
CS2034	1000	160	6	4	10	100	25
CS2090	140	95	1.5	1.5	Low	—	—
CS2099	80	60	1.3	—	—	—	—
CS2101	700	55	13	13	2	350	192
CS2108	20	30	0.7	—	—	—	—
CS2109B	200	50	4	4	—	—	—
CS2133A	2100	102	21	—	—	—	—
CS2137	530	185	3	2	—	—	—

[a] Adapted from Moore *et al.* (1978).

[b] Corrected for retardation in the heterologous rate of reassociation due to sequence divergence of repeats found in the heterologous genome.

previously obtained sea urchin plasmid clones to screen λ-sea urchin clones containing long genomic segments for related repeats, in order to examine the interspersion patterns and degrees of relatedness of individual family members. These were examined at the level of restriction mappings and Southern blot reactions. The results are complex and differ for the three cloned segments: one (2034) is found to represent a portion of a longer repeat which occurs in many small clusters, and at least one large tandem run, all repeats being closely related to one another. The second clone (2108) also identified a long (4500 ntp) repeat family which appeared to have a complex organization, being comprised of a small number of very similar members which occur singly and a much larger number of diverged repeats. The third (2109) identifies a true short repeat family (200–300 ntp), most members of which are highly diverged when compared to the mass of family members and are found to be flanked by single-copy regions. More detailed analyses, including heteroduplex visualization and direct sequencing analysis, have been performed in cloned members of these three families by Scheller *et al.* (1981) and reveal still another level of complexity. The family originally identified by probe 2108 is revealed to consist of perhaps 20 to 40 subfamilies which exhibit close relationship among their members but only distant relationship to other 2108 subfamilies. Sequence subelements within a given subfamily are arranged in colinear order over thousands of base pairs, with only a few percent sequence divergence. In other 2108 subfamilies, however, subelements are found in different order, and not all subelements are found in all subfamilies. The subfamilies appear to have relatively few members (average 25) and are themselves conserved in evolution despite their divergence from other subfamilies. The less thoroughly examined short repeat 2109 family also appears to exhibit subgroups but a spectrum of relatedness rather than discrete subfamilies exists. For 2034, however, no such substructure is seen. Further detailed nucleotide sequencing studies (Posakony *et al.*, 1981) of representatives of these and five other families have also been reported. These show that these represent families which are distinct even at the level of very short sequence comparison, but that several contain internal short direct or inverted repeats with a frequency greater than expected at random. This last study also detected a very fine-scale level of sequence rearrangement in the short repeat family 2109 (these results are summarized in Table VII).

The authors suggest that these families may represent different stages in a process in which saltations producing new subfamilies alternate with large-scale rearrangement and segment transpositions to produce isolated segment interspersion of family members. This process may be rapid, for nothing in the data appears to rule out the possibility that saltation and interposition are virtually simultaneous. Certainly these observations on individual repeat families seem to accord with the view that repeated saltational events frequently involve segment(s) of an already repeated family in sea urchin DNA, producing subfamilies

TABLE VII

SUMMARY TABLE FOR CHARACTERISTICS OF THREE *S. purpuratus* REPEATED FAMILIES STUDIED WITH CLONED REPEATS[a]

Family[b]	Length of colinear segments in genome	Interspersion[c]	Intrafamilial sequence divergence[d]
2034	Long	Grouped in one large and many small clusters throughout genome	Little sequence divergence
2108	Long for individual subfamilies—4500 ntp Short and variable for colinear portions of related subfamilies which share only some subelements	Both individual segments of subfamilies and elements of different subfamilies occur as single interspersed segments	Discontinuous—subfamily members show little divergence, high divergence between different subfamilies, which allows rearrangement of subelements
2109	Short—200–300 ntp	Single interspersed segments	Spectrum of divergence, average repeat considerably diverged from other family members. Some subelement rearrangement

[a] Adapted from data of Anderson *et al.* (1981), Scheller *et al.* (1981), and Posakony *et al.* (1981).

[b] Number designation derived from original repeat segment clone used to identify homologous segments in cloned library (see Table VI for earlier data on these families).

[c] Estimated from observed recovery of positive plaques in a λ genome library versus expected numbers if repeats occurred singly. More detailed information also obtained from selected λ clones.

[d] Based on remelting data for cloned probes reannealed with whole DNA, augmented by heteroduplex analysis, restriction map data, and primary sequence information for some individual repeat segments identified in genomic clones.

of related sequences. Interestingly, evidence for such rearrangements, at least at the level of subunits and long repeats, has also been found in cloned repeats from chick (Eden *et al.*, 1980; Eden, 1980). The further elucidation of this remarkable evolutionary process will be of great interest.

D. "Fluid" Moderate Repeats in *Drosophila*

Considerable information concerning certain moderately repetitive DNA segments has been obtained for *Drosophila melanogaster* in the last several years. The unique advantages afforded by this organism's amenability to genetic manipulation, relatively low DNA content, and possession of laterally amplified polytene chromosomes have been exploited in several studies with evolutionary implications. The results of these studies suggest that the long-term interspersed repeated segments of this organism may consist largely of elements which differ qualitatively in their pattern of sequence organization and mechanism of arriving at their locations from the short repeat segments of other organisms.

Particularly heavy research work has been focused on three moderately repetitive families, identified through cloned recombinant plasmids and known respectively as *copia*, 412, and 297 (Finnegan *et al.*, 1978), which appear to have many of the properties of bacterial transposible elements. The organization of these elements has recently been reviewed in detail (Spradling and Rubin, 1981). Here we describe the results of several studies with particular evolutionary implications and make note of others with similar implications.

Evidence suggesting that these sequences have the ability to change in number and position in the genome under conditions of relaxed selection has been obtained by Potter *et al.* (1979) in a study comparing their number and flanking sequence environment in *Drosophila* cultured cell-line genomes with the patterns observed in embryonic nuclear DNA. The rationale for this approach was that the demands made on the integrity of the genome to maintain single cells in culture, as opposed to producing a developmentally complex organism, would be far less stringent, providing a more favorable environment for frequent transposition. This study will be described in some detail because its methodology is a prototype for that used in other papers mentioned below.

When the copy number of each family was compared in embryo DNA versus Schneider's Line 2 and the Echalier Kc_0 line DNA at the study's outset, the prediction of rapid change made by Potter *et al.* was borne out. The *copia* family, for example, was found to be 5.1-fold more abundant in the Schneider's cell line than in embryonic DNA. Copy number for the other families was independently modulated, 412 is represented about the same number of times in Schneider's and embryo DNA, while 297 is increased 2.7-fold. The general trend, however, is toward increase of copy number in the cell lines. When total cell line or embryonic DNA was cut with restriction enzymes defining internal fragments of the cloned repeats, subjected to electrophoresis and Southern blotting, and hybridized with probes made from the same purified fragments, only one band of the predicted size was found with a given fragment. Since this band shows the relative intensity for each DNA predicted by its kinetically determined copy number, it appears that the additional repeats represent intact copies of the basic unit. Genomic DNAs were also digested with a restriction enzyme giving one cut within the typical repeat and the second somewhere in the flanking region, and blots of these digests were hybridized with a probe representing a region starting from the same internal site but terminating at another restriction enzyme site still within the repeated sequence. In these experiments, a highly heterogeneous pattern was always observed, and this showed numerous qualitative and quantitative differences for the various genomic DNAs, indicating the transposition had placed the new repeats adjacent to novel flanking sequences. Strong confirmation for this inference was provided by cloning a region containing both a *copia* element and a unique DNA sequence from Schneider cell DNA, and employing probe made from the unique segment to examine the size of the homologous fragment in embryonic DNA. This was found to be 4.5 ± 0.8 kb shorter, and therefore lacked the *copia* element.

Evidence that the elements of these families can also vary in location in organismic DNA was provided by Strobel *et al.* (1979). These workers examined the genomes of different laboratory strains of *Drosophila* (Oregon R, Canton S, Seto, and Swedish C), using as probes fragments from cloned repeats in a manner analogous to that employed in the cell-line studies and obtained similar results. In addition, they compared the actual chromosomal locations of the complementary sequences by performing *in situ* hybridization of clone fragments to polytene chromosomes of the various strains. The latter experiment, despite the apparent homosequentiality of all the strains, revealed that the locations of the repeat families were almost entirely divergent (e.g., less than 20% of 297 sites or 10% of *copia* sites are common to all four strains used).Closer examination revealed that considerable variability in repeat location could also be detected among individuals within a strain, and even between the paternal and maternal chromosomes of single individuals when these were compared at squash-induced asynaptic regions. Essentially equivalent results have been obtained for independently identified repeated segments of this type by Soviet investigators (Tchurikov *et al.*, 1978, 1981; Georgiev *et al.*, 1981), suggesting that such transposon-like repeated elements are an important component of the *Drosophila* genome.

The dual characteristics of conservation of repeat sequence set against plasticity of location are not confined to these well-characterized repeat families. In survey experiments, Young (1979) and Young and Schwartz (1981) analyzed Southern blots of whole restriction digested genomic DNA with probes representing 80 randomly chosen clones from a *Drosophila* library and found 18 contained dispersed repeats. Seventeen of the 18 were examined by *in situ* hybridization to the polytene chromosomes of two *D. melanogaster* laboratory strains which have been isolated for over 50 years. These strains appear to be homosequential with respect to banding pattern, but in all cases the locations of regions complementary to a given clone were highly divergent. All clones were mapped precisely on chromosome region 3R, and three clones were mapped on all chromosomes. In all cases, the number of common sites was in the minority. Based on the estimate of Manning *et al.* (1975) of the average repeat size, this survey represents 17 of the approximately 70 families which comprise the total moderately repetitive DNA class in *Drosophila,* suggesting that fluidity of location is a fairly general property of this component. Interestingly, while the location of the members of each family was very different in the two strains, the number of repeats for any given family appears to be conserved.

This ability to move about in the genomic DNA and become stably integrated in a variety of sites is a characteristic which these *Drosophila* repeat families share with prokaryotic transposons and some phages (Calos and Miller, 1980), as well as integrated proviruses in vertebrates (Hughes *et al.,* 1978; Shimotohnu *et al.,* 1980) and the mobile element Ty$_1$ of yeast (Cameron *et al.,* 1979). Additional parallels have been found by examination of the sequences of *copia,* 412, and 297. These include possession of direct terminal repeats which are identical for a given repeat though they may differ slightly between repeats of a family (Levis *et al.,* 1980) and the generation of a short base sequence duplication in the host DNA upon integration (Dunsmuir *et al.,* 1980). There does not appear to be any sequence homology to integration sites or among sites where elements integrate, suggesting that the bulk of the genomic DNA may well be susceptible to integration (Rubin *et al.,* 1981).

Though the pattern of long repeat segments apparently analogous to the transposons of other organisms is clearly responsible for a substantial portion of *Drosophila* moderately repetitive DNA, other types of repeats exist. Potter *et al.* (1980) have recently examined the sequence heterogeneity and lability in genomic location of several repetitive segments containing terminal inverted repeats. While these elements also demonstrate locational polymorphisms indicating transposon character, they differ from the *copia*-like class not only in possessing inverted terminal repeats, but also in being characterized by a greater heterogeneity in sequence between family members.

A radically different mode of repeat element organization has also been identified in a class of cloned segments by Wensink *et al.* (1979). These segments were originally identified by screening a large number of randomly cloned genomic sequences with probe from a preexisting repeat-segment clone, pDm1, and selecting four plasmids with large inserts from among those which gave a strong hybridization signal. The four were then analyzed by the technique called "Southern cross" hybridization (Sato *et al.,* 1977) for cross-homology between different restriction fragments of their own DNAs and each of the other three DNAs. The results of these internal cross-hybridizations indicate that each of the cloned segments contains a number of shorter repeat elements, often occurring more than once on different fragments of the same segments and generally interspersed among different repeats on other segments. An estimated 52 elements probably averaging less than 0.5 kb are necessary to account for all the cross-homologies observed on just the four cloned segments examined.

Despite these exceptions, it would appear that much of the moderately repetitive DNA component of *D. melanogaster* consists of the large "nomadic" segments. The variability in location of these elements seems unequivocal, but it is useful to bear in mind that their true rate of transposition in the organismic genome is still indeterminant. In discussions dealing with the heterogeneity in location of these repeats within and between laboratory strains, there is some tendency to assume that heterogeneity seen within a strain has entirely accumulated since that strain's establishment in the laboratory. In fact, older strains such as Oregon R, Swedish b and c, and Canton S were originally established from stocks obtained in the wild, and thus encompassed a considerable sample of the genetic variation present in their ancestral demes (Bridges and Brehme, 1944). Even strains established from single inseminated females will initially contain a minimum of four haploid genomes. While the tendency for strains to become more homozygous with time may be expected to reduce initial variability, the rate at which this will occur for an entity segregating at scores of loci which can undergo continual recombination is low. These arguments are not meant to suggest that elements such as *copia* do not transpose within the *Drosophila melanogaster* genome over the course of evolutionary time, since the evidence for this seems persuasive; they are merely included to point out that the rate and frequency of such events may be less than is sometimes supposed for many repeated elements when their substrate is the genome of the intact organism. A modest rate of transposition, when coupled with recombination, is quite capable of producing tremendous heterogeneity: in a population polymorphic for only 10 transposon plus and minus sites, recombination alone can generate an array of 2^{10}, or 1024, different combinations of transposon locations. One case where the recombination source of intrastrain variability has been precluded suggests that the rate of accumulation of strictly transposition-induced polymorphisms may indeed be lower than sometimes thought. This case is the X chromosome of strain g-1, which was made homozygous at the time the stock was formed. Young and Schwartz (1981) have mapped the localization of *copia* elements on the X chromosomes of 21 different individuals from this strain over 4 years and found only one instance of a larva showing a different *copia* site, a change which could have occurred in the stock at any time in the past 50 years. Certain other elements do appear to transpose at experimentally detectable levels, such as the large TE element carrying the *white* gene (Ising and Block, 1981) and the inferred transposible "P" factor of hybrid dysgenesis (Engels, 1981).

The microevolutionary variability indicated by the polymorphisms of location seen for nomadic repeats in *D. melanogaster* raises the possibility that these elements might be poorly conserved through the macroevolutionary changes and genetic bottlenecks characterizing successive speciation events. This idea is supported by other data of Young and Schwartz (1981), who examined the DNA of *D. melanogaster's* sibling species, *D. simulans,* for the presence of 15 nomadic sequences. Only two were found in amounts comparable to those seen in *D. melanogaster,* while two-thirds were present in concentrations ranging from one-fourth down to virtually undetectable levels. Tchurikov *et al.* (1981) hybridized probes for four *D. melanogaster* repeated sequences to DNA and chromosomes of *D. virilis,* a representative of a different species group, and found no detectable cross-reaction at a level and stringency where single-copy cross-hybridizations were clearly discernible. Unpublished results of Martin and Schedel described by Spradling and Rubin (1981) also indicate that repeats identified in *D. melanogaster* are variously restricted in their evolutionary distribution, ranging down to some found only in the *melanogaster* group.

Much work remains to be done to elucidate the norms and exceptions of *D. melanogaster* repeat organization and evolution. It is nevertheless evident that much of this organism's moderate-repeat class consists of rather special elements which appear to have the ability to transpose and maintain themselves within the genome autonomously. At the present time it is by no means manifest that these entities can be in any way equated with the short-period interspersed repeats more characteristic of many eukaryote genomes, or even with the long period repeats which appear to have arisen independently in other evolutionary lineages. Furthermore, their variability both in location within the genome of a species and occurrence or nonoccurrence within the genus suggests that they may have little immediate relevance to the genetic organization of the Drosophilids except as an occasional disruptive force.

E. Intragenomic Repeat Family Homology and the Origins of New Families

The interspecific comparison studies reviewed above suggest that the kinetically complex moderately repetitive components found in a variety of organisms' genomes undergo saltational augmentation through evolutionary time. The source of the sequences which supply these new repeats may differ from group to group. For example, in *Xenopus,* and particularly the heavily studied sea urchin, we have noted that the moderate-repeat families dominating extant genomes seem to be derived from a class of families which have already reached considerable levels of repetition in ancestral genomes, though the choice of which sub-group of families will become more highly repeated seems to be made independently in each descending lineage. On the other hand, in the larger genomes of salamanders as well as cereals and other higher plants, which are often dominated by their moderately repetitive components, the repeated families added during evolution appear to be nonrepeated in other lineages, at least at the levels of analysis employed. These intergenomic comparisons suggest that the genomes of different groups may vary as to the extent with which previously repeated sequences tend to be involved in additional rounds of saltation. We will now examine several types of *intra*genomic repeat homology studies performed on the large moderate-repeat components of certain vascular plants which also suggest that saltations tend to involve preexisting detectable repeats in some genomes but not in others.

The conceptual model for all of the approaches to be described is as follows. If saltation events frequently involve preexisting repetitive families, then the moderate-repetitive class as a whole will consist of related groups of families, though separate groups will be unrelated to one another. Such a situation may of course be most readily detected through the use of cloned repeats, as has now been done in sea urchin. But it should also manifest itself in ways which are detectable in experiments with the bulk moderate-repeat DNA. Under the basic assumption that a saltation represents a relatively discrete event in evolutionary time, a new repeated family should initially be comprised of virtually identical member sequences. With the passage of time, however, continual random base changes in all of the members will result in their progressive divergence from the original uniform ancestral sequence and from one another. Later occurrence of a secondary saltation involving one repeat of this original family will create a new secondary family showing little sequence difference among its own repeats, but differing from repeats of the ancestral primary family by an amount equal to their average sequence divergence among themselves, and the two families together will comprise a related family group. Frequent occurrence of preexisting repeat saltation will mean that the moderate-repeat class will be dominated by groups comprised of several families with differing extents of internal sequence divergence (heterogeneous family organization). At the other extreme, if the incidence of saltation is low relative to the rate of divergence of repeats, or tends largely to involve previously single-copy segments, most saltations will involve either true single-copy segments or repeat families whose members are so diverged in sequence that they no longer reanneal with one another and will not cross-react with the new family which has been derived from one of their cryptic members. In this case, the moderately repetitive component will be dominated by discrete families which do not cross-react with other families showing different degrees of sequence divergence (homogeneous family organization).

An early effort to discriminate between these situations was that of Bouchard and Swift (1977), in a study of the large moderately repetitive component of the fern *Thelypteris normalis*. In these experiments, radioactive *Thelypteris* DNA was reassociated through the moderately repetitive range and thermally fractionated on an HAP column into three thermal stability classes based on fidelity of repeat sequence pairing in the reassociation. These thermal stability class tracers were then individually mixed with a large excess of whole DNA, reassociated a second time, and remelted on HAP. Under these conditions, repeats pairing with a particular fidelity in the first reassociation have a full spectrum of potential pairing partners available during the second reassociation. Nevertheless, upon remelting, each fraction remelted over a much narrower range than seen for the total moderately repetitive class; the low thermal stability component again re-melted with low thermal stability, the intermediate with intermediate, and the high with high. Optical remelts of cold DNA which had been thermally fraction-ated and reassociated were comparable. These experiments involved broad stabil-ity subfractions, each containing many repeated families, but they do suggest that families characterized by a particular extent of divergence in *Thelypteris* DNA are generally not related to abundant families with very different extents of divergence.

At about the same time, a kinetic method of analysis was applied by Bendich and Anderson (1977) in a study of the moderate-repeat classes of two other ferns (*Cryptogramma crispa* and *Blechnum spicam*), as well as daffodil and barley. Their approach was based on the idea that, if most moderate repeats were capable of reassociating with partners representing a range of sequence di-vergence, changes in the temperature of reassociation (or criterion) would alter the number of available pairing partners and therefore change the observed kinetics of reassociation. The reassociation experiments in this study were monitored optically, and the data therefore required careful correction for temperature-dependent differences in collapse hypochromicity of the large non-reacting single-stranded component in each reaction, as well as the known effect of temperature alone on reassociation kinetics. After adjusting their data for these factors, Bendich and Anderson saw little or no change in kinetics of reassociation for the moderate-repeat classes they monitored, despite substantial alteration of the proportion of the genome reassociating in this class. They therefore con-cluded that moderately repetitive families are also discrete entities in these genomes, each characterized by its own characteristic and fairly homogeneous level of member sequence divergence, and unrelated to families with different extents of divergence.

Results which could point to a somewhat different conclusion were obtained for the genome of the primitive fern *Osmunda cinnamonia* by Stein *et al.* (1979). DNA of this fern was reassociated at three different criteria (T_m -10, T_m -21, and T_m -30), bound to HAP, and thermally eluted. While almost the same proportion of DNA bound at the two lower criteria, the duplexes formed at T_m

-30 remelted with lower fidelity than those formed at T_m -21, suggesting that the less stringent conditions had allowed sequences to cross-hybridize with more diverged partners. It should be pointed out that these results are somewhat confounded by the fact that reassociation was carried to a $C_0 t$ of 2000, where a substantial portion of fragments with single-copy kinetics should have reassociated. Because of this, members of diverged repeat families with reassociated thermal stabilities in the range 21 to 30°C below native could have reassociated in part as poorly paired repeats in the low criterion reaction, while many of the same repeats would pair with their single precise complementary strand as well-paired "single copy" in the higher criterion reaction. However, interspecific comparisons given in the same paper may also be interpreted as suggesting that most of the extant moderate repeats in *O. cinnamonia* are derived from recurrent saltations acting upon preexisting ancient repeated families. Therefore the moderately repetitive component of the genome of *Osmunda* may well be characterized by a substantial degree of heterogeneous family organization.

In two recent studies, Preisler and Thompson have compared the small (0.48 pg) genome of mung bean (*Vigna radiata*) with the larger genome (4.9 pg) of garden pea (*Pisum sativum*) and suggested that the moderate-repeat component of the former is characterized by homogeneous family organization, whereas the latter is highly heterogeneous. In their first paper (Preisler and Thompson, 1981a), they employed the kinetic approach of Bendich and Anderson for mung bean with unexpected results, in that the rate of reassociation for repeats still able to pair *increased* with increasing reassociation temperature. Their explanation for this observation is that the least diverged families are also those characterized by the highest numbers of member repeats, since they have also had the least time to experience member losses through deletion. For pea, on the other hand, they observed little change in reassociation rate with differences in temperature, but placed an interpretation on this opposite to that suggested by Bendich and Anderson, arguing that the increasing number of cross-reactable repeats appearing in a heterogeneous moderate-repeat component as temperature is lowered will tend to keep the overall rate of reassociation the same. By this interpretation, in pea the moderate-repeat class is heterogeneous. In order to test these conclusions more directly (Preisler and Thompson, 1981b), they optically examined the portion of the moderately repetitive component which formed high thermal stability duplexes under different stringency conditions to see whether these repeats could also reanneal with more diverged components. Their results indicate that the high thermal stability repeats in mung bean are homogeneous and unrelated to the more diverged families, whereas in pea a proportion of them can cross-anneal to diverged sequences when given the opportunity to do so, and are therefore members of heterogeneous family groups. The results presented by Preisler and Thompson suggest that pea and mung bean, which are members of the same subfamily (Faboidae), differ both with respect to their repeat interspersion patterns (see above) and the immediate origin of these repeats. These differences are also reflected at the level of the apparent "single-copy" component of these genomes (Murray *et al.*, 1981). At lower stringency reassociation, much of an ostensible fractionated single-copy tracer from pea DNA can be shown to be highly diverged, or "fossil," repeat sequence, while such diverged repeats are a minor fraction of mung bean "single-copy" tracers. The rate of fossil repeat reassociation in the pea "single-copy" tracer shows a major increase with reduction in reassociation temperature, indicating the effective number of these repeats is increased by lowering the criterion, exactly as one would expect under the model of heterogeneous family organization.

The notions of "homogeneous" versus "heterogeneous" repeat family organization (or inter- versus intrafamilial heterogeneity in Bouchard and Swift) describe conceptual extremes, much as do the "short-term" versus "long-term" models of interspersion, and instances of genomes falling along a continuum between them are quite likely to be found. The studies on intragenomic repeat relatedness discussed here are moreover subject to difficulties inherent in all examinations of mass repeat component behavior. They nevertheless tend to agree with the implications of interspecific studies in suggesting that saltations tend to involve preexisting repeats in some organisms and not in others. We will return to this point in the concluding section.

V. Summary and Speculation

As we have seen, a considerable amount of information bearing on moderately repetitive DNA in eukaryote genome evolution has been accumulated in the past decade, based on mass studies and more recently on recombinant DNA techniques as well. In this final section, we will attempt to consider the implications of our present understanding concerning the apparent patterns of moderate-repeat appearance and increase in different genomes and lineages, as these shed light on various notions concerning the possible origins and functions of these sequences. We will also present a brief consideration of ways in which important features of organismic evolution may themselves influence the evolution of moderately repetitive DNAs in ways independent of the molecular character of these repeats.

A. MULTIPLE PATTERNS OF MODERATE REPEAT ORGANIZATION AND EVOLUTION

It seems fairly evident from the results surveyed that moderately repetitive DNA elements can no longer be considered a uniform class of DNA sequences. The results for the spectrum of organisms reviewed in Section III indicate that different evolutionary lineages, and even different genomes within the same major taxon, can be characterized by quite distinct patterns of repeat segment length, repetitive component complexity, and pattern of interspersion between single-copy segments and repeats. In addition, as noted in Section IV, the character and pattern of appearance and maintenance of some moderate repeats appear to differ qualitatively. Two general categories appear to be distinguishable.

1. Long Copia-like Moderate Repeats

The large *copia*-like sequences which dominate the moderately repetitive components of *Drosophila* (see Section IV,D) and the Ty elements of *Saccheromyces* (see Section III,E) appear to constitute a distinct class, characterized by a highly sequence-dependent mode of maintenance and propagation. It appears likely that their generalized interspersion in the genome is a product of their ability to change their position autonomously at intervals rather than of any positive function relevant to the phenotype of the animal. This impression is strengthened by their combination of sequence constraint with evolutionary plasticity.

The available information concerning the distribution of such elements in lineages of the genus *Drosophila* appears to indicate that various examples of these complex entities have entered the lineage leading to *D. melanogaster* at distinct points in its evolution and cannot be detected even at single-copy levels in genomes which diverged prior to those points. This pattern is readily explicable if these entities are not derived from DNA sequences which were originally part of the ancestral *Drosophila* genome at all, but were instead invading entities which have now become coadapted and established as part of the hereditary component of the species in which they are found, albeit of no fixed abode. If the *copia*-like repeats are in fact the descendants of infective agents rather than monsters arising within the *Drosophila*-lineal genome, they may certainly be regarded as parasitic DNAs, but it seems somewhat unfair to describe them as "selfish." Indeed, to the degree that they may have become coadapted to their host genome, they are probably less detrimental than initially. They also may now be viewed as constituting a distinctive type of genetic variability which, to the extent that individual elements may come to influence adjacent genes affecting phenotype, forms part of the future evolutionary potential of *Drosophila*. Any such potential would however be a consequence of the gradual coadaptive embedding of *copia*-like elements in the genome, not the cause.

Evidence as to the origin of Ty elements in yeast is not available, though their proclivity for rapid transposition and for being lost from the genomic DNA of some strains suggests that they are somewhat tenuously established (see Section III,E). On the other hand, their ability to produce mutations and place the expression of adjacent genes under the control of the mating-type (MAT) locus gives them a functional potential (Errede *et al.*, 1981). Whether this potential has been utilized in the evolution of yeast or merely of yeast genetics is a question of considerable interest.

A clear example of the establishment of foreign sequences in genomes appears to be provided by the vertebrate retroviruses, whose many resemblances to *copia*-like elements are so frequently cited (Calos and Miller, 1980; Finnegan, 1981). While it has been argued that such infectious elements may have evolved from eukaryote transposons (Shimotohno and Temin, 1981), it would seem at least equally logical to assume, as has been done above, that the direction of evolution runs the other way. It has recently been suggested that the mouse VL30 sequences may be an example of a long-established retrovirus-like element which has lost some of its viral character (Courtney *et al.*, 1981). The presence of various retrovirus-type sequences in the mouse genome, which is dominated by a different type of moderate repeat, is in any event suggestive. It may well be that all genomes periodically acquire similar entities, but that they have as yet only been observed in small genomes where they are the dominant type of established repeat (*Drosophila* and *Saccheromyces*) or are still recognizable as invaders (vertebrate retroviruses). The phenomenology associated with maize controlling elements such as the ac/ds system (McClintock, 1965) could be explained by postulating families of transposible entities which have become highly embedded in the maize genome. Under this model, some elements have lost the ability to mobilize transposition though they can still respond to mobilization, while others which retain the capacity to mobilize themselves and others are normally under coadaptive repression but can be detached from this control by severe cytogenetic upsets such as led to their original identification. The transposon-like behavior of the P-element of *Drosophila* hybrid genesis (Engels, 1981) might also be explained as an example of an element susceptible to rapid coevolutionary embedding, so that it normally only manifests itself when placed against a noncoadapted genomic background.

2. Short Repeats and Other Non-Copia-like Repeats

The bulk of the moderately repetitive DNA sequence components found in most organisms do not show obvious affinities to *copia*-like elements, and the available data for several examples suggests that they may exhibit distinctly different properties in evolution.

We have reviewed evidence for the intensely studied sea urchin and *Xenopus* genomes, implying that the currently predominant moderate repeats are the products of recurrent cycles of saltation over long evolutionary time spans (see Section IV,C). For sea urchin in particular, enough direct sequencing data now exist from cloned segments to show that the sequence family members which are the end products of this recurrent process often show considerable sequence heterogeneity with respect to other family members and share no obvious structural features in common either with other repeated families in the same genome or known transposon-like elements (Posakony *et al.*, 1981). As we have also noted, the derivation of predominant moderate repeat families in these genomes shows features suggesting both some degree of sequence dependence (saltations seem to involve segments already somewhat repetitive) and considerable sequence independence (the subgroup of somewhat repetitive families which undergoes repeated saltation is different in different lineages). One other point to be noted is that the modal size of these genomes' repeats, and the predominant class of many other ''short-term'' genomes, is often as low as 300 ntp, far shorter than that of known transposon elements or even the terminal repeats which bracket these entities.

The state of affairs for larger genomes, often dominated by moderate-repeat components which are made up of somewhat longer repeat segments showing complex interspersion, is less well understood. At the level of detection used in the studies on *Plethodon* and the cereal genomes (Section IV,B), the bulk of moderate-repeat families appears to have entered that class during the evolution of the lineages studied. They are generally presumed to be derived from low- or single-copy elements already present in the ancestral genomes. Since only sequences present at similar repetition frequencies in the heterologous genomes could have cross-reacted under the condition used, such homologous sequences present in other lineages at these low levels would have gone undetected. The studies on intragenomic repeat homology in several other large repeat component plant genomes suggest that in some of these cases saltations also continually add ''new'' moderately repetitive families to the major class, while in others they frequently produce further amplification of families which are already detectably part of this class. In all these cases, little or no sequence specificity is indicated. In genomes such as that of pea, where much more than half the weight of the genome is found in the detectable moderately repetitive component, even a random saltation mechanism will produce secondary amplification of a segment already found in this class in a majority of instances. The continual addition of new moderate-repeat families in the cereal genomes in particular would suggest that the saltational phenomenon in this group may be rather sequence independent, or ''ignorant,'' in its action.

The ''long-term'' repeat genomes of birds, despite their apparent similarities in organization to *Drosophila,* appear to be different in terms of the type of segments forming the long repeat component. We have noted that a substantial portion of the repeats in chicken are of great antiquity in avian genomes, 23–27% being shared with the distantly related ostrich. In addition, the recent studies of cloned repeats (Eden *et al.,* 1980; Eden, 1980) have so far not revealed any obvious transposon-like organization.

Where information exists, it would appear that the saltations which produce new families of repeats and the interpolations which intersperse them widely among other segments may be closely spaced in evolutionary time. Measurements of the modal size of undiverged repeat segments as compared to all repeat segments in wheat (Flavell and Smith, 1976), rye (Smith and Flavell, 1977), and *Lilium* (Bouchard and Stern, 1980) suggest that the undiverged repeats in these plant genomes form S_1 resistant duplexes of approximately the same length as all repeats, and thus are not frequently found in long tandem arrays. In *Xenopus* and sea urchin, on the other hand, there is an inverse correlation between S_1 defined segment length and sequence divergence, and it has often been suggested that the nondiverged repeats are found in long tandem arrays on the basis of their exclusion from agarose columns (Britten *et al.*, 1976). More recently, however, data on cloned repeat segments from families showing little sequence divergence and long enough to be found in this class have shown that member segments frequently exist in isolation or in short clusters of a few repeats (Anderson *et al.*, 1981). Sequence comparisons of cloned members of a diverged short repeat family indicate that the divergence of segments is accompanied by sequence rearrangements within the segments as well as single base changes (Posakony *et al.*, 1981); the accumulation of such rearrangements may well account for some of the decline in the average length of colinear pairable duplex with divergence.

The few ubiquitous and very abundant short repeat families which seem to dominate the moderately repetitive components of mammals (see Section III,A), particularly the primate *Alu* family, may represent a different situation. It has been suggested, based on their partial sequence similarity to mammalian replicons and their abundance in the processed portions of hnRNAs, that they may be involved in one or more of these cellular functions (Jelinek *et al.*, 1980). Alternatively, if the still poorly understood cellular machinery which performs these functions also plays a part in the saltational phenomenon, these ubiquitous repeats would represent elements which have evolved in such a way as to become the overwhelmingly preferred saltational substrates in mammalian genomes. In any event, they seem to be long-established components of the lineages in which they are found and do not show obvious similarities to the known eukaryote transposon elements, including the retroviruses which inhabit some of the same genomes at much lower copy numbers. The pachytene repeats of *Lilium* and other seed plants (Section IV,C) appear to constitute a set of established conserved repeats for which a function may be inferred, though their different abundance in some groups indicates that they may still be subject to saltational change in numbers.

Overall, the bulk of eukaryote moderately repetitive elements can at present only be said to be united in their apparent lack of resemblance to transposon-like elements. Clearly, any class which is comprised of the residue of entities not assignable to another class is likely to prove heterogeneous upon closer examination. Thus some entities now placed in it may eventually prove to be autonomous transposons or replicators, while the others may be generated and interspersed by more than one set of phenomena from more than one source. The enormous phylogenetic spectrum over which such repeats are found nevertheless suggests a phenomenological unity, inasmuch as genomes of many eukaryote lineages are characterized by a continual tendency to produce and interpolate moderately repetitive sequence families in a saltational fashion. In some cases, this phenomenon shows a moderate degree of sequence dependence, in that new saltations are likely to involve a subset of sequences which have been previously amplified to some degree, while in others it appears to act in a more sequence-independent fashion.

The mechanism or mechanisms responsible for these phenomena are completely unknown at present, allowing free play for speculation. They may be an incidental pleiotropic byproduct of the complex metabolic systems responsible for replication and recombination in the eukaryote nucleus, which in most instances cannot be purged by natural selection without disrupting some aspect of the paramount normal functioning of the system (see Dover, 1980). An alternative proposal which could be put forward is that saltational phenomena result from the synergistic interaction of intrinsic nuclear DNA metabolism with other functions initially introduced from outside the genome. Possibly attractive candidates for the introduction of such functions would be the transposon-type elements themselves. Despite a considerable body of information concerning the events at the level of DNA sequence which accompany insertion of these elements, the actual molecular mechanisms producing their movement are obscure, and most probably involve the appropriation of host nuclear functions as well as functions intrinsic to the elements themselves. It would not be unreasonable if such a coordinated system should occasionally misfunction and act more or less randomly on a segment of host DNA, particularly in the case of degenerate or coadaptively embedded elements which have become relatively fixed in the genome and express their functions only imperfectly or intermittently. Regardless of such speculations, however, the simple existence of phenomena which continually produce and interpolate repeated segments forms a source of genetic variability on which the forces of evolution can act, either because of or regardless of their functions. We will conclude with a consideration of potential aspects of this evolutionary situation.

B. MODELS OF MODERATELY REPETITIVE DNA IN EVOLUTION

Various models, as noted in the Introduction, have been presented to account for the accumulation and persistence of moderately repetitive DNA components in eukaryote genomes. These models may be classified according to the level at which natural selection is presumed to have operated in producing moderate repeats and whether this operation is supposed to be sequence-dependent or sequence-independent. The various functionalist models, espoused particularly by Davidson, Britten, and their colleagues, suggest a form of sequence-dependent positive selection for moderate-repeat elements as parts of the genotype, through various postulated beneficial and indeed central effects in the developmental program which produces the evolving phenotype. The more recent sequence dependent or "selfish" evolutionary models of Orgel and Crick (1980) or Doolittle and Sapienza (1980), by contrast, decouple selection for moderate repeats almost entirely from that acting on the phenotype. These arguments suggest that moderate-repeat accumulation occurs through direct selection for those segments which demonstrate the greatest facility as parasites of the replication machinery of the genome, with phenotypic selection acting only to check the accumulation when it begins to affect severely the workings of the genome. The highly sequence independent "ignorant" model of Dover (1980) also calls for phenotypic selection only as a check, but postulates no direct selection on the repeats at all; they are rather occasional randomly generated side products of the genomic machinery for DNA metabolism. Here natural selection is involved only as it has produced this tendency in the cell machinery as a pleiotropic byproduct of postitive selection for the actual functioning of its components. The nucleotype model, espoused especially by Cavalier-Smith (1978, 1980), does argue for a linkage between selection on aspects of the phenotype and accumulation of moderate repeats, but this is also postulated to be sequence-independent. Here, selection acts indirectly, and only on overall DNA content as this determines the physical volume of the nucleus, and through this nucleus to cytoplasm transport, cell size, cell-cycle time, or other selectively significant aspects of phenotype determined by nuclear and cellular size. The spectrum of actual cases which we have reviewed suggests that all of these modes of evolution may occur, but that different genomes may be subject to them to different degrees. It may in fact be suggested that this should be the case, since the several models actually focus on different points in the overall phenomenon of moderate-repeat accumulation in evolution.

This point may be best understood by reference to a schema which breaks down the overall phenomenon into interconnected hierarchical levels. In doing this, we may draw a primary distinction between phenomena bearing on the genome's tendency to produce or facilitate saltational amplification and interpolation of sequences, which I shall term its *predisposition* to produce moderate repeats, and those aspects of the genome's overall biology which determine the rate at which it will accept such events, that is its *tolerance* to moderate-repeat accumulation.

It would appear that the question of predisposition is most directly addressed by that aspect of the Dover ignorant DNA model which suggests that the production of periodic saltations is an unavoidable byproduct of the complex machinery generated by natural selection for DNA metabolism in eukaryote nuclei. This postulate does not seem unreasonable to anyone familiar with the opportunistic tendency in the evolution of all eukaryote characteristics, including those of the genetic system. As Darlington (cf. 1958, 1971) has cogently argued, characteristics which have an overall selective advantage tend to be adopted in evolution even if they also carry disadvantages of lesser magnitude, and once they are thoroughly incorporated into the complex functions of the essential hereditary system, they can be quite difficult to change or replace. We may also include at the level of predisposition the suggestion of Jain (1980) to the effect that there may have been positive selection for some level of saltational tendency as a potential source of evolutionary novelty not provided by point mutations. The tendency toward predisposition, and even the extent to which it produces random vs sequence-independent saltations, may be expected to vary from one genome and lineage to another under both models, based on both the present demands of selection and those which have been made during the lineage's history. Under the "ignorant" model, however, this variation should be more random since it is produced as a byproduct of selection acting on the entirely distinct direct functions of the system of DNA metabolism.

The possibility of selectively modulated differences in tolerance is most clearly implied in the many ingenious suggestions concerning possibilities for direct yet sequence-independent nucleotypic selection in different groups presented by Cavalier-Smith (1978). Occasional amplifications and interspersions of sequences with actual genetic function is a separate issue, since these will be tolerated in accordance with their positive or negative effects as directly viewed by selection in a sequence-dependent manner. One may also evoke vaguely such possibilities as varying susceptibility of the recombinational system to distruption, or energetic considerations, though Loomis (1973) has argued that the replication of the entire DNA complement is a trivial component of the cell's energy budget. In addition, as I shall argue at the close of this section, aspects of population structure and the processes of speciation may also influence the accumulation of moderate repeats through evolution, and are thus also elements contributing to tolerance.

The position of "selfish" transposon-like elements and "preferred replicator" sequences in this schema depends on their source. If they are ultimately derived from infectious entities of external origin, they represent a separate source of repeats from genomic predisposition, but are still subject to selection at the level of tolerance. If there is some tendency for genomes to produce such segments from sequences of intrinsic origin, however, then their accumulation can also be governed by aspects of predisposition. In may be safest to postulate that both sources can play a role until more information is available.

In this discussion, the role assigned to possible functions of moderate repeats as arbiters of their accumulation is small. This is not to argue, however, that saltation/interpolation events may never produce changes with potential functional impact in the genome. If a genome is indeed subject to regular rounds of such events over evolutionary time, it is quite reasonable to argue (Britten and Davidson, 1971) that they may sometimes involve sequences with regulatory potency which alter the potential expression in developmental programs of genes into whose vicinity individual segments interpolate. Certainly such events could occasionally have great selective significance. Saltation/interpolation episodes which involve segments with recombinational activity (Bouchard and Stern, 1980) or of a type recently observed which remove actual gene segments from the domain of their normal regulatory environment (Childs *et al.*, 1981) are other examples of events which may eventually be fixed because of their selective impact. They are unlikely to occur with any frequency, however, except in lineages which are characterized by an overall tendency to moderate-repeat generation and accumulation.

Under the unified scheme proposed in Table VIII, the potential for accumulation of moderate repeats within a given genome will have been governed by the extent of predisposition, receipt from external sources, and tolerance which its lineal ancestors have exhibited over evolutionary time. The amount and characteristics of the moderate-repeat components observed at present will be a function of the passive rate of accumulation over time (balanced to an unknown degree by probably passive deletion). In addition, some portion of the accumulated repeats may have become fixed due to having possessed or acquired a selectively advantageous function or because they possess autonomous tendencies (e.g., transposons).

This entirely speculative model is presented simply to summarize how a hierarchy of effects acting over a period of time may govern the type of moderate-repeat component observed in a genome at present. This model of course makes no claim to include all relevant effects or explain all possible observations. It is nevertheless interesting to observe that the assignment of different weights to various components in the schema accords with several of the patterns observed in actual genomes.

We may note first of all that a low level of tolerance should result in a strong bias against sequence-independent accumulation. Sequence-dependent accumulators, particularly transposon-elements in cases where predisposition is also low, should therefore predominate since these can in effect counter the negative force of selection with their own autonomous tendency to accumulate, so that the genome is forced to adopt to them and they to it if both are to survive. The *Drosophila* situation would then be the product of a long history of low tolerance and predisposition, while that seen in most fungi would imply an even lower level of tolerance, such that any repeat accumulation is either confined to a few sectors or precluded entirely. Cavalier-Smith (1978) has suggested that the filiment growth of higher fungi, which requires nuclei to squeeze continually through minute pores, has selected for a minimal nuclear size produced by

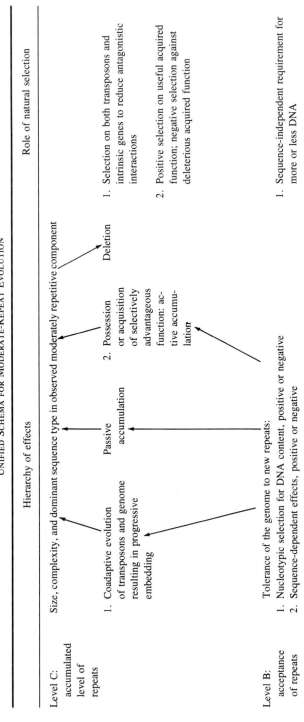

TABLE VIII

UNIFIED SCHEMA FOR MODERATE-REPEAT EVOLUTION

Hierarchy of effects	Role of natural selection
Level C: accumulated level of repeats	
Size, complexity, and dominant sequence type in observed moderately repetitive component	1. Selection on both transposons and intrinsic genes to reduce antagonistic interactions
Passive accumulation → 2. Possession or acquisition of selectively advantageous function: active accumulation → Deletion	2. Positive selection on useful acquired function; negative selection against deleterious acquired function
1. Coadaptive evolution of transposons and genome resulting in progressive embedding	
Level B: acceptance of repeats	
Tolerance of the genome to new repeats: 1. Nucleotypic selection for DNA content, positive or negative 2. Sequence-dependent effects, positive or negative	1. Sequence-independent requirement for more or less DNA

180

3. Undefined disruptive or beneficial effects on recombination
4. Population structure and events of speciation

2. Direct positive or negative selection
3. Negative or positive selection
4. Sequence-independent consequences of evolution at superorganismic levels

Level A: sources of repeats

1. Genomic predisposition toward saltational amplification and interspersion of sequences in either sequence-dependent or sequence-independent fashion: predisposition to evolve transposon elements from intrinsic sequences

2. Transposons derived from invading extrinsic infectious agents

1. Pleiotropic creation of either sequence-independent or moderately sequence-dependent saltational tendency in DNA metabolism through selection of other functions. Possible positive selection for certain level of saltational tendency. Selection of autonomous transposons with parasitic behavior
2. Evolution of parasitic DNA elements with broad host specificities, selected for ability to invade genomes and maintain themselves autonomously

181

haploid dikaryosis and low DNA content; the present tolerance of yeast for transposons would then be due to their reversion to a nonthallose mode of existence. The situation seen in sea urchins and perhaps *Xenopus* could be due to an intermediate level of predisposition and a moderate level of tolerance. In such a case, saltationally generated repeats may be expected to come into existence. However, there can be selection on the repeats for acquisition of a degree of sequence dependence if there is any inherent tendency of the saltational mechanism to prefer some sequences over others. After the first accidental saltation of a sequence which is slightly preferred, its probability of being chosen again is immediately increased by as many times as it is now repeated. The accumulation of sequence changes in the member sequences of families produced in this way can then alter the original slightly preferred sequence to states having greater or lesser liability to saltation as individual segments. This increasing variance will be distributed both within and between families, so that in different lineages different sets of families would evolve toward a higher likelihood of resaltation, and some of them will eventually experience it. Their increased membership will then again increase the immediate likelihood of additional saltations, and further refinement in preference can also occur among these added members to increase it further still. By contrast, a relatively high tolerance with intermediate predisposition could produce the larger homogeneous family moderate-repeat components seen in some plants, such as mung bean, while very high tolerance and intermediate predisposition may characterize cereal and salamander genomes, and high levels of both could account for a genome like that of pea, which is almost entirely dominated by a heterogeneous family moderate-repeat component.

Whether it is actually possible or fruitful to construct a sort of Hertzprung–Russell diagram of genomes on Tolerance versus Predisposition coordinates is open to question, particularly given our present lack of specific information on the make-up of most moderately repetitive components and the complexity of the phenomena subsumed under these two terms. The correct position of several of the examples used here would also be debatable. They are provided only to suggest that some conceptual separation between phenomena which provide potential sources of new repeats and phenomena which may govern their incorporation into the genome may be useful in attempting to interpret the variety of natural situations encountered. We would now like to close this discussion with an examination of certain aspects of eukaryote evolutionary biology which may have a bearing on the level of incorporation, or tolerance, shown by particular groups at levels which transcend genomic or organismic biology.

In the various models we have discussed, genome and species have been treated as archtypal entities. While this is a convenient form of logical shorthand, it neglects an important reality of evolution. Just as *the species* actually consists of the living array of genetically distinct individuals distributed among local populations in the environment, *the genome* is not an entity but an abstraction standing for all the genetic information of all individuals. Thus for a diploid species there exists at any given time n individuals containing $2n$ active genomes and an almost infinite number of potential genomes in their gametes, from which the realized genomes in the next generation will be derived. Since new moderate repeats, like other genetic novelties, arise in this context, we may speculate that their fixation and evolution should be governed by the laws of population genetics and the phenomena attendant on speciation.

Unless the saltation and interpolation of a particular segment are entirely orthogenetic at particular points in evolution, it may be considered to be a form of mutation like any other. A saltation event will therefore occur, not in *the genome*, but within an individual genome within an individual organism which is a member of a local population. The immediate fate of the products of this saltation will be tied to the fate of the descendants of the portions of this genome in subsequent generations.

If saltation and interpolation are essentially cotemporal, the fate of each new repeat element will be partly independent of its brethren. Provided the original genome experiencing the saltation survives long enough to leave offspring, segments on different chromosomes will assort independently, while those on the same chromosome will be gradually broken up with additional time. They may thus be treated as a set of plus-alleles distributed through the karyotype, while the corresponding unsaltated chromosome segments can be considered to carry null-alleles at the corresponding locations. If the initial saltation event is presumed to generate a long tandem array of segments, which are later broken up and interpolated by some secondary process, the same situation would eventually be arrived at, but the initial chances for survival of the saltated segments would be far lower, since they could all be lost from genomes as a unit until this had occurred. We will therefore deal only with the fate of segments which have successfully survived both saltation and interpolation, whatever the temporal spacing of these events.

We will make the assumption for this discussion that the vast majority of saltation/interpolation-produced segments are either disruptive or neutral in their original effect, whatever positive function some of them may later be supposed to acquire after they have become a fixed component of all genomes in some descendant lineage. We will further discount the disruptive class as likely to be purged eventually, and certainly unlikely to be fixed. If these premises are accepted, the accumulation of saltationally generated repeats then reduces largely to the fixation and accumulation of relatively neutral interpolations in evolution.

The labyrinthian mathematical theory of neutral allele evolution is outside the scope of this discussion, which will confine itself to certain major aspects treated by Lewontin (1974). Under the steady-state stochastic assumptions of this theory, the probability of fixation for an individual neutral site will be low and will decline with increasing effective population size according to the formula, $P_0 = 1/2N$. A larger proportion may rise to temporary abundance in the population before again declining and ultimately being lost. The chances for individual segments rise somewhat when the effects of linkage are taken into account, since the presence of linked loci carrying alleles undergoing positive selection can allow them to hitchhike to relative abundance before the linkage disequilibrium is broken up.

With respect to interpolated repeats, this suggests that the probability of fixation (which is required if they are subsequently to become part of *the genome*) is very low for each individual segment if these are treated independently. In a local population with relatively small effective size $N = 200$ which remains genetically isolated for some time, for example, a given segment has only a 0.25% chance of becoming fixed at random. Any significant genetic flow between this small population and the rest of the gene pool will vastly increase the effective N, with a corresponding decrease in the individual segment likelihood of fixation. Under linkage some small proportion of interpolations may be expected to enter chromosome segments undergoing positive selective increase early enough in their history to be passively carried into some abundance, and those few entering the immediate sequence environment of a new functional allele which will eventually undergo fixation may be carried to virtual fixation with it. Nevertheless, under these assumptions, the fate of virtually all products of all saltations which occur within populations of a species should be immediate or eventual dilution and extinction in the species gene pool.

The probable fate may be very different, however, for those saltations which have occurred in the recent history of isolated populations which go on to give rise to new species. It is a long-established tenet of evolutionary biology that speciation events frequently represent "genetic bottlenecks," in which the gene pool of the new species takes its departure from the restricted basis of a small ancestral population representing only a sample of the genetic variability of the parent species (Wright, 1940a,b; White, 1959; Grant, 1963; Dobzhansky, 1970). During this period, genetic drift plays a prominent role and results in the frequent fixation of alleles and gene combinations which are relatively rare in the parent species but which happen to have been temporarily abundant in the local population. Cases of near or complete fixation of specific rare chromosome segments or isozyme alleles accompanying population and subspecies isolation have been observed in the laboratory (Dobzhansky and Pavlovsky, 1957) and in nature (Prakash et al., 1969; Carson et al., 1970 Selander et al., 1969). In addition, the formation of species through such genetic sampling is presumed to both permit and impel a considerable amount of rapid adaptive genetic change, an actual acceleration of the evolutionary tempo (Simpson, 1944; Carson, 1959), involving positive and negative selection on aspects of the starting pool of variability. Such selection can involve both immediate changes in the average selection acting on individual alleles in their altered genetic background (Mayr, 1958) and directional selection for changes in the overall phenotype and therefore its underlying genotype as the incipient species adjusts and refines its adaptations to its new niche (Stebbins, 1950, 1970; Grant, 1963; Mayr, 1966). In its extreme form, this has been postulated to involve a genetic revolution, as in the "punctuated equilibrium" model of Eldredge and Gould (1972). Whether one is willing to go this far or not, there is a reasonable consensus that the disjunctions produced in evolutionary lineages by speciation episodes are accompanied by concomitant increases in opportunities for drift and by the potential for rapid selective changes in abundance of regions genetically linked to genes involved in aspects of the new species' adaptations. It therefore seems quite reasonable to suggest that the products of recent saltations in such lineages, some of which will be either temporarily abundant or linked to genes which will become fixed in the new species, have a greatly increased chance of fixation during the events accompanying the formation of the new species' genome. Once a set of families has been established in a lineage by such an event, the families should tend to persist for long periods and through many subsequent bifurcations of the lineage, since while establishment has occurred for many segments at once, deletion (unless it is also saltational, which would require remarkably precise sequence dependence) should occur for individual members of families only on a segment-by-segment basis. Such deletion would be particularly difficult for segments which had successfully interpolated into the immediate domains of genetically functional regions, since many random deletions in such regions would remove adjacent regions required for activity of the gene and thus be selectively disfavored, while successful deletion of the repeat would have little or no positive selective value. We would then predict that subsequent speciation events, devolving from particular descendant populations, would tend to conserve the ancestral families, though perhaps lowering their repetition frequency somewhat, while also adding new sets of families representing the subsample then prominent in the population of all sequences undergong saltation in the total genome pool at the time. Different daughter species, descending from different populations, would be expected to contain largely nonoverlapping sets of new families but to share most families found in the ancestral genome. While this entirely speculative model can in no way be said to "explain" the fairly extensive observations of precisely this sort of situation in examples including *Plethodon*, the cereals, and sea urchin, it is certainly consistent with these observations.

The various models we have dealt with all rightly evoke the concept of hierarchy in considering the evolution of moderate repeats. In general, they have drawn their distinctions between evolutionary changes in DNA sequences at the genomic level from those produced by selection at the organismic level of phenotype, stressing various possibilities as to the coupling or lack of coupling between evolution at one level and the other. We have added the last few paragraphs only to point out that additional levels of hierarchy exist in the processes of evolution and that the dynamics of events occurring at these levels may influence the outcome of molecular changes, as well as the reverse.

C. Prospects

The study of moderately repetitive DNA in evolution has already proven to be a fruitful undertaking. The gradual accumulation of information on genome organization and evolutionary patterns found in different lineages now makes possible an appreciation of the diversity and complexity of actual situations which underlie the basic observation that most eukaryote genomes contain moderately repetitive components. Comparisons of the differences in overall modes and details of moderate-repeat organization among groups also serve to point up inadequacies in models propounded on the basis of detailed observations for particular genomes. Future studies of general sequence organization may be expected to fill in some of the present large gaps in our picture of genome structure in different phylogenetic lineages. Meanwhile, the current explosion in recombinant DNA cloning may be expected to add rapidly to our knowledge concerning specific sequence families. Much information of this type will of course be derived from studies aimed directly at problems of moderate-repeat evolution. An additional source will be provided as a byproduct of detailed studies of cloned long stretches of genomic DNA containing genes of experimental interest and long stretches of adjacent sequence, including individual moderate-repeat family members. While a long list of potentially interesting questions could be drawn up, several with evolutionary dimensions seem particularly pertinent at present:

1. How closely spaced in time are events of saltation and interpolation, and does the temporal spacing of these events differ for different families? As this article was being completed, Moore et al. (1981) published results suggesting that families which have undergone recent saltation in the sea urchin genome may in fact differ in the extent to which individual segments have become interpolated. More detailed information on the proportions which undergo rapid versus retarded interpolation in this or other genomes will be of great interest in attempting to understand details of the processes governing interpolated repeat accumulation in the evolving genome.

2. How abundant are interpolated repeats which retain an ability to exchange information? The accumulation of sequence divergence in moderate-repeat classes suggests that dispersed segments generally evolve in a somewhat independent manner. However, the pachytene repeats appear to undergo concerted sequence change in evolution, and recent data of Brown and Dover (1981) suggest that the EcoRI–BglII family of mouse does also (see Section III,A,

mammals). Further discoveries of such families will be most noteworthy, since they may be expected to represent a distinct evolutionary component.

3. Do the individual short repeat segments which are interpolated near genes in some genomes have a role in their function? Comparative sequence information on the sequence environments of given genes in different organisms should speak to this question as it accumulates. Assays of the functionality for expression of such long cloned stretches in *Xenopus* oocyte microinjection experiments and the increasingly sophisticated *in vitro* test systems with and without alterations or deletions of the repeated elements should offer a more direct test. Comparisons with respect to functionality between the gene contiguous repeats of sea urchin and especially *Xenopus,* which represent hundreds of distinct sequence families, and those of primates and other mammals, which represent very few, will be of particular significance. If functions requiring a large array of different sequences in some deuterostome and lower vertebrate genomes have become the provenance of only a few in the mammals, this would seem to represent a major redesigning of the genome's regulatory system somewhere at the root of the mammalian lineage.

It is quite conceivable that some new and unforeseen observations or the accumulation of enough detailed genome organization and specific sequence information will again reveal aspects of unity and simplicity underlying the apparent diversity which now seems to characterize moderately repetitive DNA in evolution. At present, however, it seems equally likely that further analyses will reveal more diversity. This is indeed to be expected if the accumulation of moderate repeats is a complex process in which different sources participate in their generation and a variety of types and levels of selection influence their retention, with aspects of all these components being uniquely and separately determined by the present biology and evolutionary history of each lineage. It may well be that in considering moderately repetitive DNA, we confront a point at which the unity of biochemistry and the diversity of organic evolution meet.

REFERENCES

Adams, J. W., Kaufaman, R. E., Kretschmer, P. J., Harrison, M., and Nienliuis, A. W. (1980). *Nucleic Acids Res.* **8,** 6113–6128.
Allen, J. R., Roberts, T. M., Loeblich, A. R., III, and Klotz, L. C. (1975). *Cell* **6,** 161–169.
Ammermann, D., Steinbrück, G., Von Berger, L., and Hennig, W. (1974). *Chromosoma* **45,** 401–429.
Anderson, D. M., Scheller, R. H., Posakony, J. W., McAllister, L. B., Trabert, S. G., Beall, C., Britten, R. J., and Davidson, E. H. (1981). *J. Mol. Biol.* **145,** 4–28.
Angerer, R. C., Davidson, E. H., and Britten, R. J. (1975). *Cell* **6,** 29–39.
Arthur, R. R., and Straus, N. A. (1978). *Can. J. Biochem.* **56,** 257–263.
Bachmann, K., and Price, H. J. (1977). *Chromosma* **61,** 267–275.
Baldari, C. T., and Amaldi, F. (1976). *Chromosoma* **59,** 13–22.
Beauchamp, R. S., Pasternak, J., and Straus, N. A. (1979). *Biochemistry* **18,** 245–251.

Bedbrook, J. R., O'Dell, M., and Flavell, R. B. (1980). *Nature (London)* **288,** 133–137.

Bendich, A. J., and Anderson, R. S. (1977). *Biochemistry* **16,** 4655–4663.

Bennett, M. D., and Smith, J. B. (1976). *Phil. Trans. R. Soc. London Ser. B* **274,** 227–274.

Bonner, J., Garrard, W. T., Gottesfeld, J., Holmes, D. S., Sevall, J. S., and Wilkes, M. (1974). *Cold Spring Harbor Symp. Quant. Biol.* **38,** 303–310.

Borror, D. J., DeLong, D. M., and Triplehorn, C. A. (1976). "An Introduction to the Study of Insects," rev. ed. Holt, New York.

Bouchard, R. A., and Stern, H. (1980). *Chromosoma* **81,** 349–363.

Bouchard, R. A., and Swift, H. (1977). *Chromosoma* **61,** 317–333.

Bozzoni, I., and Beccari, E. (1978). *Biochim. Biophys. Acta* **520,** 246–252.

Bridges, C. B., and Brehme, K. S. (1944). "The Mutants of *Drosophila melanogaster.*" Carnegie Inst. Wash. Publ. 552, Washington, D.C.

Britten, R. J., and Davidson, E. H. (1969). *Science* **165,** 349–357.

Britten, R. J., and Davidson, E. H. (1971). *Q. Rev. Biol.* **46,** 111–138.

Britten, R. J., and Kohne, D. E. (1968). *Science* **161,** 529–540.

Britten, R. J., Graham, D. E., and Neufeld, B. R. (1974). *In* "Methods in Enzymology XXIX" (L. Grossman and K. Moldave, eds.). Academic Press, New York.

Britten, R. J., Graham, D. E., Eden, F. C., Painchard, D. M., and Davidson, E. H. (1976). *J. Mol. Evol.* **9,** 1–23.

Brown, S. D. M., and Dover, G. (1981). *J. Mol. Biol.* **150,** 441–466.

Calos, M. P., and Miller, J. M. (1980). *Cell* **20,** 579–595.

Cameron, J. R., Loh, E. Y., and Davis, R. W. (1979). *Cell* **16,** 739–751.

Carson, H. (1959). *Cold Spring Harbor Symp. Quant. Biol.* **24,** 87–105.

Carson, H. L., Hardy, D. E., Spieth, H. T., and Stone, W. S. (1970). *In* "Essays in Evolution and Genetics in Honor of Theodosius Dobzhansky" (M. K. Hecht and W. C. Steere, eds.), pp. 437–543. Appleton, New York.

Cavalier-Smith, T. (1978). *J. Cell Sci.* **34,** 247–278.

Cavalier-Smith, T. (1980). *Nature (London)* **285,** 617–618.

Cech, T. R., and Hearst, J. E. (1976). *J. Mol. Biol.* **100,** 227–256.

Chamberlin, M. B., Britten, R. J., and Davidson, E. H. (1975). *J. Mol. Biol.* **96,** 317–333.

Childs, G., Maxson, R., Cohn, R. H., and Kedes, L. (1981). *Cell* **23,** 651–663.

Chooi, W. Y. (1971). *Genetics* **68,** 213–230.

Constantini, F. D., Scheller, R. H., Britten, R. J., and Davidson, E. H. (1978). *Cell* **15,** 173–187.

Constantini, F. D., Britten, R. J., and Davidson, E. H. (1980). *Nature (London)* **287,** 111–117.

Courtney, M. G., Steffen, D. L., and Getz, M. J. (1981). *J. Cell Biol.* **91,** 140a.

Crain, W. R., Eden, F. C., Pearson, W. R., Davidson, E. H., and Britten, R. J. (1976a). *Chromosoma* **56,** 309–326.

Crain, W. R., Davidson, E. H., and Britten, R. J. (1976b). *Chromosoma* **59,** 1–12.

Cullis, C. A., and Schweizer, D. (1974). *Chromosoma* **44,** 417–421.

Darlington, C. D. (1958). "Evolution of Genetic Systems," 2nd ed. Oliver & Boyd, Edinburgh.

Darlington, C. D. (1971). *Nature (London)* **234,** 521–525.

Davidson, E. H., and Britten, R. J. (1973). *Q. Rev. Biol.* **48,** 565–613.

Davidson, E. H., and Britten, R. J. (1979). *Science* **204,** 1052–1059.

Davidson, E. H., Hough, B. R., Amenson, C. S., and Britten, R. J. (1973). *J. Mol. Biol.* **77,** 1–23.

Davidson, E. H., Graham, D. E., Neufeld, B. R., Chamberlin, M. E., Amenson, C. S., Hough, B. R., and Britten, R. J. (1974). *Cold Spring Harbor Symp. Quant. Biol.* **38,** 295–302.

Davidson, E. H., Galau, G. A., Angerer, R. C., and Britten, R. J. (1975a). *Chromosoma* **51,** 253–259.

Davidson, E. H., Hough, B. R., Klein, W. H., and Britten, R. J. (1975b). *Cell* **4,** 217–238.

Davidson, E. H., Klein, W. H., and Britten, R. J. (1977). *Dev. Biol.* **55,** 69–84.

Deininger, P. L., and Schmid, C. W. (1979). *J. Mol. Biol.* **127,** 437–460.

Deininger, P., Jolly, D., Friedmann, T., Rubin, C., Houck, C., and Schmid, C. (1980). *In* "Mechanistic Studies of DNA Replication and Genetic Recombination" (B. Alberts and C. F. Fox, eds.), pp. 369–378. Academic Press, New York.

Deininger, P. L., Jolly, D. J., Rubin, C. M., Friedmann, T., and Schmid, C. W. (1981). *J. Mol. Biol.* **151,** 17–33.

Dobzhansky, T. (1970). "Genetics of the Evolutionary Process." Columbia Univ. Press, New York.

Dobzhansky, T., and Pavlovsky, O. (1957). *Evolution* **11,** 311–319.

Doolittle, W. F., and Sapienza, C. (1980). *Nature (London)* **284,** 601–603.

Dover, G. (1980). *Nature (London)* **285,** 618–620.

Dover, G., and Doolittle, W. F. (1980). *Nature (London)* **288,** 646–647.

Dunsmuir, P., Brorein, W. J., Jr., Simon, M. A., and Rubin, G. M. (1980). *Cell* **21,** 575–579.

Eden, F. C. (1980). *J. Biol. Chem.* **255,** 4854–4863.

Eden, F. C., and Hendrick, J. P. (1978). *Biochemistry* **17,** 5838–5844.

Eden, F. C., Hendrick, J. P., and Gottlieb, S. S. (1978). *Biochemistry* **17,** 5113–5121.

Eden, F. C., Burns, A. T. H., and Goldberger, R. F. (1980). *J. Biol. Chem.* **255,** 4843–4853.

Efstratiadis, A., Crain, W. R., Britten, R. J., Davidson, E. H., and Kafatos, F. C. (1976). *Proc. Natl. Acad. Sci. U.S.A.* **73,** 2289–2293.

Eibel, H., Gafner, J., Stotz, A., and Philippsen, P. (1981). *Cold Spring Harbor Symp. Quant. Biol.* **45,** 609–617.

Eldredge, N., and Gould, S. J. (1972). *In* "Models in Paleobiology" (T. J. M. Schopf, ed.), pp. 82–115. Freeman, San Francisco, California.

Emmons, S. W., Klass, M. R., and Hirsh, D. (1979). *Proc. Natl. Acad. Sci. U.S.A.* **76,** 1333–1337.

Emmons, S. W., Rosenzweig, B., and Hirsh, D. (1980). *J. Mol. Biol.* **144,** 481–500.

Engels, W. R. (1981). *Cold Spring Harbor Symp. Quant. Biol.* **45,** 561–565.

Epplen, J. T., Leipoldt, M., Engel, W., and Schmidtke, J. (1978). *Chromosoma* **69,** 307–321.

Epplen, J. T., Diedrich, U., Wagenmann, M., Schmidtke, J., and Engel, W. (1979). *Chromosoma* **75,** 199–214.

Errede, B., Cardillo, T. S., Wever, G., and Sherman, F. (1981). *Cold Spring Harbor Symp. Quant. Biol.* **45,** 593–607.

Fedoroff, N. (1979). *Cell* **16,** 697–710.

Fink, G., Farabaugh, P., Roeder, G., and Chaleff, D. (1981). *Cold Spring Harbor Symp. Quant. Biol.* **45,** 575–580.

Finnegan, D. J. (1981). *Nature (London)* **292,** 800–801.

Finnegan, D. J., Rubin, G. M., Young, M. W., and Hogness, D. S. (1978). *Cold Spring Harbor Symp. Quant. Biol.* **42,** 1053–1063.

Firtel, R. A., and Kindle, K. (1975). *Cell* **5,** 401–411.

Flavell, R. B., and Smith, D. B. (1976). *Heredity* **37,** 231–252.

Flavell, R. B., Bennett, M. D., Smith, J. B., and Smith, D. B. (1974). *Biochem. Genet.* **12,** 257–269.

Flavell, R. B., Rimpau, J., and Smith, D. B. (1977). *Chromosoma* **63,** 205–222.

Flavell, R. B., O'Dell, M., and Hutchinson, J. (1981). *Cold Spring Harbor Symp. Quant. Biol.* **45,** 501–508.

Free, S. J., Rice, P. W., and Metzenberg, R. L. (1979). *J. Bacteriol.* **137,** 1219–1226.

French, C. K., and Manning, J. E. (1980). *J. Mol. Evol.* **15,** 277–289.

Fritsch, E. F., Shen, C. K. J., Lawn, R. M., and Maniatis, T. (1981). *Cold Spring Harbor Symp. Quant. Biol.* **45,** 761–775.

Galau, G. A., Chamberlin, M. E., Hough, B. R., Britten, R. J., and Davidson, E. H. (1976). *In* "Molecular Evolution" (F. J. Ayala, ed.), pp. 200–224. Sinauer, Sunderland, Massachusetts.

Georgiev, G. P., Ilyin, Y. V., Chmeliauskaite, V. G., Ryskov, A. P., Kramerov, D. A., Skryabin,

K. G., Krayev, A. S., Lukanidin, E. M., and Grigoryan, M. S. (1981). *Cold Spring Harbor Symp. Quant. Biol.* **45**, 641–654.

Goldberg, R. B. (1978). *Biochem. Genet.* **16**, 45–68.

Goldberg, R. B., Crain, W. R., Ruderman, J. V., Moore, G. P., Barnett, T. R., Higgins, R. C., Gelfand, R. A., Galau, G. A., Britten, R. J., and Davidson, E. H. (1975). *Chromosoma* **51**, 225–251.

Graham, D. E., Neufeld, B. R., Davidson, E. H., and Britten, R. J. (1974). *Cell* **1**, 127–137.

Grant, V. (1963). "The Origin of Adaptations." Columbia Univ. Press, New. York.

Hake, S., and Walbot, V. (1980). *Chromosoma* **79**, 251–270.

Ham, R. G., and Veomett, M. J. (1980). "Mechanisms of Development," Appendix A, pp. 759–766. Mosby, St. Louis, Missouri.

Harpold, M. M., and Craig, S. P. (1977). *Nucleic Acids Res.* **4**, 4425–4437.

Harshey, R. M., Jayaram, M., and Chamberlin, M. E. (1979). *Chromosoma* **73**, 143–151.

Heller, R., and Arnheim, N. (1980). *Nucleic Acids Res.* **8**, 5031–5042.

Hinegardner, R. (1976). *In* "Molecular Evolution" (F. J. Ayala, ed.), pp. 179–199. Sinauer, Sunderland, Massachusetts.

Holland, C. A., and Skinner, D. M. (1977). *Chromosoma* **63**, 223–240.

Hotta, Y., and Stern, H. (1975). *In* "The Eukaryote Chromosome" (W. S. Peacock and R. D. Brock, eds.), pp. 283–300. Aust. N. Univ. Press, Canberra.

Houck, C. M., and Schmid, C. W. (1981). *J. Mol. Evol.* **17**, 148–155.

Houck, C. M., Rinehart, F. P., and Schmid, C. W. (1978). *Biochim. Biophys. Acta* **518**, 37–52.

Houck, C. M., Rinehart, F. P., and Schmid, C. W. (1979). *J. Mol. Biol.* **132**, 289–306.

Hudspeth, M. E. S., Timberlake, W. E., and Goldberg, R. B. (1977). *Proc. Natl. Acad. Sci. U.S.A.* **74**, 4332–4336.

Hughes, S. H., Shank, R. P., Spector, D. H., King, H. J., Bishop, J. M., Varmus, H. E., Vogt, P. K., and Breitman, M. L. (1978). *Cell* **15**, 1397–1410.

Hutchinson, J., Rees, H., and Seal, A. G. (1979). *Heredity* **43**, 411–421.

Hutchinson, J., Narayan, R. K. J., and Rees, H. (1980). *Chromosoma* **78**, 137–145.

Ising, G., and Block, K. (1981). *Cold Spring Harbor Symp. Quant. Biol.* **45**, 527–544.

Jain, H. K. (1980). *Nature (London)* **288**, 647–648.

Jelinek, W., Molloy, G., Fernandez-Monoz, R., Salditt, M., and Darnell, J. E. (1974). *J. Mol. Biol.* **82**, 361–370.

Jelinek, W., Evans, R., Wilson, M., Salditt-Georgieff, M., and Darnell, J. E. (1978). *Biochemistry* **17**, 2776–2783.

Jelinek, W. R., Toomey, T. P., Leinwand, L., Duncan, C. H., Biro, P. A., Coudray, P. V., Weisman, S. M., Rubin, C. M., Houck, C. M., Deininger, P. L., and Schmid, C. W. (1980). *Proc. Natl. Acad. Sci. U.S.A.* **77**, 1398–1402.

Kimmel, A. R., and Firtel, R. A. (1979). *Cell* **16**, 787–796.

Kindle, K. L., and Firtel, R. A. (1979). *Nucleic Acids Res.* **6**, 2403–2422.

Klein, W. H., Thomas, T. L., Lai, C., Scheller, R. H., Britten, R. J., and Davidson, E. H. (1978). *Cell* **14**, 889–900.

Krayev, A. S., Kramerov, D. A., Skryabin, K. G., Ryskov, A. P., Bayev, A. A., and Georgiev, G. P. (1980). *Nucleic Acids Res.* **8**, 1201–1215.

Krumlauf, R., and Marzluf, G. A. (1979). *Biochemistry* **18**, 3705–3713.

Laird, C. D., and McCarthy, B. J. (1969). *Genetics* **63**, 865–882.

Laird, C. D., Chooi, W. Y., Cohen, E. H., Dickson, E., Hutchinson, N., and Turner, S. H. (1974). *Cold Spring Harbor Symp. Quant. Biol.* **38**, 311–327.

Lauth, M. R., Spear, B. B., Heumann, J., and Prescott, D. M. (1976). *Cell* **7**, 67–74.

Lawn, R. M., Heumann, J. M., Herrick, G., and Prescott, D. M. (1978). *Cold Spring Harbor Symp. Quant. Biol.* **42**, 483–492.

Lee, A. S., Britten, R. J., and Davidson, E. H. (1978). *Cold Spring Harbor Symp. Quant. Biol.* **42,** 1065–1076.

Levis, R., Dunsmuir, P., and Rubin, G. M. (1980). *Cell* **21,** 581–588.

Lewin, B. (1974). "Gene Expression," Vol. 2: "Eucaryotic Chromosomes." Wiley, New York.

Lewontin, R. C. (1974). "The Genetic Basis of Evolutionary Change." Columbia Univ. Press, New York.

Loomis, W. F. (1973). *Dev. Biol.* **30,** f-3, f-4.

McClintock, B. (1965). *Brookhaven Symp. Biol.* **18,** 162–184.

McTavish, C., and Sommerville, J. (1980). *Chromosoma* **78,** 147–164.

Malyshev, S. I. (1968). "Genesis of the Hymenoptera and the Phases of their Evolution." Methuen, London.

Manning, J. E., Schmid, C. W., and Davidson, N. (1975). *Cell* **4,** 141–156.

Mayfield, J. E., McKenna, J. F., and Lessa, B. S. (1980). *Chromosoma* **76,** 277–294.

Mayr. E. (1958). *In* "Evolution as a Process" (J. Huxley, ed.), 2nd Ed., pp. 188–213. Allen & Unwin, London.

Mayr, E. (1966). "Animal Species and Evolution." Harvard Univ. Press, Cambridge, Massachusetts.

Miksche, J. P., and Hotta, Y. (1973). *Chromosoma* **41,** 29–36.

Mizuno, S., and Macgregor, H. C. (1974). *Chromosoma* **48,** 239–296.

Mizuno, S., Andrews, C., and Macgregor, H. C. (1976). *Chromosoma* **58,** 1–31.

Moore, G. P., Scheller, R. H., Davidson, E. H., and Britten, R. J. (1978). *Cell* **15,** 649–660.

Moore, G. P., Constantini, F. D., Posakony, J. W., Britten, R. J., and Davidson, E. H. (1980). *Science* **208,** 1046–1048.

Moore, G. P., Pearson, W. R., Davidson, E. H., and Britten, R. J. (1981). *Chromosoma* **84,** 19–32.

Moreschalli, A., and Serra, V. (1972). *Experientia* **30,** 487–489.

Moyzis, R., Bonnet, J., and Ts'o, P. O. P. (1977). *J. Cell Biol.* **75,** 130a.

Murray, M. G., Cuellar, R. E., and Thompson, W. F. (1978). *Biochemistry* **17,** 5781–5790.

Murray, M. G., Palmer, J. D., Cuellar, R. E., and Thompson, W. F. (1979). *Biochemistry* **18,** 5259–5266.

Murray, M. G., Peters, D. L., and Thompson, W. F. (1981). *J. Mol. Evol.* **17,** 31–42.

Narayan, R. K. J., and Rees, H. (1976). *Chromosoma* **54,** 141–154.

Narayan, R. K. J., and Rees, H. (1977). *Chromosoma* **63,** 101–107.

Orgel, L. E., and Crick, F. H. C. (1980). *Nature (London)* **284,** 604–607.

Orgel, L. E., Crick, F. H. C., and Sapienza, C. (1980). *Nature (London)* **288,** 645–646.

Posakony, J. W., Scheller, R. H., Anderson, D. M., Britten, R. J., and Davidson, E. H. (1981). *J. Mol. Biol.* **149,** 41–67.

Potter, S. S., Brorein, W. J., Jr., Dunsmuir, P., and Rubin, G. M. (1979). *Cell* **17,** 415–427.

Potter, S. S., Truett, M., Phillips, M., and Maher, A. (1980). *Cell* **20,** 639–647.

Prakash, S. R., Lewontin, R. C., and Hubby, J. L. (1969). *Genetics* **61,** 841–858.

Preisler, R. S., and Thompson, W. F. (1981a). *J. Mol. Evol.* **17,** 78–84.

Preisler, R. S., and Thompson, W. F. (1981b). *J. Mol. Evol.* **17,** 85–93.

Proscott, D. M., and Murti, K. G. (1974). *Cold Spring Harbor Symp. Quant. Biol.* **38,** 609–618.

Rees, H., and Jones, G. H. (1972). *Int. Rev. Cytol.* **32,** 53–92.

Rimpau, J., Smith, D. B., and Flavell, R. B. (1978). *J. Mol. Biol.* **123,** 327–359.

Rimpau, J., Smith, D. B., and Flavell, R. B. (1980). *Heredity* **44,** 131–149.

Romer, A. S. (1959). "The Vertebrate Story." Univ. of Chicago Press, Chicago, Illinois.

Roth, G. E. (1979). *Chromosoma* **74,** 355–371.

Rubin, C. M., Houck, C. M., Deininger, P. L., Friedmann, T., and Schmid, C. W. (1980). *Nature (London)* **284,** 372–374.

Rubin, G. M. (1978). *Cold Spring Harbor Symp. Quant. Biol.* **42,** 1041–1046.

Rubin, G. M., Brorein, W. J., Jr., Dunsmuir, P., Flavell, A. J., Levis, R., Strobel, E., Toole, J. J., and Young, E. (1981). *Cold Spring Harbor Symp. Quant. Biol.* **45**, 619–628.

Samols, D., and Swift, H. (1979). *Chromosoma* **75**, 129–143.

Sato, S., Hutchinson, C. A., III, and Harris, J. I. (1977). *Proc. Natl. Acad. Sci. U.S.A.* **74**, 542–546.

Scagel, R. F., Bandoni, R. J., Rouse, G. E., Schofield, W. B., Stein, J. R., and Taylor, T. M. C. (1965). "An Evolutionary Survey of the Plant Kingdom." Wadsworth, Belmont, California.

Schachat, F., O'Connor, D. J., and Epstein, H. F. (1978). *Biochim. Biophys. Acta* **520**, 688–692.

Scheller, R. H., Thomas, T. L., Lee, A. S., Klein, W. H., Niles, W. D., Britten, R. J., and Davidson, E. H. (1977). *Science* **196**, 197–200.

Scheller, R. H., Constantini, F. D., Kozlowski, M. R., Britten, R. J., and Davidson, E. H. (1978). *Cell* **15**, 189–203.

Scheller, R. H., Anderson, D. M., Posakony, J. W., McAllister, L. B., Britten, R. J., and Davidson, E. H. (1981). *J. Mol. Biol.* **149**, 15–39.

Schmid, C. W., and Deininger, P. L. (1975). *Cell* **6**, 345–358.

Schmidtke, J., Epplen, J. T., and Engel, W. (1979a). *Comp. Biochem. Physiol.* **63B**, 455–458.

Schmidtke, J., Schmitt, E., Leipoldt, M., and Engel, W. (1979b). *Comp. Biochem. Physiol.* **64B**, 117–120.

Selander, R. K., Hunt, W. G., and Yang, S. Y. (1969). *Evolution* **23**, 379–390.

Shen, C.-K. J., and Maniatis, T. (1980). *Cell* **19**, 379–391.

Shimotohno, K., and Temin, H. M. (1981). *Cold Spring Harbor Symp. Quant. Biol.* **45**, 719–730.

Shimotohono, K., Mizutani, S., and Temin, H. M. (1980). *Nature (London)* **285**, 550–554.

Simpson, 1944. "Tempo and Mode in Evolution." Columbia Univ. Press, New York.

Smith, D. B., and Flavell, R. B. (1977). *Biochim. Biophys. Acta* **474**, 82–97.

Smith, D. B., Rimpau, J., and Flavell, R. B. (1976). *Nuceic Acids Res.* **3**, 2811–2825.

Smyth, D. R., and Stern, H. (1973). *Nature (London)* New Biol. **245**, 94–96.

Spradling, A. C., and Rubin, G. M. (1981). *Annu. Rev. Genet.* **15**, in press.

Stebbins, G. L. (1950). "Variation and Evolution in Plants." Columbia Univ. Press, New York.

Stebbins, G. L. (1970). In "Essays in Evolution and Genetics in Honor of Theodosius Dobzhansky" (M. K. Hecht and W. C. Steere, eds.), pp. 173–208. Appleton, New York.

Stebbins, L. G. (1971). "Chromosomal Evolution in Higher Plants." Addison-Wesley, Reading, Massachusetts.

Stein, D. B., and Thompson, W. F. (1975). *Science* **189**, 888–890.

Stein, D. B., Thompson, W. F., and Belford, H. S. (1979). *J. Mol. Evol.* **13**, 215–232.

Storer, T. I., and Usinger, R. L. (1965). "General Zoology," 4th ed. McGraw-Hill, New York.

Straus, N. A. (1971). *Proc. Natl. Acad. Sci. U.S.A.* **68**, 799–802.

Straus, N. (1972). *Carnegie Inst. Year Book* **71**, 257–259.

Strobel, E., Dunsmuir, P., and Rubin, G. M. (1979). *Cell* **17**, 429–439.

Tchurikov, N. A., Zelentsova, E. S., and Georgiev, G. P. (1978). *Nucleic Acids Res.* **5**, 2169.

Tchurikov, N. A., Ilyin, Y. V., Skryabin, K. G., Ananiev, E. V., Bayev, A. A., Jr., Krayev, A. S., Zelentsova, E. S., Kulguskin, V. V., Lyubomirskaya, N. V., and Georgiev, G. P. (1981). *Cold Spring Harbor Symp. Quant. Biol.* **45**, 655–665.

Timberlake, W. E. (1978). *Science* **202**, 973–975.

Timberlake, W. E., Shumard, D. S., and Goldberg, R. B. (1977). *Cell* **10**, 623–632.

Ullrich, R. E., Kohorn, B. D., and Specht, C. A. (1980). *Chromosoma* **81**, 371–378.

Walbot, V., and Dure, L. S., III. (1976). *J. Mol. Biol.* **101**, 503–536.

Walbot, V., and Goldberg, R. (1979). In "Nucleic Acids in Plants" (T. C. Hall and J. W. Davies, eds.), Vol. 1, pp. 3–40. CRC Press, Boca Raton, Florida.

Wells, R., Royer, H.-D., and Hollenberg, C. P. (1976). *Mol. Gen. Genet.* **147**, 45–51.

Wesink, P. C. (1978). *Cold Spring Harbor Symp. Quant. Biol.* **42**, 1033–1039.

Wensink, P. C., Tabata, S., and Pachl, C. (1979). *Cell* **18,** 1231–1246.
Wetmur, J. G., and Davidson, N. (1968). *J. Mol Biol.* **31,** 349–370.
White, M. J. D. (1959). *Aust. J. Sci.* **22,** 32–39.
Wilmore, J. P., and Brown, A. R. (1975). *Chromosoma* **51,** 337–345.
Wright, S. (1940a). *In* "The New Systematics" (J. Huxley, ed.), pp. 161–183. Oxford Univ. Press, London.
Wright, S. (1940b). *Am. Nat.* **74,** 232–248.
Wu, J.-R., Pearson, W. R., Posakony, J. W., and Bonner, J. (1977). *Proc. Natl. Acad. Sci. U.S.A.* **74,** 4382–4386.
Yao, M.-C., and Gorovsky, M. A. (1974). *Chromosoma* **48,** 1–18.
Young, M. W. (1979). *Proc. Natl. Acad. Sci. U.S.A.* **76,** 6274–6278.
Young, M. W., and Schwartz, H. E. (1981). *Cold Spring Harbor Symp. Quant. Biol.* **45,** 629–640.
Zuckerkandl, E. J. (1976). *J. Mol. Evol.* **9,** 73–122.

Structural Attributes of Membranous Organelles in Bacteria[1]

Charles C. Remsen

Center for Great Lakes Studies and Department of Zoology, University of Wisconsin–Milwaukee, Milwaukee, Wisconsin

I. Introduction

The past 25 years has seen some major changes occur in our perception of the structural organization in prokaryotic cells. Whereas the bacterium was once thought of as a simple bag of enzymes, it has now been shown to be a complex, highly ordered cell. In some instances, the bacterial cell has even shown some attempt at cellular differentiation or compartmentalization (Watson *et al.*, 1971). This change in perception is primarily the result of improved technology, particularly in high resolution transmission electron microscopes, but it is also due to refinements in the methodologies of fixation and embedding of specimens as well as the development of new techniques for visualizing cell structures (Moor *et al.*, 1961; Branton, 1966; Remsen and Watson, 1972).

The large volume of information that is available concerning the ultrastructure of prokaryotes now permits us to make some judgments that heretofore were impossible to make. For example, the ultrastructural organization of a bacterial cell can now be used for the taxonomic separation of genera and even species

---

[1]Contribution No. 233, Center for Great Lakes Studies, University of Wisconsin–Milwaukee, Milwaukee, Wisconsin 53201.

(Buchanan and Gibbons, 1974; Davies and Whittenbury, 1970; Romanovskaya *et al.*, 1978).

The concept of discrete organelles in prokaryotic cells has historically been rejected by biologists in general and by bacteriologists and microbiologists in particular. Traditionally, this characteristic of prokaryotes—the absence of membrane-bound cytoplasmic organelles—has been used as one of several major differences between eukaryotic and prokaryotic cells (cf. Pfenning, 1978).

As our ability to resolve greater detail within the prokaryotic cell has increased, and the association between structure and function has become more clearly defined, the presence of discrete organelles in bacteria is once again the topic of much discussion. There can no longer be any doubt that certain bacteria have evolved extremely specialized structures designed to carry out, in some instances, a single function. Furthermore, the discovery of cytoplasmic structures bounded by a nonunit membrane (Wullenweber *et al.*, 1977; Staehelin *et al.*, 1978) brings into question certain assumptions that have traditionally been accepted as fact. No longer can we say that "the cytoplasmic membrane . . . is the only major bounding agent which can be structurally defined" (Stanier and van Niel, 1962).

The purpose of this article is to examine, as the title suggests, the structural attributes of membranous organelles in bacteria. The use of the term organelle is not intended to be misleading, rather it is intended to challenge some historical concepts that have perhaps outlived their usefulness.

II. Membranous Organelles in Bacteria

Internal membrane systems have been described for a wide variety of microorganisms. Perhaps the earliest to be described, and lately perhaps the most controversial, is the mesosome (Fitz-James, 1960), or chondriod (Kellenberger *et al.*, 1958) or plasmalemmasome (Edwards and Stevens, 1963) or, more recently, the technikosome (Fooke-Achterrath *et al.*, 1974). Mesosomes, as they will be referred to in this article, have a characteristic appearance and have been found to be widely distributed among the bacteria although they are more commonly associated with the gram-positive bacteria. In addition to the mesosomes, intracytoplasmic membranes are found quite extensively in a variety of autotrophic bacteria including the phototrophic bacteria (Cohen-Bazire, 1963), the nitrifying bacteria (Murray and Watson, 1965), the methane-oxidizing bacteria (Davies and Whittenbury, 1970), and the hydrogen-oxidizing bacteria (Zeikus and Wolfe, 1973; Mayer *et al.*, 1977; Doddema *et al.*, 1979).

For the purpose of this discussion, intracytoplasmic membrane systems or membranous organelles will be defined structurally as consisting of lamellar, vesicular, or tubular membranes of the "unit-membrane" type (Robertson,

1959) which are usually but not necessarily still associated with the cytoplasmic or cell membrane. Furthermore, they will be defined functionally as being the site of specific, energy-yielding reactions.

A. Mesosomes

Mesosomes have intrigued microscopists and microbiologists from the time they were first described as "peripheral bodies" by Chapman and Hillier (1953) right up to the present where their structure and function are suspect (Fooke-Achterrath et al., 1974). Since a great deal of time and effort over the past 20 years has been devoted to the mesosome, and a number of significant reviews have been produced (Reusch and Burger, 1973; Ghosh, 1974; Greenawalt and Whiteside, 1975; Salton and Owen, 1976; Salton, 1978), this article will not attempt to cover the mesosome in any detail. Rather, the mesosome is included only because it has historical precedence in its possible role as a membranous organelle.

The significance of the mesosome is that its discovery, and subsequent associa-

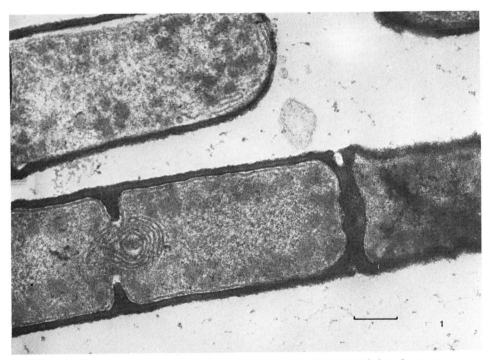

Fig. 1. Thin section of chemically fixed *Bacillus subtilis* showing the association of a mesosome and the forming division septum. Bar represents 0.2 μm (Remsen, unpublished micrograph).

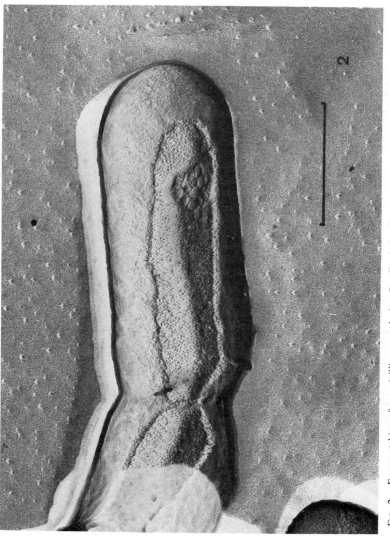

FIG. 2.   Freeze-etching of a penicillinase-producing *Bacillus licheniformis* showing a cluster of vesicles (mesosomes) in the periplasmic space of the cell wall. Bar represents 0.5 μm (Remsen, Ghosh, and Lampen, unpublished micrograph).

tion with most major functions of the bacterial cell, suggested that it could be classified as a membrane-bounded organelle. It appeared to be associated with cell division (Fig. 1) and sporulation (Chapman and Hillier, 1953; Fitz-James, 1960; Ryter, 1965; Remsen, 1966), nuclear division (Giesbrecht, 1960; Van Iterson, 1961; Remsen, 1968), sites of oxidation and reduction (Vanderwinkle and Murray, 1962; Van Iterson and Leene, 1964a, b), and the release of extracellular enzymes (Ghosh et al., 1969) (Fig. 2). However, when these apparent functions were examined more critically by applying cell fractionation techniques and measuring enzymatic activities, no specific functional role could be shown to be applicable (Owen and Freer, 1972; Reusch and Burger, 1973; Salton and Owen, 1976).

Higgins and co-workers (Higgins and Daneo-Moore, 1974; Higgins et al., 1976) have examined the effects of fixation and other physical insults on bacteria, and the resulting effect on mesosomes. It is their feeling, and earlier that of Nanninga (1971), that such insults affect a "mesosome precursor" present in naturally growing cells, that permits the visualization of a mesosome after such things as fixation and freeze-fracturing.

While this feeling appears to have gained widespread support (and consequently a lack of interest in studying mesosomes) there remain those investigators who still feel that the mesosome represents a specialized structure of some importance. Van Iterson et al. (1975) and Van Iterson and Aten (1976) provide striking micrographs of *Bacillus subtilis* illustrating "coordinated behavior between nucleoids and mesosomes." As they have stated, their work "sustains the contention that, during nuclear division, the mesosome interacts with the nucleoplasm to make what is probably a significant contribution to replication."

B. Photosynthetic Intracytoplasmic Membranes

As indicated earlier, this article is not intended to critically review the body of literature dealing with fine structure of the prokaryotic photosynthetic apparatus. The reader is encouraged to refer to other reviews where this task has been admirably done (Fuller et al., 1963; Stanier, 1963; Cohen-Bazire and Sistrom, 1966; Pfenning, 1967; Oelge and Drews, 1972; Remsen, 1978).

Photosynthetically active intracytoplasmic membranes have been described in all members of the families Rhodospirillaceae and Chromatiaceae. In most instances, these membranes appear to arise by the invagination of the cytoplasmic membrane. These invaginations can then form a variety of arrangements ranging from simple intrusions into the cytoplasm as in *Rhodopseudomonas spheroides* (Peters and Cellarius, 1972) to extremely ordered lamellar stacks as in *Ectothiorhodospira mobilis* (Raymond and Sistrom, 1967; Remsen et al., 1968;

Holt *et al.*, 1968), and tubular intracytoplasmic membranes as in *Thiocapsa pfennigii* (Eimhjellen *et al.*, 1967; Eimhjellen, 1970).

Typically, the cytoplasmic membrane differentiates by producing vesicular or tubular invaginations which extend inwardly in the shape of connected tubes. As growth of these invaginations proceeds, vesicles become tightly packed within the cell. In sectioned (Fig. 3) and frozen-etched (Fig. 4) preparations, the tight packing of vesicles is easily seen; often it becomes difficult to resolve the individual vesicle membranes.

While most investigators believe that the intracytoplasmic membranes observed in photosynthetic bacteria (especially the Rhodospirillaceae) remain attached to, and are continuous with, the cytoplasmic membrane, some workers feel that this is not always true. In at least some members of the Chromatiaceae, it is possible that vesicles may separate from the cytoplasmic membrane even though they almost certainly arise from it. It has even been speculated that such free, "chromatophore" vesicles are closely associated with the formation of intracellular sulfur granules, and the membrane associated with these storage granules (Remsen, 1978).

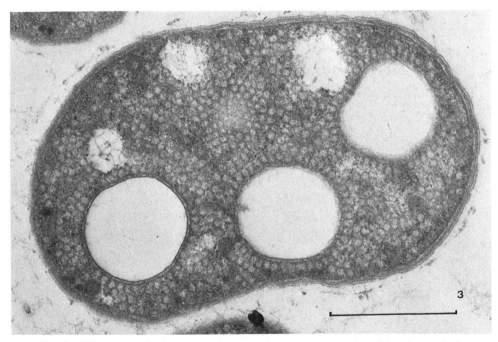

FIG. 3. Thin section of *Thiocapsa violacea* showing the dense packing of the photosynthetic intracytoplasmic membrane system. The large, membrane-bound transparent regions are sulfur granules. Bar represents 0.5 μm (Remsen, unpublished micrograph).

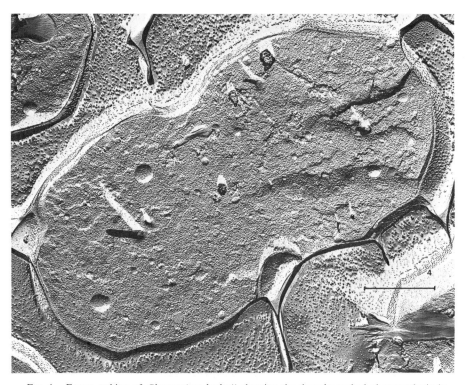

FIG. 4. Freeze-etching of *Chromatium buderii* showing the densely packed photosynthetic intracytoplasmic membranes. Bar represents 1.0 μm (Remsen, unpublished micrograph).

C. Nonphotosynthetic Intracytoplasmic Membranes

When one examines the fairly extensive literature dealing with the fine structure of bacteria, it soon becomes apparent that the presence of extensive intracytoplasmic membranes is limited to those species of bacteria that utilize inorganic or single carbon energy sources. The sulfur/iron oxidizing thiobacilli (Remsen and Lundgren, 1966; Mahoney and Edwards, 1966; Shively *et al.*, 1970) seem to be the one exception, but they are able to synthesize a major part of their ATP by substrate phosphorylations which do not involve electron transport (Peck, 1968). There may also be some methane-oxidizing bacteria that do not have cytomembrane systems.

1. The Nitrifying Bacteria

The complex arrangement of intracytoplasmic membranes in the nitrifying bacteria was first described in 1965 by Murray and Watson. They compared the ultrastructure of *Nitrosocystis oceanus,* a newly isolated marine ammonia-oxidizing bacterium, with the two classical nitrifying bacteria, *Nitrosomonas europeae* and *Nitrobacter agilis.* They were able to show that all three had characteristically arranged "cytomembrane systems," a discovery perhaps not so surprising since energy yielding reactions such as those carried out by these bacteria are most often membrane-linked (Murray, 1963).

Further studies on both ammonia-oxidizing (Remsen *et al.,* 1967; Watson *et*

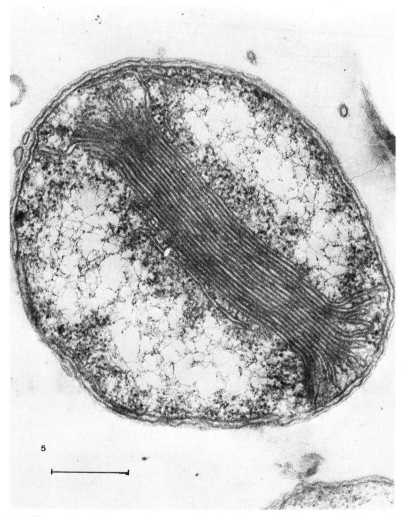

FIG. 5.   Thin section of *Nitrosocystis oceanus* showing the extensive intracytoplasimic membrane system within the cell. Bar represents 0.5 μm (Watson and Remsen, unpublished micrograph).

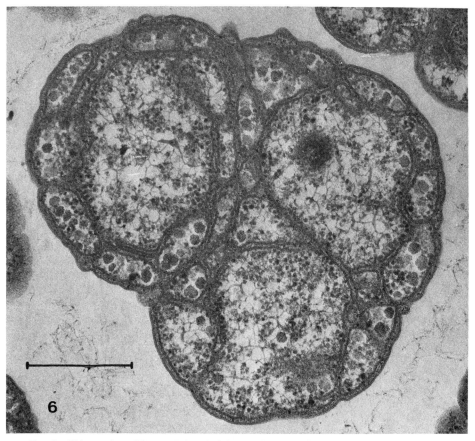

FIG. 6. Thin section of *Nitrosolobus multiformis* showing lobular nature and the presence of membrane-bound compartments. Bar represents 0.5 μm (Watson *et al.*, 1971; by permission of Springer-Verlag, Berlin and New York).

al., 1971) and nitrite-oxidizing bacteria (Watson and Waterbury, 1971) confirmed the earlier observations that these chemolithotrophic bacteria all possessed extensive intracytoplasmic membrane systems. While the majority of these bacteria have membranes arranged in characteristic pairs and/or stacks (Fig. 5) arising by an invagination of the plasma membrane (e.g., *Nitrosocystis oceanus*, Murray and Watson, 1965; Remsen *et al.*, 1967; and *Nitrobacter agilis*, Murray and Watson, 1965), there are at least two notable exceptions. In the first case, the ammonia-oxidizing bacterium, *Nitrosolobus multiformis* (Watson *et al.*, 1971) has pairs of cytomembranes, originating from the cytoplasmic membrane, separated by the cell wall mucopeptide layer, which intrude deeply

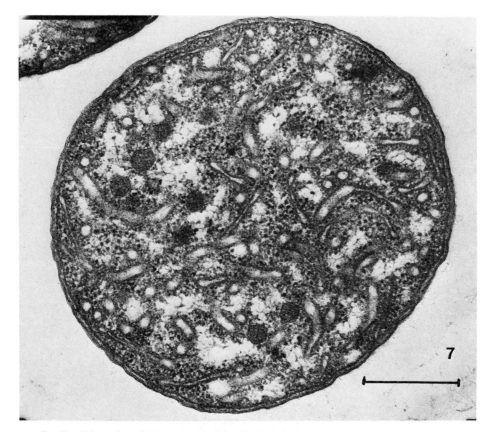

FIG. 7. Thin section of *Nitrococcus mobilis* showing tubular intracytoplasmic membrances. Note also the presence of polyhedral bodies (carboxysomes) in the cytoplasm. Bar represents 0.5 μm (Waterbury, Watson, and Remsen, unpublished micrograph).

into the cytoplasm resulting in an apparent compartmentalization of the cell (Fig. 6). The second exception is represented by the nitrite-oxidizing bacterium, *Nitrococcus mobilis* (Watson and Waterbury, 1971) which has a unique tubular cytomembrane system (Fig. 7).

2. The Methane-Oxidizing Bacteria

According to the work of several investigators (Proctor *et al.*, 1969; Davies and Whittenbury, 1970; Ribbons *et al.*, 1970; Hazeu, 1975; Romanovskaya *et al.*, 1978) the methane-oxidizing bacteria can be classified according to the arrangement of their intracytoplasmic membranes. Those bacteria that have membranes arranged in characteristic stacks and extend into the cell away from the cytoplasmic membrane belong to Type I (Fig. 8). Those that have membranes arranged as peripheral layers running parallel to the cytoplasmic membrane belong to Type II (Figs. 9 and 10) (Smith and Ribbons, 1970; Weaver and Dugan, 1975).

As in the case of the nitrifying bacteria, the intracytoplasmic membranes described in the methane-oxidizing bacteria consist of closely packed pairs of membranes resulting in a triplet structure approximately 10 nm thick. In almost all cases, thin sections have shown a definite connection between the cytoplasmic membrane and the intracytoplasmic membranes. This, of course, suggests that invaginations of the cytoplasmic membrane give rise to the complex intracytoplasmic membranes.

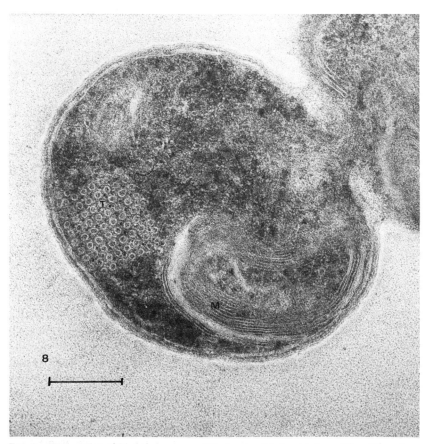

FIG. 8.   Thin section of *Methylococcum mobilis* showing tubular structures (T) and Type I membranes. Bar represents 0.2 μm (from Hazeu *et al.,* 1980; by permission of Springer-Verlag, Berlin and New York).

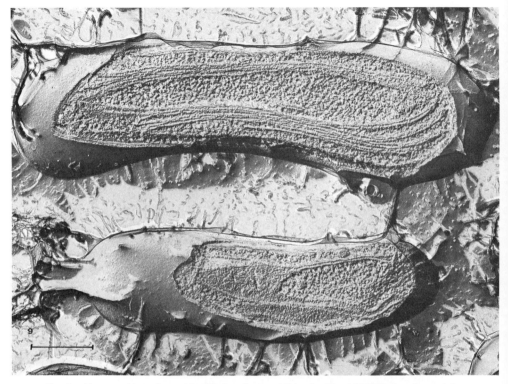

FIG. 9. Freeze-etching of an unidentified methane-oxidizing bacillus (#OB56) showing peripherally located, Type II, intracytoplasmic membranes. Bar represents 0.5 μm (Watson, Remsen, and Whittenbury, unpublished micrograph).

III. Atypical Membranous Organelles

In recent years there have been numerous scientific reports describing structures within the bacterial cell that are bounded by a single or non-"unit membrane." These structures include such things as gas vesicles, carboxysomes, storage products such as sulfur and PHB granules, and most recently, magnetosomes and chlorosomes. These atypical membranous organelles appear to have one thing in common, they are bounded by a single membrane (Table I) varying, according to the literature, in thickness from 2.5 to 5.0 nm. In the following discussion, a brief description of the structure of each of these atypical membranous organelles will be presented.

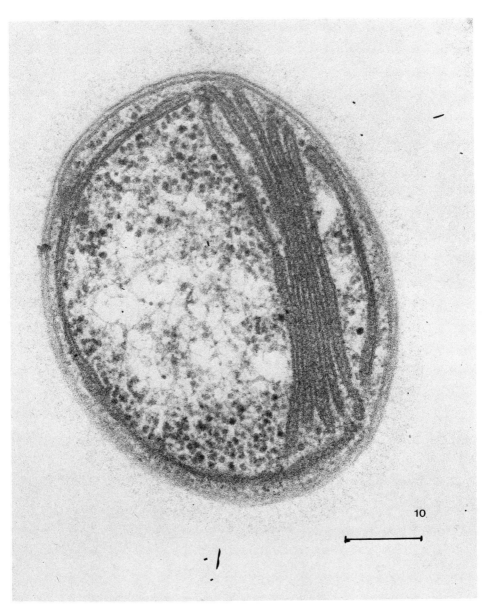

FIG. 10. Thin section of an unidentified methane-oxidizing bacillus (#OB56). Cross section shows Type II, intracytoplasmic membranes. Bar represents 0.2 μm (Watson, Remsen, and Whittenbury, unpublished micrograph).

TABLE I

Bacterial Structures Bounded by a Single, Nonunit Membrane

Structure	Characteristics of the atypical membrane	References
Gas vesicle	2.5 ± 0.4 nm thick	Jost and Jones (1980)
	1-layered structure, 30 Å thick	Larsen et al. (1967)
	Single membranes only 2 nm thick	Bowen and Jensen (1965a,b)
PHB	3.0–3.5 nm	Lundgren et al. (1964); Boatman (1964)
	Bounded by a single, electron dense layer	Balkwill et al. (1980)
Sulfur granules	"Proteinaceous membrane," 2.5 nm components	Nicholson and Schmidt (1971)
	"Encased by a nonunit membrane"	Wirsen and Jannasch (1978)
Carboxysomes	(3.0–4.0 nm)	Shively et al. (1970)
	"monolayer envelope about 4 nm thick"	Wullenweber et al. (1977)
Magnetosomes	1.4-nm-thick electron-dense layer plus a 1.6-nm electron transparent layer	Balkwill et al. (1980)
Chlorosomes	A nonunit (single layered) membrane 2–3 nm thick	Staehelin et al. (1978)
	Thin, electron-dense membrane, 5 nm thick	Cohen-Bazire et al. (1964)
	A nonunit membrane, 2 nm or less thick (2.0–5.0 nm)	Pfenning and Cohen-Bazire (1967)

A. Gas Vesicles

While the existence of gas vesicles had been speculated many years ago (Kolkwitz, 1928), it was not until 1965 when Bowen and Jensen (1965a,b), using thin sections of blue-green algae, were able to show that the gas vesicles consisted of stacks of electron-transparent cylinders which they termed "gas vesicles." They were described as hollow cylinders, approximately 75 nm in diameter and varied in length from 200 to 1000 nm. In addition, they appeared to be bounded by a single membrane only 2 nm thick.

Subsequent to this early description of gas vesicles in blue-green algae, gas vesicles have been isolated and described in many of the Cyanobacteria as well as bacteria including the photosynthetic green and purple sulfur bacteria (Walsby, 1972). As a result, the structure of the membrane surrounding the gas vesicle has been described in great detail.

According to Walsby (1972) in his review on gas vesicles, the gas vesicle membrane is considerably thinner than the typical "unit-membrane" or cytoplasmic membrane characteristic of all cells. It measures only 2 nm wide and has striations, consisting of ribs 4.5 nm wide, "running at right angles to the long axis of the structure; and the ribs seem to be made up of particles." These particles are approximately 2.8–3.5 nm in diameter. Chemical analyses of these isolated membranes show them to consist entirely of a single protein having a molecular weight of approximately 14,000–15,000 (Jones and Jost, 1970; Jost and Jones, 1970).

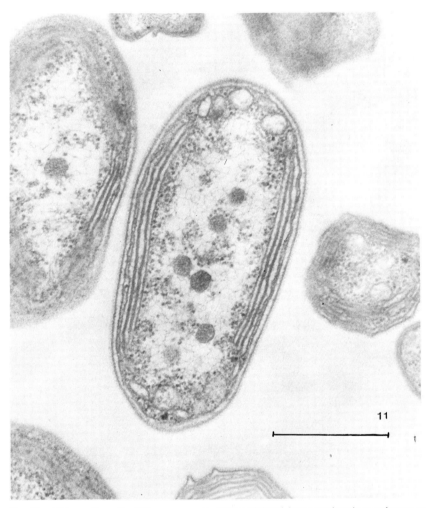

11

FIG. 11. Thin section of a *Nitrosomonas* sp. with peripheral intracytoplasmic membranes and several polyhedral bodies (carboxysomes). Note the thin membrane surrounding these polyhedral bodies. Bar represents 0.5 μm (Remsen and Watson, unpublished micrograph).

B. Carboxysomes

Polyhedral or hexagonal bodies (Fig. 11), approximately 100 nm in diameter, have been reported in a wide variety of prokaryotic cells. They have been described in the nitrifying bacteria (Murray and Watson, 1965; Pope *et al.*, 1969; Van Gool *et al.*, 1969; Watson and Waterbury, 1971), cyanobacteria (Gantt and Conti, 1969; Hoare *et al.*, 1971; Wolk, 1973), and in the thiobacilli (Mahoney and Edwards, 1966; Wang and Lundgren, 1969; Shively *et al.*, 1970, 1973a).

As described by Watson and Waterbury (1971), these 100 nm bodies "consisted of 6–8 nm subunits surrounded and enclosed by a single layered membrane." The membrane appears to be approximately 3.0 to 4.0 nm thick in bacteria (Pankratz and Bowen, 1963; Watson and Waterbury, 1971; Shively *et al.*, 1973a) but is reported to be thicker, about 5.5 nm, in some of the cyanobacteria (Wildman, 1969).

Purified carboxysomes, isolated from *Thiobacillus neapolitanus* (Shively *et al.*, 1973a), were found to consist of a mass of 10-nm particles surrounded by a 3.5-nm-thick single membrane. The particles were identified as the enzyme, ribulose-1,5-diphosphate carboxylase, hence the name given to the polyhedral bodies.

C. Chlorosomes

The term "chlorosome" has been proposed to identify vesicular structures, attached to the cytoplasmic membrane of members of the Chlorobiaceae and Chloroflexaceae, that contain bacteriochlorophyll *c* and *d*. This proposal (Staehelin *et al.*, 1978) comes nearly 15 years after Cohen-Bazire (1963) examined the fine structure of green sulfur bacteria and found them to have a complex fine structure unlike other photosynthetic bacteria. In this study, she proposed the term "chlorobium vesicle" to describe oblong structures adjacent to the cytoplasmic membrane that were surrounded by a thin, electron-dense membrane, approximately 5 nm thick. In a later study (Cohen-Bazire *et al.*, 1964), the photosynthetic pigments were shown to be associated primarily with the chlorobium vesicle fraction. Thus, the reader is asked to equate the term chlorobium vesicle with the new term, chlorosome; they are one-in-the-same structure.

Due to its close proximity to the cytoplasmic membrane, the chlorosome is difficult to resolve in thin sections. Figure 12a and 12b show thin sections of *Chloroflexus aurantiacus* with chlorosomes (arrows) closely associated with the cytoplasmic membrane (Staehelin *et al.*, 1978). Freeze-fracturing, on the other hand, reveals the structure of the chlorosome with much greater detail (Fig. 12c and d). Not only can one obtain information on the shape and distribution of these structures, but their intimate association with the cytoplasmic membrane is

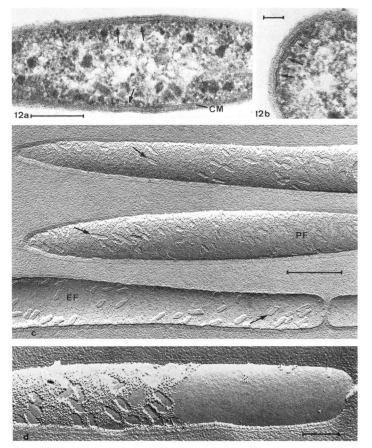

FIG. 12.   (a) Thin section of *Chloroflexus aurantiacus* showing chlorosomes lying adjacent to the cytoplasmic membrane (arrows) (CM). Bar represents 0.2 μm (Staehelin *et al.*, 1978; by permission of Springer-Verlag, Berlin and New York). (b) Thin section of *Chloroflexus aurantiacus* showing chlorosomes (arrows) lying next to the cytoplasmic membrane. Bar represents 0.05 μm (Staehelin *et al.*, 1978; by permission of Springer-Verlag, Berlin and New York). (c) Freeze-fracture of *Chloroflexus auranticus*. Where the chlorosomes are attached to the cytoplasmic membrane (arrows), the fracture plane is deflected into the vesicles. Bar represents 0.5 μm (Staehelin *et al.*, 1978; by permission of Springer Verlag, Berlin and New York). (d) Freeze-fracture of *Chloroflexus aurantiacus* showing yet another view of the chlorosomes attached to the cytoplasmic membrane. Bar represents 0.2 μm (Staehelm *et al.*, 1978; by permission of Springer-Verlag, Berlin and New York).

readily apparent. According to Staehelin *et al.* (1978), the average chlorosome is 106 ± 24 nm long, 32 ± 10 nm wide, and 12 ± 2nm thick. Upon closer examination (cf. Fig. 11 in Staehelin *et al.*, 1978) the chlorosome exhibits ''tightly packed, thin, rod-shaped elements (approximately 5.2 nm in diameter) that are oriented parallel to the long axis of the vesicles.'' These rod elements can be resolved even further into ''rows of globular particles with a periodicity of approximately 6 nm.''

Isolated and purified chlorosomes show a similar structure when examined by electron microscopy as negatively stained preparations (Fig. 13; Schmidt *et al.*, 1980). The periodic structure so clearly illustrated by Staehelin *et al.* (1978) can also be occasionally seen in these negatively stained preparations.

While the chlorosome now seems to be accepted as the photosynthetic apparatus in the green sulfur bacteria, recent studies have indicated that there may be some differences between the Chlorobiaceae and the Chloroflexaceae. Schmidt (1980) has shown ''that there are some differences in structural organization and pigment-protein-complex, especially with respect to the organization of that part of the photosynthetic apparatus which is localized in the cytoplasmic membrane.''

D. STORAGE PRODUCTS

1. *Poly-β-hydroxybutyrate Granules (PHB)*

Early studies on the fine structure of chemically fixed bacterial cells (Boatman, 1964), on carbon replicas of isolated PHB granules (Lundgren *et al.*, 1964), and

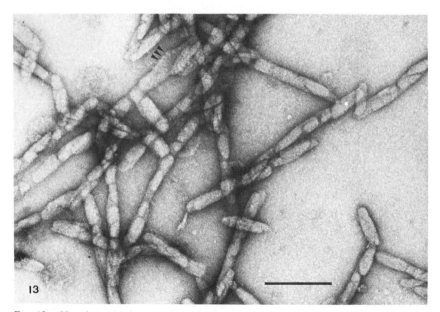

FIG. 13.   Negative stained preparation of chlorosomes isolated from *Chloroflexus aurantiacus*. Arrows indicate an area of periodic structure on a chlorosome. Bar represents 200 nm (Schmidt, 1980; by permission of Springer-Verlag, Berlin and New York).

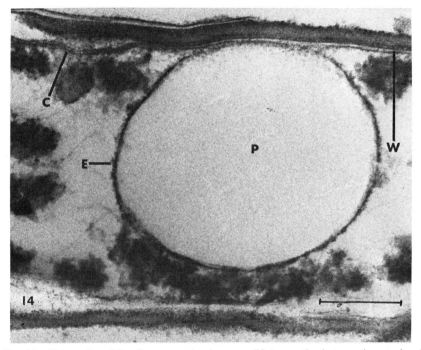

Fig. 14. Thin section of *Bacillus cereus* showing a possible connection between the cytoplasmic membrane (C) and the coating (E) surrounding the PHB (P) granule. W, Cell wall. Bar represents 0.2 μm (Pfister *et al.*, unpublished micrograph).

on freeze-etched preparations of *Bacillus cereus* (Dunlop and Robards, 1973) revealed that some sort of membrane separated the polymer, PHB, from the cytoplasm of the cell. Later studies consistently showed the presence of a membrane surrounding PHB granules (Wang and Lundgren, 1969; Jensen and Sicko, 1973).

The membrane (Fig. 14) appears as a single membrane approximately 2.0–4.0 nm thick (Jensen and Sicko, 1971) and at least one study (Pfister *et al.*, 1969) has suggested that it may arise from the cytoplasmic membrane. Finally, the polymerizing enzyme, PHB synthetase, as well as all of the depolymerizing complex are associated with the granule, most probably the membrane (Griebel *et al.*, 1968; Griebel and Merrick, 1971).

2. Sulfur Granules

Previous studies (de Boer *et al.*, 1961; Kran *et al.*, 1963; Hageage *et al.*, 1969; Schmidt and Kamen, 1970) have shown that sulfur was stored as granules which appeared to have a limiting membrane. This limiting membrane, described as a ''proteinaceous membrane'' composed of 2.5 nm subunits (Nicholson and Schmidt, 1971), can be clearly seen in thin sections of chemically fixed photosynthetic sulfur bacteria (Remsen and Truper, 1973; Remsen, 1978) or nonphotosynthetic sulfur bacteria (Wirsen and Jannasch, 1978). Figure 15 shows the ''nonunit membrane'' surrounding sulfur granules in thin sections of *Thiovulum* sp., a large colorless sulfur bacterium.

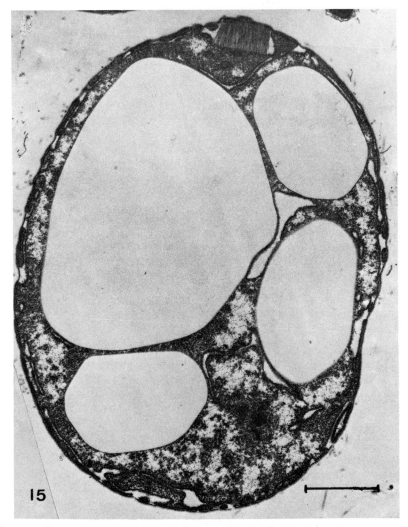

FIG. 15.   Thin section of *Thiovulum* sp. showing several sulfur inclusions encased by a nonunit membrane. Bar represents 1.0 μm (Wirsen and Jannasch, 1978; by permission from the American Society for Microbiology, Bethesda, Maryland).

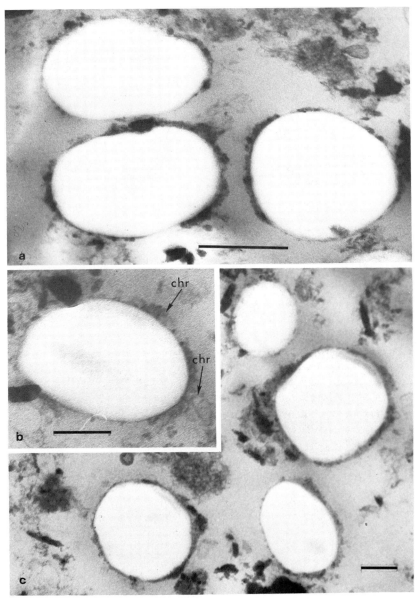

FIG. 16. (a) Example of a purified sulfur granule preparation of *Chromatium buderi*. Sulfur granules were fixed with glutaraldehyde-osmium prior to sectioning. Note nonunit membrane and attached chromatophore material. Bar represents 0.5 μm (Gonye, Schroeder, and Remsen, unpublished micrograph). (b) Thin section showing intact chromatophores (chr) visible on and near the surface of the purified sulfur granule. Bar represents 0.2 μm (Gonye, Schroeder, and Remsen, unpublished micrograph). (c) Thin section of sulfur granule preparation. Note electron-dense areas at outer edge of sulfur granules. Bar represents 0.2 μm (Gonye, Schroeder, and Remsen, unpublished micrograph).

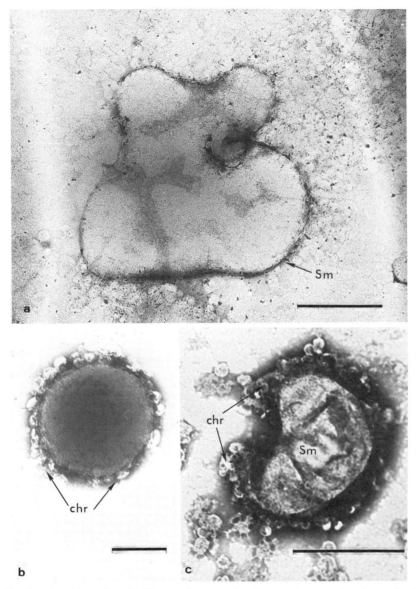

FIG. 17.   (a) Thin section of sulfur granule preparation showing the coalescing of several sulfur granules to form an intact sulfur storage granule. Note continuous nonunit membrane (Sm). Bar represents 0.5 μm (Gonye, Schroeder, and Remsen, unpublished micrograph). (b) Electron micrograph of negatively stained (2% LTA) sulfur granules with intact chromatophores (chr) visible at the surface of the sulfur granule membrane. Bar represents 0.2 μm (Gonye et al., unpublished micrograph). (c) Electron micrograph of negatively stained (2% LTA) sulfur granule with attached chromatophores (chr). Note bug-like appearance of the sulfur granule. Bar represents 0.5 μm (Gonye et al., unpublished micrograph).

When sulfur granules are isolated from photosynthetic bacteria and concentrated as purified sulfur granule preparations (Gonye, Schroeder, and Remsen, unpublished data), the nonunit membrane surrounding them is quite evident (Figs. 16a–c and 17a) in thin sections. Measurements of this atypical membrane showed a thickness of 3–4 nm. One interesting observation was the relationship between sulfur granules and chromatophores (see photosynthetic intracytoplasmic membranes) and the suggestion of some direct link between the chromatophore membrane and the membrane surrounding the sulfur granules (Remsen, 1978). This association can be seen in negatively stained preparations of sulfur granules isolated from *Chromatium buderii* (Fig. 17b and c; Gonye *et al.*, unpublished data).

The single membrane associated with sulfur granules is composed of a single protein having a molecular weight of about 13,500 (Schmidt *et al.*, 1971) and it has been speculated that these proteins may represent, in part, enzymes responsible for sulfur metabolism (Schmidt *et al.*, 1971; Remsen, 1978).

E. Magnetosomes

One of the more recent and fascinating studies on bacterial ultrastructure and its relationship to function centers around the discovery of magnetotactic bacteria. Bacteria collected from salt marshes of Cape Cod, Massachusetts and shown to respond to a magnetic field were examined with the transmission electron microscope. Chains, consisting of "approximately five to ten electron-opaque crystal-like particles, characterized magnetotactic cells" (Blakemore, 1975).

In a later study (Balkwill *et al.*, 1980), these electron-dense particles were examined in greater detail. Thin sections of a magnetotactic spirillum showed a series of electron-dense particles, usually located in close proximity to the inner surface of the cytoplasmic membrane (Fig. 18a and b). When examined in detail, the particles were generally cubic in shape (Fig. 18c) and varied from 25 to 55 nm in width. Each particle was surrounded by a membrane consisting of a 1.4-nm-thick electron-dense layer separated from the particle surface by a 1.6-nm electron transparent layer (Fig. 18d). The electron-dense layer surrounding the particles is consistent with the protein "membranes" found on other bacterial inclusion bodies (Shively, 1974), however, the additional 1.6-nm electron transparent layer is not. As the authors suggest, this could represent a "true biological membrane (i.e. lipid bilayer)" but the evidence to support that suggestion is lacking.

Since these electron-dense particles have been shown by energy dispersive X-ray analysis to be the site of cellular iron resembling magnetite, there is no question that they are responsible for the "passive alignment of magnetotactic cells in magnetic fields." In addition, they have a boundary, consisting of a "membrane" other than the cytoplasmic membrane (a typical biological mem-

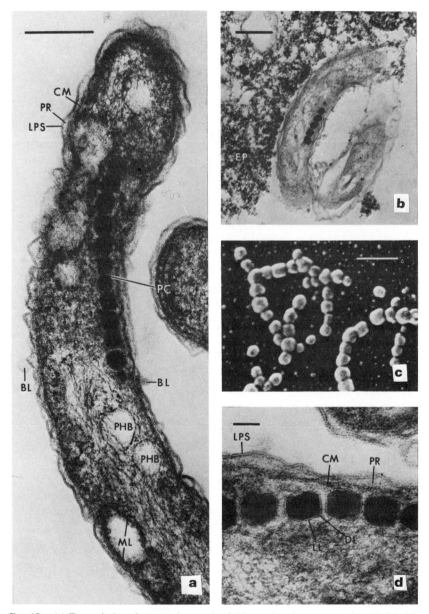

FIG. 18. (a) Transmission electron micrograph of thin-sectioned magnetotactic cell. BL, Membranous blobs extending from cell surface; CM, cytoplasmic membrane; ML, membranous loop; OM, outer membrane; PC, electron-dense particle chain; PHB, structures resembling PHB granules; PR, periplasmic region. Bar represents 0.25 μm (Balkwill *et al.*, 1980; by permission of the American Society for Microbiology, Bethesda, Maryland). (b) Transmission electron micrograph of

brane). It would appear that the magnetosome meets the basic criteria of an organelle.

IV. Summary—Do Bacteria Have Discrete Organelles?

In order to answer the question posed above, it is first necessary to understand the definition of an organelle, and at the same time, to question some of the assumptions that have become so well established that they are often considered as fact. *Webster's New Collegiate Dictionary* (1980) defines organelle in the following way: "a specialized cellular part (as a mitochondrion) that is analogous to an organ." An organ is defined as: "a differentiated structure consisting of cells and tissue and performing some specific function in an organism." If we take these two definitions and combine them (since an organelle is analogous to an organ), we get the following definition: a specialized cellular part performing some specific function in an organism.

Let us now examine some of the assumptions that have represented scientific dogma for so many years. As any college student who has taken an introductory Microbiology course will tell you, one of the major differences between a bacterial (or prokaryotic) cell and a plant or animal (or eukaryotic) cell, is the absence of discrete organelles in the bacterial cell. This view has traditionally been based on the absence of any evidence that would suggest that the major cellular functions of the bacterium are structurally compartmentalized (Pfenning, 1967). Stanier and van Niel (1962) in their essay on the concept of a bacterium, state that "Within the enclosing cytoplasmic membrane of the eukaryotic cell, certain smaller structures, which house subunits of cellular function, are themselves surrounded by individual membranes, imposing a barrier between them and other internal regions of the cell. In the prokaryotic cell, there is no equivalent structural separation of major subunits of cellular functions; the cytoplasmic membrane itself is the only major bounding element which can be structurally defined." Furthermore, they continue "in the eukaryotic cell, the enzymatic machinery of respiration and of photosynthesis is housed in specific organelles enclosed by membranes, the mitochondrion and chloroplast, respectively.

---

thin-sectioned magnetotactic cell showing peripheral electron-dense chain (magnetosomes). Bar represents 0.25 μm (Balkwill *et al.*, 1980; by permission of the American Society for Microbiology, Bethesda, Maryland). (c) Scanning electron micrograph of electron-dense particles released from the magnetotactic variant of strain MS-1. Bar represents 0.15 μm (Balkwill *et al.*, 1980; by permission of the American Society for Microbiology, Bethesda, Maryland). (d) Transmission electron micrograph of the electron dense particle chain (magnetosomes) in a thin sectioned magetotatic cell. CM, Cytoplasmic membrane; DL, electron dense layer surrounding each particle; LL, electron transparent layer; OM, outer membrane; PR, periplasmic region. Bar represents 50 nm (Balkwill *et al.*, 1980; by permission of the American Society for Microbiology, Bethesda, Maryland).

Homologous, membrane-bounded organelles responsible for the performance of these two metabolic functions have not been found in the prokaryotic cell.''

In reflecting on the statements that have just been quoted, it is important to recognize that first of all, a great deal more is known today in terms of prokaryotic structure and function than was known even as little as 20 years ago. Second, as scientists we should be aware that we often limit our thoughts by constructing such artificial boundaries as assuming that certain scientific dogma are fact.

As a case in point, it is suggested that the absence of organelles in prokaryotic cells was an assumption that has now outlived its usefulness; a suggestion, by the way that is not unique, but has been pointed out by others as well (cf. Murray, 1963). An acceptable definition of organelles must be revised to include such prokaryotic structures as chlorosomes, carboxysomes, magnetosomes, etc. These structures, as Stanier and van Niel (1962) have classically stated, "are themselves surrounded by individual membranes, imposing a barrier between them and other internal regions of the cell.''

In point of fact, these aytpical membranous structures represent true bacterial organelles. They are bounded by a membrane, have a specific function, and often are the sites of specific metabolic functions. The intracytoplasmic membrane systems typified by the photosynthetic membranes, may be no more than specialized extensions of the cytoplasmic membrane and therefore not discrete organelles in themselves. On the other hand, if one stretches a point, it should be possible to consider the cytoplasmic membrane with its specialized extensions as a single organelle.

ACKNOWLEDGMENTS

This work was supported in part by the Center for Great Lakes Studies, University of Wisconsin–Milwaukee.

The author acknowledges the scientific achievements of many individuals without whose efforts this review would have been impossible. The assistance of Delpfine Welch in the preparation of the manuscript is greatly appreciated.

REFERENCES

Balkwill, D. L., Maratea, D., and Blakemore, R. P. (1980. *J. Bacteriol.* **141**, 1399.
Blakemore, R. P. (1975). *Science* **190**, 377.
Boatman, E. S. (1964). *J. Cell Biol.* **20**, 297.
Bowen, C. C., and Jensen, T. E. (1965a). *Science* **147**, 1460.
Bowen, C. C., and Jensen, T. E. (1965b). *Am. J. Bot.* **52**, 641.
Branton, D. (1966). *Proc. Natl. Acad. Sci. U.S.A.* **55**, 1048.
Buchanan, R. E., and Gibbons, N. E. (1974). *In* "Bergey's Manual of Determinative Bacteriology," 8th Ed. Williams & Wilkins, Baltimore, Maryland.

Chapman, G. B., and Hillier, J. (1953). *J. Bacteriol.* **66**, 362.

Cohen-Bazire, G. (1963). *In* "Bacterial Photosynthesis" (H. Gest, A. San Pietro, and L. P. Vernon, eds.), pp. 89–114. Antioch Press, Yellow Springs, Ohio.

Cohen-Bazire, G., and Sistrom, W. R. (1966). *In* "The Chlorophylls" (L. P. Vernon and G. R. Seely, eds.), pp. 313–341. Academic Press, New York.

Cohen-Bazire, G., Pfenning, N., and Kunisawa, 4. (1964). *J. Cell Biol.* **22**, 207.

Davies, S. L., and Whittenbury, R. (1970). *J. Gen. Microbiol.* **61**, 227.

de Boer, W. E., LaRiviere, J. W. M., and Houwink, A. L. (1961). *Antonie van Leeuwenhoek* **27**, 447.

Doddema, H. J., van der Drift, C., Vogels, G. D., and Beenhuis, M. (1979). *J. Bacteriol.* **140**, 1081.

Dunlop, W. F., and Robards, A. W. (1973). *J. Bacteriol.* **114**, 1271.

Edwards, M. R., and Stevens, R. W. (1963). *J. Bacteriol.* **86**, 414.

Eimhjellen, K. E. (1970). *Arch. Mikrobiol.* **73**, 193.

Eimhjellen, K. E., Steensland, H., and Tratteberg, J. (1967). *Arch. Mikrobiol.* **59**, 62.

Fitz-James, P. C. (1960). *J. Biophys. Biochem. Cytol.* **8**, 507.

Fooke-Achterrath, M., Lickfeld, K. G., Reusch, V. M., Jr., Aebi, W., Tschöpe, W., and Menge, B. (1974). *J. Ultrastruct. Res.* **49**, 270.

Fuller, R. C., Conti, S. F., and Mellin, D. B. (1963). *In* "Bacterial Photosynthesis" (H. Gest, A. San Pietro, and L. P. Vernon, eds.), pp. 71–87. Antioch Press, Yellow Springs, Ohio.

Gantt, E., and Conti, S. F. (1969). *J. Bacteriol.* **97**, 1486.

Ghosh, B. K. (1974). *Sub-Cell Biochem.* **3**, 311.

Ghosh, B. K., Lampen, J. O., and Remsen, C. C. (1969). *J. Bacteriol.* **100**, 1002.

Giesbrecht, P. (1960). *Zentralbl. Bakteriol. Parasitenkd. Infektionskr. Hyg. Abt. 1 Orig.* **179**, 538.

Greenawalt, J. W., and Whiteside, T. L. (1975). *Bacteriol. Rev.* **39**, 405.

Griebel, R. J., and Merrick, J. M. (1971). *J. Bacteriol.* **108**, 782.

Griebel, R. J., Smith, Z., and Merrick, J. M. (1968). *Biochemistry* **7**, 3676.

Hageage, G. L., Jr., Eanes, E. D., and Gherna, R. L. (1969). *Bacteriol. Proc.* **69**, 38.

Hazeu, W. (1975). *Antonie van Leeuwenhoek* **41**, 121.

Hazeu, W., Batenburg-van der Vegte, W., and de Bruyn, J. C. (1980). *Arch. Microbiol.* **123**, 211.

Higgins, M. L., and Daneo-Moore, L. (1974). *J. Cell Biol.* **61**, 288.

Higgins, M. L., Tsien, H. C., and Daneo-Moore, L. (1976). *J. Bacteriol.* **127**, 1519.

Hoare, D. S., Ingram, L. O., Thurston, E. L., and Walkup, R. (1971). *Arch. Mikrobiol.* **78**, 310.

Holt, S. C., Truper, H. C., and Takaes, B. J. (1968). *Arch. Mikrobiol.* **62**, 111.

Jensen, T. E., and Sicko, L. M. (1971). *J. Bacteriol.* **106**, 683.

Jensen, T. E., and Sicko, L. M. (1973). *Cytologia* **38**, 381.

Jones, D. D., and Jost, M. (1970). *Arch. Mikrobiol.* **70**, 43.

Jost, M., and Jones, D. D. (1970). *Can. J. Microbiol.* **16**, 159.

Kellenberger, E., Ryter, A., and Séchaud, J. (1958). *J. Biophys. Biochem. Cytol.* **4**, 671.

Kolkwitz, R. (1928). *Ber. Dtscho Bot. Ges.* **46**, 29.

Kran, von, G., Schlote, F. W., and Schlegel, H. G. (1963). *Naturwissenschaft* **50**, 128.

Larsen, H., Omang, S., and Steensland, H. (1967). *Arch. Mikrobiol.* **59**, 197.

Lundgren, D. G., Pfister, R. M., and Merrick, J. M. (1964). *J. Gen. Microbiol.* **34**, 441.

Mahoney, R. P., and Edwards, M. R. (1966). *J. Bacteriol.* **92**, 487.

Mayer, F., Lurg, R., and Schoberth, S. (1977). *Arch. Microbiol.* **115**, 207.

Moor, H., Waldner, H., and Frey-Wyssling, A. (1961). *J. Biophys. Biochem. Cytol.* **10**, 1.

Murray, R. G. E. (1963). *In* "The General Physiology of Cell Specialization" (D. Mazia and A. Tyler, eds.), p. 28. McGraw-Hill, New York.

Murray, R. G. E., and Watson, S. W. (1965). *J. Bacteriol.* **89**, 1594.

Nanninga, N. (1971). *J. Cell Biol.* **48**, 219.

Nicholson, G. I., and Schmidt, G. L. (1971). *J. Bacteriol.* **105,** 1142.

Oelge, J., and Drews, G. (1972). *Biochim. Biophys. Acta* **265,** 209.

Owen, P., and Freer, J. H. (1972). *Biochem. J.* **129,** 907.

Pankratz, H. S., and Bowen, C. C. (1963). *Am. J. Bot.* **50,** 387.

Peck, H. D. (1968). *Annu. Rev. Microbiol.* **22,** 489.

Peters, G. A., and Cellarius, R. A. (1972). *Bioenergetics* **3,** 345.

Pfenning, N. (1967). *Annu. Rev. Microbiol.* **21,** 285.

Pfenning, N. (1978). *In* ''The Photosynthetic Bacteria'' (R. K. Clayton and W. R. Sistrom, eds.), p. 3. Plenum, New York.

Pfenning, N., and Cohen-Bazire, G. (1967). *Arch. Mikrobiol.* **59,** 226.

Pfister, R. M., Lundgren, D. G., and Remsen, C. C. (1969). *Bacteriol. Proc.* **69,** 38.

Pope, L. M., Hoare, D. S., and Smith, A. J. (1969). *J. Bacteriol.* **97,** 936.

Proctor, H. M., Norris, J. R., and Ribbons, D. W. (1969). *J. Appl. Bacteriol.* **32,** 118.

Raymond, J. C., and Sistrom, W. R. (1967). *Arch. Mikrobiol.* **59,** 255.

Remsen, C. C. (1966). *Arch. Mikrobiol.* **54,** 266.

Remsen, C. C. (1968). *Arch. Mikrobiol.* **61,** 40.

Remsen, C. C. (1978). *In* ''The Photosynthetic Bacteria'' (R. K. Clayton and W. R. Sistrom, eds.), p. 31. Plenum, New York.

Remsen, C. C., and Lundgren, D. G. (1966). *J. Bacteriol.* **92,** 1765.

Remsen, C. C., and Truper, H. G. (1973). *Arch. Mikrobiol.* **90,** 269.

Remsen, C. C., and Watson, S. W. (1972). *Int. Rev. Cytol.* **33,** 253.

Remsen, C. C., Valois, F. W., and Watson, S. W. (1967). *J. Bacteriol.* **94,** 422.

Remsen, C. C., Watson, S. W., Waterbury, J. B., and Truper, H. G. (1968). *J. Bacteriol.* **95,** 2374.

Reusch, V. M., Jr. and Burger, M. M. (1973). *Biochim. Biophys. Acta* **300,** 79.

Reusch, V. M., Jr. and Burger, M. M. (1974). *J. Biol. Chem.* **249,** 5337.

Ribbons, D. W., Harrison, J. E., and Wadzinski, A. M. (1970). *Annu. Rev. Microbiol.* **24,** 135.

Robertson, J. D. (1959). *Biochem. Symp.* **16,** 3.

Romanovskaya, V. A., Malashenko, Yu. R., and Bogachenko, V. N. (1978). *Microbiology* **47,** 96.

Ryter, A. (1965). *Ann. Inst. Pasteur Paris* **108,** 40.

Salton, M. R. J. (1978). *Symp. Soc. Gen. Microbiol.* **28,** 202.

Salton, M. R. J., and Owen, P. (1976). *Annu. Rev. Microbiol.* **30,** 451.

Schmidt, G. L., and Damen, M. D. (1970). *Arch. Mikrobiol.* **73,** 1.

Schmidt, G. L., Nicholson, G. L., and Kamen, M. D. (1971). *J. Bacteriol.* **105,** 1137.

Schmidt, K. (1980). *Arch. Microbiol.* **124,** 21.

Schmidt, K., Maarzahl, M., and Mayer, F. (1980). *Arch. Microbiol.* **127,** 87.

Shively, J. M. (1974). *Annu. Rev. Microbiol.* **28,** 167.

Shively, J. M., Decker, G. L., and Greenawalt, J. W. (1970). *J. Bacteriol.* **101,** 618.

Shively, J. M., Ball, L., and Kline, B. W. (1973a). *J. Bacteriol.* **116,** 1405.

Shively, J. M., Ball, L., Brown, D. H., and Saunders, R. E. (1973b). *Science* **182,** 584.

Smith, U., and Ribbons, D. W. (1970). *Arch. Mikrobiol.* **74,** 116.

Staehelin, L. A., Golecki, J. R., Fuller, R. C., and Drews, G. (1978). *Arch. Microbiol.* **119,** 269.

Stanier, R. Y. (1963). *In* ''The General Physiology of Cell Specialization'' (D. Mazia and A. Tyler, eds.), pp. 242–252. McGraw-Hill, New York.

Stanier, R. Y., and van Niel, C. B. (1962). *Arch. Mikrobiol.* **42,** 17.

Vanderwinkle, E., and Murray, R. G. E. (1962). *J. Ultrastruct. Res.* **7,** 185.

Van Gool, A. P., Lambert, R., and Landelot, H. (1969). *Arch. Mikrobiol.* **69,** 281.

Van Iterson, W. (1961). *J. Biophys. Biochem. Cytol.* **9,** 183.

Van Iterson, W., and Aten, J. A. (1976). *J. Bacteriol.* **26,** 384.

Van Iterson, W., and Leene, W. (1964a). *J. Cell Biol.* **20,** 361.

Van Iterson, W., and Leene, W. (1964b). *J. Cell Biol.* **20,** 377.

Van Iterson, W., Michels, P. A. M., Vyth-Dreese, F., and Aten, J. A. (1975). *J. Bacteriol.* **121,** 1189.

Qalsby, A. E. (1972). *Bacteriol. Rev.* **36,** 1.

Wang, W. S., and Lundgren, D. G. (1969). *J. Bacteriol.* **97,** 947.

Watson, S. W., and Waterbury, J. B. (1971). *Arch. Mikrobiol.* **77,** 203.

Watson, S. W., Graham, L. B., Remsen, C. C., and Valois, F. W. (1971). *Arch. Mikrobiol.* **76,** 183.

Weaver, T. L., and Dugan, P. R. (1975). *J. Bacteriol.* **121,** 704.

Wildman, R. B. (1969). Ph.D. Thesis, Iowa State University, Ames.

Wirsen, C. O., and Jannasch, H. W. (1978). *J. Bacteriol.* **136,** 765.

Wolk, C. P. (1973). *Bacteriol. Rev.* **37,** 32.

Wullenweber, M., Koops, H.-P., and Harms, H. (1977). *Arch. Microbiol.* **112,** 69.

Zeikus, J. G., and Wolfe, R. S. (1973). *J. Bacteriol.* **113,** 461.

Zeikus, J. G., Hegge, P. W., and Anderson, M. A. (1979). *Arch. Microbiol.* **122,** 41.

INTERNATIONAL REVIEW OF CYTOLOGY, VOL. 76

Separated Anterior Pituitary Cells and Their Response to Hypophysiotropic Hormones

CARL DENEF, LUC SWENNEN, AND MARIA ANDRIES

Laboratory of Cell Pharmacology, Department of Pharmacology, School of Medicine, Campus Gasthuisberg, Katholieke Universiteit Leuven, Leuven, Belgium

I. Introduction

The mammalian anterior pituitary gland is a neuroendocrine tissue composed of many different cell types each producing different peptide hormones (Herlant, 1964; Farquhar *et al.*, 1975). This cellular heterogeneity raises many problems of interpreting the responses to the hypothalamic hypophysiotropic hormones and their mechanism of action. For example, many contradictory reports have been published with regard to whether or not a change in adenylate cyclase activity mediates the primary response to hypothalamic releasing or inhibiting hormones (see Barnes *et al.*, 1978). A significant change in cyclic AMP (cAMP) levels in the target cell type may be obscured by the background levels in other cell types particularly when the proportional number of the target cells is low. Moreover, if one regulatory hormone affects more than one cell type, such as thyrotropin

225

releasing hormone (TRH) which stimulates both thyrotropin (TSH) and prolactin (PRL) secretion (Blackwell and Guillemin, 1973), it cannot be shown in which cell type(s) cAMP levels change. Another problem is that the topographical distribution of the different cell types in the pituitary gland is not random. It is well documented that certain cell types are more abundant in some areas in the gland while others occur more frequently in other regions (see Siperstein *et al.*, 1954; Girod, 1976). If the intact pituitary is used in an *in vitro* test system, mainly cells located at the periphery of the tissue will be influenced by diffusible substances in the medium (Farquhar *et al.*, 1975). As it is not known whether all cell types located in the periphery display the same functional characteristics as the same cell type in deeper layers, a particular response to a specific physiological stimulus may not be respresentative for the overall response occurring *in vivo*.

It is also most important to bear in mind that different pituitary cell types may functionally interact among each other and some evidence has already been reported (Denef, 1981). In the case a substance can affect the release of more than one pituitary hormone, it is difficult if not impossible to distinguish between primary responses and responses due to intercellular communication.

All these data clearly indicate the need to purify the various cell types of the anterior pituitary. Today this goal has not been fully achieved but, as will be reviewed in the present article, significant sorting-out of certain cell types has been realized, yielding populations of 60 to 90% purity. The cells retain their typical ultrastructural features, are able to respond to various physiological stimuli, and can be established in primary culture over prolonged periods of time.

II. Separation of Anterior Pituitary Cell Types

A. Velocity Sedimentation at Unit Gravity

Enzymatically dispersed pituitary cells are allowed to sediment by gravital force through a linear gradient of bovine serum albumin (BSA) (usually 0.3–2.4%) in tissue culture medium (such as Eagle's medium, Medium 199, Dulbecco's modified Eagle's medium) (Hymer, 1975). Cells separate mainly according to size but also according to density (Miller, 1973; Lloyd and McShan, 1973; Denef *et al.*, 1980). It was found that the average size and density were relatively specific for each cell type so that significantly enriched populations can be obtained. Table I gives an overview of the achievements obtained today using rat anterior pituitary cells. As can be seen, the best results are obtained if an animal model is used in which the proportional number of the desired cell type is already high in the initial cell suspension or if its size differs markedly from the other cell types. This is the case for somatotrophs from adult male rats, gonadotrophs from 14-day-old female rats, and lactotrophs from adult female rats.

Somatotrophs from adult male rats are larger than those of immature animals (Denef *et al.*, 1978a) and sediment into the deeper layers of the gradient attaining up to 80–87% purity (Hymer *et al.*, 1973; Lloyd and McShan, 1973; Denef *et al.*, 1978a).

The proportional number of the gonadotrophs is very low in adult rats but in 14-day-old females, their number is 3 to 4 times higher (Denef *et al.*, 1978a). Moreover, at that age part of the gonadotrophs are enlarged (Siperstein *et al.*, 1954; Denef *et al.*, 1976), probably as a consequence of the very high follicle stimulating hormone (FSH) and luteinizing hormone (LH) secretion rates *in vivo* (see below). These large gonadotrophs sediment into the bottom of the gradient reaching a purity of near 80% (Denef *et al.*, 1976, 1978a; Snyder *et al.*, 1980).

In male rats, the number of mammotrophs is only about 1% of the pituitary cell population and almost no enrichment can be obtained by unit gravity sedimentation (Lloyd and McShan, 1973). However, in female rats they represent 33% of the initial cell suspension. They sediment into the middle layers of the gradient where they reach about 70% of the population (Hymer *et al.*, 1974).

Enrichment of thyrotrophs from normal adult rats has not been convincingly shown. Lloyd and McShan (1973) did not report any enrichment although not all fractions of the gradient were examined. Barnes *et al.* (1978) have shown significant banding of immunoassayble TSH in a unit gravity sedimentation gradient but no data on the actual cell number were given. Significant banding of immunoassayable TSH (but no data on cell number) was also found by Leuschen *et al.* (1978) but the distance of sedimentation did not correspond to that reported by Barnes *et al.* (1978). An important finding by Leuschen and associates, however, was that after 1 week in primary culture thyrotrophs reached 60–80% of the population of cells derived from the top of the gradient. The authors suggested that the selective enrichment of thyrotrophs during 1 week culturing might be due to differences in survival time or plating efficiency of the different cell types. Some enrichment of thyrotrophs has been obtained when using 14-day-old male rat pituitaries (Denef *et al.*, 1978a). Again, in the latter the proportional number of thyrotrophs is considerably higher than in adult rat pituitaries.

Experimental manipulations can also favor the purification of certain cell types. Hymer *et al.* (1973) succeeded in obtaining a 60% pure thyrotroph population from chemically thyroidectomized rats. Thyroidectomy indeed markedly increases number and size of the thyrotrophs whereas it diminishes the number of somatotrophs (Surks and De Fesi, 1977). Thyrotrophs from thyroidectomized rats, therefore, sedimented into the bottom of the gradient which otherwise would be occupied by somatotrophs.

Perhaps because of their very low proportional number, it has been very

TABLE I

SEPARATION OF RAT ANTERIOR PITUITARY CELL TYPES BY GRADIENT SEDIMENTATION

Cell type	Donor pituitary	Proportional number (%)			Main contaminant (%)	Method of identification[e]	Reference
		Before separation	In most enriched fraction	In poorest fraction			
Somatotrophs	36- to 45-day-old ♂	35	60	2	Basophils (10)	H	Hymer et al. (1973)
	Adult ♂	52	87	8	Gonadotrophs (1.2) Thyrotrophs (1.7)	EM	Lloyd and McShan (1973)
	Adult ♂	n.d.	80	2	Gonadotrophs (8)	H	Denef et al. (1978a)
	Adult ♀	32	58	1	Mammotrophs (11)	H,RIA	Hymer et al. (1974)
	Adult ♂	38	85[a]	15	Basophils (7)	H,EM,RIA	Snyder and Hymer (1975)
	Adult ♂	34	90[b]	<2	Basophils (5)	H,EM,RIA	Hymer et al. (1972); Kraicer and Hymer (1974)
Mammotrophs	Diestrus ♀	33	70	<10	n.d.[a]	H,IP,IF,EM,RIA	Hymer et al. (1974)
	Estrus ♀	33	65	<16	n.d.	H,IP,IF,EM,RIA	Hymer et al. (1974)
	Random cycle ♀	33	66	<6	Somatotrophs (21)	H,IP,IF,EM,RIA	Hymer et al. (1974)

	Lactating ♀ (pups removed)	n.d.	61	12	Various cell types (6–8)	H,RIA	Hymer et al. (1973, 1974)
Gonadotrophs	Adult ♂	1.3	1.2	0.7	Somatotrophs (87)	EM	Lloyd and McShan (1973)
	Adult ♂	8	42	1	Somatotrophs (33)	EM,RIA	Lloyd and McShan (1973)
	Adult ♂	n.d.	17	<2	Chromophobes (25) Somatotrophs (34)	IP,H,RIA	Denef et al. (1978a)
	36- to 45-day-old ♂	<4	<15	2	Chromophobes (43)	H	Hymer et al. (1973)
	Random cycle ♀	<5	<25	n.d.	n.d.	H	Hymer et al. (1974)
	14-day-old ♀	17	78	<5	Thyrotrophs (12)	H,IP,EM,RIA	Denef et al. (1976, 1978a) Denef (1980); Snyder et al. (1980)
Thyrotrophs	Adult ♂	3	3	1.7	Somatotrophs (38)	EM	Lloyd and McShan (1973)
	Adult ♂	5–10	80[c]	<10	n.d.	IP,EM	Leuschen et al. (1978)
	Thyroidectomized adult ♂	n.d.	60	11	Gonadotrophs (5)	H	Hymer et al. (1973)
	14-day-old ♂	17	30	<10	Chromophobes (65)	IP	Denef et al. (1978a)
	Adult ♂	n.d.	n.d.	n.d.	n.d.	RIA	Barnes et al. (1978)

[a] Obtained by density gradient sedimentation.

[b] Obtained by density gradient sedimentation of a somatotroph-enriched fraction separated on a unit-gravity sedimentation gradient.

[c] Proportional number estimated after a 7-day culture period.

[d] n.d., No data.

[e] H, Histochemistry; EM, electron microscopy; RIA, radioimmunoassay; IP, immunoperoxidase staining; IF, immunofluorescence.

difficult to assess whether or not some enrichment of corticotrophs can be obtained (Lloyd and McShan, 1973). Rotsztejn et al. (1980), however, found considerable banding of immunoassayable ACTH after unit gravity sedimentation but did not present data on the actual number of corticotrophs. It should be borne in mind that measuring the amount of intracellular hormone or hormone release is not a reliable index of the number of cells present. Over- or underestimations can be made since smaller cells might have a lower hormone content than larger cells and hormone release is not necessarily in direct proportion to intracellular hormone content (Leuschen et al., 1978; Denef et al., 1980).

Enriched populations of various cell types have also been prepared from human pituitaries (Hymer et al., 1976). As with rat pituitaries a 2- to 3-fold enrichment was obtained.

The reliability of the unit gravity sedimentation method has been found most satisfactory (Hymer, 1975; Denef et al., 1978a). The proportional distribution of the various cell types in the initial dispersed cell suspension as well as after gradient separation reflects the distribution in the intact pituitary (Hymer, 1975; Surks and De Fesi, 1977; Denef et al., 1978a). Cells recovered from the gradient are viable for over 95% (trypan blue exclusion test). Plating efficiency on culture dishes coated (Denef et al., 1976, 1978b) or not (Denef, unpublished data) with polylysine was also very high and similar for all gradient fractions. All studies have reported excellent ultrastructural preservation. Overall recovery of the cells ranges between 65 and 80% and the proportional number of cells in each collected gradient fraction is highly reproducible. However, cells may lose part of their stored hormone during the dispersion and sedimentation procedures. These losses, however, are not massive and in case two hormones are stored in the same cell (FSH and LH) losses are of similar magnitude for both hormones. Loss of PRL can be diminished by adding dopamine to the media (Hymer et al., 1974). Finally, there are no selective losses of cell types.

B. DENSITY GRADIENT CENTRIFUGATION

A few studies have reported successful enrichement of pituitary cell types after centrifugation through a density gradient. Different cell types separate on the basis of differencies in density. Ishikawa (1969) enriched several types of anterior pituitary cells by centrifugation of pituitary cell suspensions through discontinuous gradients of Dextran A. Hymer et al. (1972) (Table I) have purified somatotrophs from adult male rat pituitaries to a purity of 90% by centrifugation of an enriched population of somatotrophs, obtained by unit gravity sedimentation, on a BSA density gradient. Later it was shown that an almost equally purified population of somatotrophs could be prepared in half the time required previously by directly applying the initial pituitary cell suspension on a discontinuous density gradient (Snyder and Hymer, 1975) (Table I). After centrifugation in isopicnic BSA gradients, Hirsch *et al.* (1979) obtained a 7- to 16-fold enrichment of gonadotrophs, thyrotrophs, and lactotrophs from bovine anterior pituitary.

C. Affinity Binding to Ligands Covalently Bound on Solid Supports

This method is based on the principles of affinity chromatography (Edelman *et al.*, 1971). A hypophysiotropic hormone is bound covalently to nylon fibers, sepharose, or glass beads and target cells are expected to bind to these beads when a receptor for the covalently bound hormone is present on their plasma membrane. Only a few studies have reported specific binding of target cells to these immobilized ligands. GH_3 cells (rat pituitary tumor cells secreting PRL and growth hormone) selectively bind to immobilized triiodothyronine (Venter *et al.*, 1976). Tal and associates (1978) reported a more than 16-fold enrichment of thyrotrophs from a pituitary cell suspension by binding to TRH immobilized on nylon fibers, but enrichment of mammotrophs, which are also responsive to TRH, was not seen. A significant enrichment of gonadotrophs was obtained with luteinizing hormone-releasing hormone (LHRH) immobilized on sepharose (Ishikawa *et al.*, 1978).

III. Functional Responses of Separated Pituitary Cell Types

There is general agreement that cells freshly separated by unit gravity sedimentation, density gradient centrifugation, or affinity binding look very healthy. Somatotrophs (Hymer *et al.*, 1973; Hymer, 1975; Lloyd and McShan, 1973), mammotrophs (Hymer *et al.*, 1974), gonadotrophs (Lloyd and McShan, 1973; Denef *et al.*, 1978b), and thyrotrophs (Tal *et al.*, 1978; Leuschen *et al.*, 1978) retain their typical ultrastructural features.

In general, freshly separated cells remain functional. They incorporate radioactive amino acids into protein in a linear fashion as a function of time (Hymer *et al.*, 1972, 1973). Mammotrophs are capable to incorporate these amino acids into newly synthesized PRL (Hymer *et al.*, 1974) whereas [^{3}H]glucosamine, which labels the glycoprotein hormones FSH and LH, is incorporated at a much higher rate in gonadotroph-enriched fractions than in gonadotroph-poor fractions (Lloyd and McShan, 1973). Isolated somatotrophs respond to various secretagogues such as partially purified growth hormone releasing factor, dibutyryl-cAMP, prostaglandin E_2, and isobutylmethylxanthine, a phosphodiesterase inhibitor increasing the intracellular levels of cAMP (Kraicer and Hymer, 1974; Snyder and Hymer, 1975; Spence *et al.*, 1980). In mammotroph-rich populations PRL release is stimulated by TRH (Barnes *et al.*, 1978) and vasoactive intestinal peptide (VIP) (Rotsztejn *et al.*, 1980; Samson *et al.*, 1980) whereas it is inhibited by dopamine (Hymer *et al.*, 1974; Barnes *et al.*, 1978). Gonadotroph-rich fractions secrete LH in response to LHRH (Lloyd and McShan, 1976; Andries and Denef, unpublished observations) and convert testosterone into 5α-dihydrotestosterone (Lloyd and Karavolas, 1975; Denef, 1979). Enriched populations of thyrotrophs respond to TRH by increasing TSH release (Tal *et al.*, 1978) and intracellular levels of cAMP (Barnes *et al.*, 1978).

All these different cell types have also been established in primary culture with preservation of their functional responses (Lloyd and McShan, 1973; Hymer *et al.*, 1974; Denef *et al.*, 1976, 1978b, 1980, and unpublished results; Leuschen *et al.*, 1978; Rotszejn *et al.*, 1980). Moreover, there is evidence that at least part of the physiological characteristics of the pituitary *in vivo* is retained, at least during short-term culture (Hymer *et al.*, 1974; Denef *et al.*, 1978b).

IV. The Value of Separated Pituitary Cell Preparations toward the Elucidation of the Mechanisms of Hypophysiotropic Hormone Action

A. Relationship of the Proportional Number and Size of the Cell Types with the Secretory Activity of the Pituitary *in Vivo*

It is well known that pituitary hormone secretion fluctuates dramatically under different physiological and experimental conditions and during development. It is a generally accepted concept that both hypothalamic regulatory peptides (releasing and inhibiting factors) and amines (dopamine) as well as peripheral hormones that feedback into the hypothalamo-pituitary system are responsible for these changes in pituitary hormone release. However, the cell-physiological events underlying these changes are not well understood. Changes in hormone secretion could be brought about by certain biochemical changes in the cell but also by changes in the number and size of the cell type involved. The unit of gravity sedimentation procedure is a most suitable technique to evaluate the latter phenomenon and it has been found that various physiological, experimental, or developmental changes in pituitary hormone secretion are paralleled by selective changes in the number and sedimentation rate of certain cell types.

Large mammotrophs of estrogen-primed female rats are more numerous than large mammotrophs of normal controls (Hymer *et al.*, 1974). In contrast, the number of small mammotrophs increases in ovariectomized rats (Hymer *et al.*, 1974). An impressive finding, reported by Hymer and his associates (1976), is that in patients with breast cancer, the number of large mammotrophs is increased 2-fold. These large mammotrophs contained and secreted in primary culture substantially more PRL than large mammotrophs separated from normal postmortem pituitaries.

Thyrotrophs increase several fold in number and size after thyroidectomy

(Surks and De Fesi, 1977). The proportional number of large-sized densily granulated thyrotrophs also increases (Hymer et al., 1973).

The proportional distribution of cell types also changes during development. Denef et al. (1976, 1978a) showed that the majority of large cells sedimenting into the bottom of a unit gravity sedimentation gradient were somatotrophs in adult male rats but gonadotrophs in immature (14-day-old) male and female rats. The same authors also showed that the proportional number of large as well as medium-sized gonadotrophs was higher in pituitaries from these 14-day-old rats than in those from adult rats. The proportional number of thyrotrophs was also found to be higher.

All these findings correlate well with the respective functional state of the pituitary in vivo. The increased number of large mammotrophs in estrogen-primed rats agrees well with the well-known stimulatory effect of estrogen of PRL synthesis and release (Maurer and Gorski, 1977). The proportional increase of large-sized thyrotrophs after thyroidectomy correlates with the increment of TSH release. The rise in number and size of gonadotrophs in 14-day-old females in consistent with the very high FSH and LH secretion at that age (Ojeda and Ramirez, 1972; Döhler and Wüttke, 1975).

B. Separation of Morphological and Functional Variants

1. Mammotrophs

Snyder et al. (1976) showed that mammotrophs from normal adult female rats fractionated by unit gravity sedimentation were heterogeneous in terms of baseline PRL release in primary culture. Secretion rates (expressed per 1000 PRL cells) positively correlated with cell size but the largest mammotrophs, representing about 10% of the recovered PRL cells, had a surprisingly low secretion rate. In contrast, the latter subpopulation became about 6 times more active in estrogen-primed females and exceeded the secretory activity of all other mammotrophs. Moreover, the proportional number of these mammotrophs increased about 3-fold. In ovariectomized rats, the proportional number of these large mammotrophs was similar to that of normal rats but they displayed very low secretion rates. Since it is well known that estrogen treatment in vivo increases PRL synthesis and release (Voogt et al., 1970; Maurer and Gorski, 1977), these observations show that not all PRL cells respond in the same fashion to the estrogen stimulus. It should be worthwhile to further investigate whether gradient-separated subpopulations of mammotrophs would respond similarly to estrogen given during culturing.

The existence of functional variants of mammotrophs has also been suggested by others using different techniques (Walker and Farquhar, 1980). Various subpopulations could be distinguished depending on the rate newly synthesized PRL is released. In addition, TRH did not affect newly synthesized PRL but stimulated the release of older PRL. Whether or not the latter subpopulations can be separated by gradient sedimentation remains to be investigated. However, preliminary data point toward this possibility. It has been shown that dopamine preferentially inhibits the release of newly synthesized PRL (MacLeod, 1969, 1976) and PRL secretion from slowly sedimenting mammotrophs appears to be more readily inhibited by dopamine than mammotrophs sedimenting deeper into the gradient (Denef and Swennen, unpublished).

2. Somatotrophs

Snyder and Hymer (1975) were able to isolate two morphological distinct subpopulations of somatotrophs by density gradient centrifugation. Somatotrophs of lesser density (type I) had fewer secretory granules and a more developed rough endoplasmic reticulum and Golgi region than somatotrophs of higher density (type II). Over a period of 7 days in culture (Snyder *et al.*, 1977), type I cells increased their GH production rate while type II cells did not. Hydrocortisone stimulated GH production in type I but not in type II cells. In response to dibutyryl-cAMP, type I cells also released more GH in proportion to their GH content than type II. Type I cells were also more sensitive to the GH inhibiting effect to somatostatin. In a preliminary developmental study the same authors found a physiological correlate of their findings. Type I cells were relatively more abundant during the period of rapid body growth (around 25 days of age). Whereas these studies again give strong evidence for the existence of functionally distinct subpopulations within one cell type, they also showed that after 6 days in culture the morphological differences between type I and type II cells became less prominent. Type I cells became more granulated while type II cells became smaller and contained less granules than at the beginning of culture, indicating that morphological and functional heterogeneity is not indefinitely retained during culture.

The existence of functional heterogeneity among somatotrophs has also been suggested by Farquhar and associates (1975) on the basis of differential amino acid incorporation into protein.

3. Gonadotrophs

Histochemical studies and electron microscopic examinations have revealed that the gonadotroph cell population is morphologically heterogeneous. Originally two different cell types were distinguished (Herlant, 1964; Farquhar and Rinehart, 1954; Barnes, 1962; Yoshimura and Harumiya, 1965; Kurosumi and Oota, 1968). One was characterized by the presence of two populations of secretory granules, the size of which was about 200 and 700 nm, respectively. The other was smaller and contained secretory granules of homogeneous size (~ 250 nm). There is now strong evidence that each of these cell types contains both the gonadotropins FSH and LH (Moriarty, 1975, 1976; Tixier-Vidal *et al.*, 1975b; Nakane, 1970; Tougard *et al.*, 1977; Phifer *et al.*, 1973; Herbert, 1975; Robyn *et al.*, 1973; Pelletier *et al.*, 1976). More recently, additional morphological variants have been identified which also contained both FSH and LH. In normal castrated male rats Tixier-Vidal *et al.* (1975a,b) found gonadotrophs with ultrastructural characteristics intermediate between the two cell types described originally. Still another cell type was small-sized and the predominant type during fetal life. The proportional number of the latter decreased during development in favor of the other types but some remained detectable in adult life (Tougard *et al.*, 1977; Tougard, 1980).

Of most importance was the finding that the different morphological subtypes were of different size (Tougard, 1980). One would expect therefore, that gradient sedimentation should separate these subtypes or at least produce enriched populations of them. Recently, evidence was obtained that this may indeed be so. Electron microscopic examination of 14-day-old male pituitary cells separated by unit gravity sedimentation demonstrated that most of the largest gonadotrophs corresponded to the large type gonadotrophs with secretory granules of 200 and 700 nm. In gradient fractions with small cells the latter gonadotrophs were rarely found and mainly gonadotrophs with 200 nm secretory granules as well as the intermediate forms were seen (see Fig. 1).

Evidence has been given that morphologically distinct gonadotroph cell types are also functionally different. Moriarty (1975) has shown that the ultrastructural features of the gonadotrophs and the distribution of their immunoreactive secretory granules fluctuate with the estrous cycle and are dependent on sex (Moriarty, 1975). In 1954, Siperstein and associates found that between 10 and 14 days of life, there is a very rapid increase in size, number, and stainability (which reflects hormone content) of a pool of gonadotrophs and that this is more pronounced in female than in male rats. More recently, a comparative study of 14-day-old male and female rat pituitary cells separated by unit gravity sedimentation revealed that the proportional and total number of large-sized gonadotrophs as well as their FSH and LH content is much higher in the female than in the male pituitary (Denef *et al.*, 1978a). These values also largely exceed those of adult male rat pituitary. Others have shown that female rats at 14 days of age sharply increase their plasma FSH levels to values never reached later in life and exceeding several fold the values of male littermates (Ojeda and Ramirez, 1972; Kragt and Dahlgren, 1972; Döhler and Wüttke, 1975). Fourteen-day-old females are also more responsive to LHRH than 14-day-old males (Ojeda *et al.*, 1977). At this age, female but not male rats also show periodic spikes of high LH release (Döhler and Wüttke, 1975).

All these data indirectly show that changes in the ultrastructural appearance of the gonadotrophs reflect significant changes in their cell-physiological functions. Direct *in vitro* evidence that the gonadotrophs may also be functionally

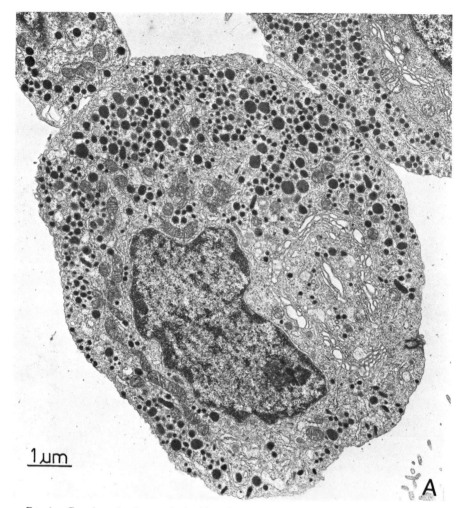

FIG. 1. Gonadotroph subtypes obtained by unit gravity gradient sedimentation of dispersed pituitary cells from 14-day-old male rats. (A) Representative gonadotroph from the bottom fraction of the gradient; the great majority of these gonadotrophs are large, have small- and large-sized secretory granules and areas of dilated endoplasmic reticulum. (B) Gondotroph from one of the upper layers

heterogeneous has also been given (Denef *et al.*, 1978b). Again pituitary cells from 14-day-old male and female rats, separated according to size by unit gravity sedimentation, were used as a model. The cells were studied after 3 or 4 days in primary culture. In 14-day-old female rats the large-sized gonadotrophs displayed a 20-fold higher secretion rate of FSH and LH in response to LHRH than the small-sized gonadotrophs. The increment above baseline levels was also

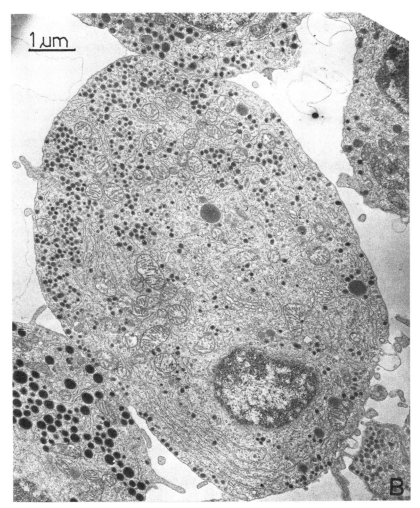

(fraction 4) of the gradient; most gonadotrophs in this fraction have only small-sized secretory granules; note the linear and parallel shape of the endoplasmic reticulum. ×15,000. Photomicrograph taken by Dr. C. Tougard, Groupe de Neuroendocrinologie Cellulaire, Collège de France, Paris, France.

much higher in the large than in the small gonadotrophs. About 70% of the overall secretory potential came from this pool of large gonadotrophs which represented only 10% of the total number of gonadotrophs. In preparations from 14-day-old males the difference in secretory potential between large and small gonadotrophs was less prominent than in those from 14-day-old females and secretion rates were more equally distributed over the different gonadotroph populations. Thus, all gonadotrophs are not identical in terms of their secretory potential in response to LHRH. It is logical to assume therefore that the overall response of the pituitary gland to LHRH will depend on the proportional number of the different gonadotroph subtypes. The elevated number (see above) of the large gonadotrophs in 14-day-old females and their high responsiveness to LHRH and secretory potential most likely favor the high FSH and (episodic) LH release of these animals *in vivo*.

Gonadotrophs also differ in terms of their relative content and release of FSH and LH. Immunocytochemical staining has shown that the majority of the gonadotrophs in 14-day-old males and females store both FSH and LH (Denef *et al.*, 1978a). However, the proportional amounts of these hormones differed according to cell size as well as between males and females. In small- and medium-sized gonadotrophs from 14-day-old males the FSH/LH ratio was only half of that in large gonadotrophs and considerably lower than in those of 14-day-old females. As shown in Fig. 2 similar differences were seen in the proportional release of these hormones from the latter pool of gonadotrophs (in confirmation of Denef *et al.*, 1978b). In small- and medium-sized gonadotrophs from 14-day-old males the release was in favor of LH. In adult males the opposite pattern was found. There, the small- and medium-sized gonadotrophs released

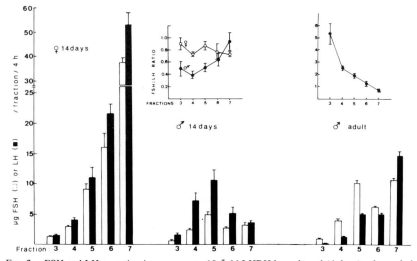

FIG. 2.   FSH and LH secretion in response to 10^{-7} *M* LHRH by cultured (4 days) subpopulations of rat anterior pituitary cells separated by unit gravity sedimentation. The figure shows the variation of the proportion of FSH and LH released by the different gradient fraction as well as the differences among animal models. Values are means ± SEM of hormone measured by radioimmunoassay (RIA) and expressed in terms of the RP-1 reference preparations of the RIA kits obtained from U.S. National Pituitary Agency, NIAMDD, National Institutes of Health, Bethesda, MD.

proportionally more FSH than LH and release in favor of LH was found in the large gonadotrophs. Thus, functional variants of gonadotrophs not only differ in terms of their responsiveness to LHRH but also in terms of differential FSH and LH release. The latter observation is important in view of the fact that LHRH stimulates both FSH and LH secretion but that the proportional release of these hormones changes with physiological, experimental, and developmental conditions (Blackwell and Guillemin, 1973). Denef *et al.* (1976, 1978b, 1980) have proposed that the high proportional release of FSH in 14-day-old females may repose on the selective proliferation of a specific pool of gonadotrophs. In these females the most active subpopulation of gonadotrophs not only releases much more FSH than the most active subpopulation of gonadotrophs in males but the release of FSH relative to that of LH is also higher in the former than in the latter (Denef, 1980, and Fig. 2).

Finally, it has been shown that the differential modulation of FSH and LH synthesis and release by androgens may also be based on the functional heterogeneity of the gonadotrophs (Denef *et al.*, 1980).

3. *Thyrotrophs*

Evidence for functional heterogeneity among enriched populations of thyrotrophs isolated from adult male rats has been given by Leuschen *et al.* (1978). By unit gravity sedimentation, two distinct thyrotroph cell types could be separated. One had a slightly lower sedimentation rate and secreted in monolayer culture much more TSH than the other.

C. Distinction between Direct Effects of Hypophysiotropic Regulatory Hormones and Effects Mediated through Intercellular Communication

As the pituitary is a heterogeneous tissue in which one stimulus may affect the release of several hormones (see Section I), the question arises whether the primary site of action of a specific stimulus which releases a specific hormone is the cell type storing that hormone or another cell type which transmits the message through an intercellular communication system (see Hertzberg *et al.*, 1981). On the other hand the response of a particular cell type may be conditioned by the neighboring cells.

Direct experimental evidence for such intercellular effects can be obtained by coculturing different cell types which have been enriched or highly purified by unit gravity sedimentation. Using this approach, evidence has been presented that there is functional communication between gonadotrophs and mammotrophs. LHRH has no effect on PRL release in mammotroph-enriched populations but if mammotrophs are cocultured with a highly purified population of gonadotrophs LHRH does stimulate PRL release (Denef, 1981, and see Fig. 3)

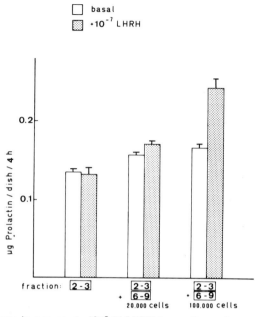

FIG. 3.   PRL release in response to 10^{-7} M LHRH in cocultures of an enriched population of mammotrophs (2–3) and a highly purified population of gonadotrophs (6–9). The cells were from 14-day-old female rats. Values are means ± SEM of hormone measured as described in Fig. 2.

and this from concentrations as low as 10^{-11}–10^{-10} M (Denef and Andries, in preparation). In addition, in such cocultures the release of LH in response to LHRH rises considerably as compared to LH released by the purified gonado-trophs cultured alone (Denef, 1980). These findings correlate with *in vivo* obser-vations. In the pituitary *in situ* gonadotrophs occur frequently in close apposition with mammotrophs (Nakane, 1970; Horvath *et al.*, 1977). On the other hand it has been reported that under certain conditions LHRH is capable of stimulating PRL release *in vivo* (Giampietro *et al.*, 1979; Yen *et al.*, 1980; Casper and Yen, 1981). Taken together these data suggest that functional communication between pituitary cell types may be a modulatory system in the control of pituitary hormone release.

D. Localization of Receptors of Hypophysiotropic Hormones and of the Stimulus–Effect Coupling Mechanisms

When studying receptors of hypophysiotropic hormones by radioligand bind-ing methods, it is necessary to know whether or not binding sites of the hormone are located in the cell type which is biologically responsive to this hormone.

The unit gravity sedimentation technique has been used with success to locate LHRH binding sites (Andries and Denef, submitted). A good correlation was found between the amount of binding and the number of gonadotrophs present in the different fractions collected from 14-day-old female and male rat pituitary. No specific binding was seen in fractions virtually devoid of gonadotrophs or in a highly purified population of somatotrophs.

Other parameters of hormone action have also been successfully located. The formation of dihydrotestosterone which occurs in many androgen-dependent tissues appears to occur primarily in gonadotrophs (Denef, 1979). The conversion was exceedingly high in a highly purified population of gonadotrophs isolated from 14-day-old female rats but very low in a mammotroph enriched or a highly purified population of somatotrophs.

Finally, sorting-out of cell types by unit gravity sedimentation has proven useful in studying stimulus–effect coupling. The hypophysiotropic hormone TRH has been shown to stimulate intracellular cAMP levels in pituitary cells but, as the peptide stimulates the release of both TSH and PRL, it remained unclear whether cAMP levels rise in thyrotrophs, in mammotrophs, or in both. Barnes *et al.* (1978) have shown the TRH raises cAMP levels in both a thyrotroph and a lactotroph-enriched population. The authors also showed that dopamine, which inhibits PRL release but not TSH release, inhibits TRH-stimulated accumulation of cAMP in the mammotroph-enriched but not in the thyrotroph-enriched population. The involvement of cAMP or cGMP in the mechanism of action of LHRH has been a matter of serious debate. Recently it was found that LHRH was not able to stimulate cAMP levels in a highly enriched population of gonadotrophs (Snyder *et al.*, 1980). On the other hand LHRH appears to stimulate cGMP in the gonadotroph-rich population but not in gonadotroph-poor populations (Snyder *et al.*, 1980).

V. Conclusion

The preparation of enriched and highly purified populations of different anterior pituitary cell types has proven to be a fruitful approach to better understanding the regulation of pituitary hormone secretion and the mechanism of action of the hypophysiotropic regulatory hormones. By using such preparations, evidence has been presented that the population of one cell type, even when producing a single hormone, consists of a continuum of variants with different ultrastructural and functional characteristics. Depending on physiological or experimental conditions selective proliferation of certain variants may occur. The latter changes might be an important cellular basis of the adaptive changes in pituitary hormone release when physiological conditions change.

The unit gravity sedimentation technique, particularly when different animal models are used, produces subpopulations of pituitary cells in which receptors, stimulus–receptor coupling, and other mechanisms of action of the hypophysiotropic hormones can be more reliably studied than in conventional *in vitro* systems.

Finally, important progress is represented by the fact that functional interaction between the various pituitary cell types can now be thoroughly investigated.

ACKNOWLEDGMENTS

The authors wish to thank Miss M. Bareau for excellent typographical work. Published and unpublished work reported in this article has been supported by grants from F.G.W.O., Onderzoeksfonds K.U.Leuven, and I.W.O.N.L.

REFERENCES

Barnes, B. G. (1962). *Endocrinology* **71**, 618–628.
Barnes, G. D., Brown, B. L., Gard, T. G., Atkinson, D., and Ekins, R. P. (1978). *Mol. Cell. Endocrinol.* **12**, 273–284.
Blackwell, R. E., and Guillemin, R. (1973). *Annu. Rev. Physiol.* **35**, 357–390.
Casper, R. F., and Yen, S. S. C. (1981). *J. Clin. Endocrinol. Metab.* **52**, 934–936.
Denef, C. (1979). *Neuroendocrinology* **29**, 132–139.
Denef, C. (1980). *In* "Synthesis and Release of Adenohypophyseal Hormones" (M. Jutisz and W. McKerns, eds.), pp. 659–676. Plenum, New York.
Denef, C. (1981). *Ann. Endocrinol.* **42**, 65–66.
Denef, C., Hautekeete, E., and Rubin, L. (1976). *Science* **194**, 848–851.
Denef, C., Hautekeete, E., De Wolf, A., and Vanderschueren, B. (1978a). *Endocrinology* **103**, 724–735.
Denef, C., Hautekeete, E., and Dewals, R. (1978b). *Endocrinology* **103**, 736–747.
Denef, C., Hautekeete, E., Dewals, R., and De Wolf, A. (1980). *Endocrinology* **106**, 724–729.
Döhler, K. D., and Wüttke, W. (1975). *Endocrinology* **97**, 898–907.
Edelman, G. M., Rutishauser, U., and Millette, C. F. (1971). *Proc. Natl. Acad. Sci. U.S.A.* **68**, 2153–2157.
Farquhar, M. G., and Rinehart, J. F. (1954). *Endocrinology* **54**, 516–541.
Farquhar, M. G., Skutelsky, E. H., and Hopkins, C. R. (1975). *In* "The Anterior Pituitary" (A. Tixier-Vidal and M. G. Farquhar, eds.), pp. 83–135. Academic Press, New York.
Gard, T. G., Atkinson, D., Brown, B. L., Tait, J. F., and Barnes, G. D. (1975). *J. Endocrinol.* **67**, 40P–41P.
Giampietro, O., Moggi, G., Chisci, R., Coluccia, A., Dalle Luche, A., Simonini, N., and Brunori, I. (1979). *J. Clin. Endocrinol. Metab.* **49**, 141–143.
Girod, C. (1976). *In* "Handbuch der Histochemie" (Suppl.), Vol. 8, pt. 4, p. 99. Fisher, Stutgart.
Herbert, D. C. (1975). *Am. J. Anat.* **144**, 379–385.
Herbert, D. C. (1976). *Endocrinology* **98**, 1554–1557.
Herlant, M. (1964). *Int. Rev. Cytol.* **17**, 299–382.
Hertzberg, E. L., Lawrence, T. S., and Gilula, N. B. (1981). *Annu. Rev. Physiol.* **43**, 479–491.
Hirsch, M. A., Lipner, H., and Leif, R. C. (1979). *Cell Biophys.* **1**, 93–115.

Horvath, E., Kovacs, K., and Ezrin, C. (1977). *IRCS Med. Sc.* **5**, 511.

Hymer, W. C. (1975). *In* "The Anterior Pituitary" (A. Tixier-Vidal and M. G. Farquhar, eds.), pp. 137–180. Academic Press, New York.

Hymer, W. C., Kraicer, J., Bencosme, S. A., and Haskill, J. S. (1972). *Proc. Soc. Exp. Biol. Med.* **141**, 966–973.

Hymer, W. C., Evans, W. H., Kraicer, J., Mastro, A., Davis, J., and Griswold, E. (1973). *Endocrinology* **92**, 275–287.

Hymer, W. C., Snyder, J., Wilfinger, W., Swanson, N., and Davis, J. A. (1974). *Endocrinology* **95**, 107–122.

Hymer, W. C., Snyder, J., Wilfinger, W., Bergland, R., Fisher, B., and Pearson, O. (1976). *J. Natl. Cancer Inst.* **57**, 995–1007.

Ishikawa, H. (1969). *Endocrinol. Jpn.* **16**, 517–529.

Ishikawa, H., Shiino, M., and Rennels, E. G. (1978). *Cell Tissue Res.* **189**, 31–40.

Kragt, C. L., and Dahlgren, J. (1972). *Neuroendocrinology* **9**, 30–40.

Kraicer, J., and Hymer, W. C. (1974). *Endocrinology* **94**, 1525–1530.

Kurosumi, K., and Oota, Y. (1968). *Z. Zellforsch.* **85**, 34–46.

Leuschen, M. P., Tobin, R. B., and Moriarty, C. M. (1978). *Endocrinology* **102**, 509–518.

Lloyd, R. V., and Karavolas, H.J. (1975). *Endocrinology* **97**, 517–526.

Lloyd, R. V., and McShan, W. H. (1973). *Endocrinology* **92**, 1639–1651.

Lloyd, R. V., and McShan, W. H. (1976). *Proc. Soc. Exp. Biol. Med.* **151**, 160–162.

MacLeod, R. M. (1969). *Endocrinology* **85**, 916–923.

MacLeod, R. M. (1976). *In* "Frontiers of Neuroendocrinology" (L. Martini and F. Ganong, eds.), Vol. 4, pp. 169–194. Raven, New York.

Maurer, R. A., and Gorski, J. (1977). *Endocrinology* **101**, 76–84.

Miller, R. G. (1973). *In* "New Techniques in Biophysics and Cell Biology" (R. H. Pain and B. J. Smith, eds.), p. 87. Wiley (Interscience), New York.

Moriarty, G. C. (1975). *Endocrinology* **97**, 1215–1225.

Moriarty, G. C. (1976). *J. Histochem. Cytochem.* **24**, 846–863.

Nakane, P. K. (1970). *J. Histochem. Cytochem.* **18**, 9–20.

Ojeda, S. R., and Ramirez, V. D. (1972). *Endocrinology* **90**, 466–472.

Ojeda, S. R., Jameson, H. E., and McCann, S. M. (1977). *Endocrinology* **100**, 440–451.

Pelletier, G., Leclerc, R., and Labrie, F. (1976). *Mol. Cell. Endocrinol.* **6**, 123–128.

Phifer, R. F., Midgley, A. R., and Spicer, S. S. (1973). *J. Clin. Endocrinol. Metab.* **36**, 125–141.

Robyn, C., Leleux, P., Vanhaelst, L., Golstein, J., Herlant, M., and Pasteels, J. L. (1973). *Acta Endocrinol.* **72**, 625–642.

Rotsztejn, W. H., Benoist, L., Besson, J., Beraud, G., Bluet-Pajot, M. T., Kordon, C., Rosselin, G., and Duval, J. (1980). *Neuroendocrinology* **31**, 282–286.

Samson, W. K., Said, S. I., Snyder, G., and McCann, S. M. (1980). *Peptides* **1**, 325–332.

Siperstein, E., Nichols, C. W., Griesbach, W. E., and Chaikoff, I. L. (1954). *Anat. Rec.* **118**, 593–619.

Snyder, G., and Hymer, W. C. (1975). *Endocrinology* **96**, 792–796.

Snyder, G., Wilfinger, W., and Hymer, W. C. (1976). *Endocrinology* **98**, 25–32.

Snyder, G., Hymer, W. C., and Snyder, J. (1977). *Endocrinology* **101**, 788–799.

Snyder, G., Naor, Z., Fawcett, C. P., and McCann, S. M. (1980). *Endocrinology* **107**, 1627–1632.

Spence, J. W., Sheppard, M. S., and Kraicer, J. (1980). *Endocrinology* **106**, 764–769.

Surks, M. I., and De Fesi, C. R. (1977). *Endocrinology* **101**, 946–958.

Tal, E., Savion, S., Hanna, N., and Abraham, M. (1978). *J. Endocrinol.* **78**, 141–146.

Tixier-Vidal, A., Gourdji, D., and Tougard, C. (1975a). *Int. Rev. Cytol.* **41**, 173–239.

Tixier-Vidal, A., Tougard, C., Kerdelhué, B., and Jutisz, M. (1975b). *Ann. N.Y. Acad. Sci.* **254**, 433–461.

Tougard, C. (1980). *In* "Synthesis and Release of Adenohypophyseal Hormones" (M. Jutisz and K. W. McKerns, eds.), pp. 15–37. Plenum, New York.

Tougard, C., Picart, R., and Tixier-Vidal, A. (1977). *Dev. Biol.* **58,** 148–163.

Venter, B. R., Venter, J. C., and Kaplan, N. O. (1976). *Proc. Natl. Acad. Sci. U.S.A.* **73,** 2013–2017.

Voogt, J. L., Chen, C. L., and Meites, J. (1970). *Am. J. Physiol.* **218,** 396–399.

Walker, A. M., and Farquhar, M. G. (1980). *Endocrinology* **107,** 1095–1104.

Yen, S. S. C., Hoff, J. D., Lesley, B. L., Casper, R. F., and Sheehan, K. (1980). *Life Sci.* **26,** 1963–1967.

Yoshimura, F., and Harumiya, K. (1965). *Endocrinol. Jpn.* **12,** 119–151.

INTERNATIONAL REVIEW OF CYTOLOGY, VOL. 76

What Is the Role of Naturally Produced Electric Current in Vertebrate Regeneration and Healing?

RICHARD B. BORGENS

*The Institute for Medical Research,
San Jose, California*

I. Introduction

When a child accidentally amputates the tip of a finger and is brought to a hospital, the conventional wisdom has been to suture the tip closed. There are several different surgical techniques for accomplishing this, however, they all serve to cover the open wound with skin—and the result is a blunt fingerstump (Illingworth, 1974). Salamanders and newts are animals well known for their regenerative ability. If they lose an entire limb or tail, they regenerate a near perfect replica in a matter of a few weeks. However this remarkable process can be inhibited completely if one sews a full thickness flap of skin across the top of the injured stump (Polezhaev and Favorina, 1935; Mescher, 1976).

245

One might reasonably ask what will occur if the fingertip stumps in children are not surgically tampered with? The answer to this question is that they grow back. From reports of Cynthia Illingworth (1974), B. S. Douglas (1972), and others (see also Farrel *et al.*, 1977), it is now quite clear that a cosmetically perfect replacement, complete with fingerprint and fingernail, will grow back within 4 months. The amputation must be no more proximal than the joint between the middle phalange and the distal phalange, and the amputation surface cannot be sewn shut.

I have recently experimented with 4-week-old mice and have observed this same regrowth phenomenon. In mice (as is probably the case with humans), the origin of these replacement tissues is not the same as in amphibian limbs (also see Section V)—but these processes do share some things in common: one being that regenerating limbs and finger stumps that will regrow all drive a substantial amount of electric current out of the injury. Is this current flow relevant to the regrowth?

II. Current Flow in Biological Systems

What is meant by a steady flow of electric current? In biological systems, electric current is carried through tissues, body fluids, and cytoplasm by ions and not electrons. So in reality, what we are discussing is a steady ionic current. The convention for describing the direction of this current flow is that it is defined as moving in the direction in which positive charge moves. For the net flow of cations, the direction of current flow is the actual direction in which these ions are moving. For anions, it will be opposite. For example, if a localized area of cell membrane allows Na^+ and Ca^{2+} to specifically leak into the cytoplasm, this could be described as a net incurrent. If Cl^-, on the other hand, was pumped out of this hypothetical cell, this would also be described as a net incurrent. Electrically speaking, negative charge moving *out* is the same as positive charge moving *in*. If charges are moving in or out of a cell against their electrochemical gradient, this requires the metabolic machinery of the membrane. Thus, these ions are "pumped." If an ion flows down its electrochemical gradient, we will refer to this as a "leak." Of course, the net flow of current (charge) moving into a cell from the medium, must be balanced by a net movement out of the same cell. I will discuss the two most obvious consequences, of this flow of charge: (1) ionic concentrations can be changed, especially inside of cells exposed to a steady current flow. This may result in actual ionic gradients in cytoplasm, depending on the mobility of the particular ionic component of a transcellular current. Furthermore, changes in the ionic composition of cytoplasm can profoundly alter the physiology, even the structure, of the cell.

(2) Cytoplasm, body fluids, skin, both hard and soft tissue, etc., all have

characteristic resistances to the flow of current. Current being driven across such areas will be associated with an electrical field. Therefore, we should expect *voltage-mediated processes* (such as electrophoresis) to be associated with developmental currents as well as ionically mediated processes.

Electrical events associated with injury may perhaps be one of the most ancient signals of trauma. One of the necessary properties of primal cells was the ability to separate their internal milieu from the external world by a membrane. Evolving cells isolated their interiors; they developed mechanisms for the elimination of waste materials, and they separated (and qualitatively altered) the sea water on their insides from their sea water environment. This was accomplished by the development of metabolically powered ionic "pumps" which in the process of maintaining these differences, also helped to produce a large potential difference across the membrane itself; therefore it is characteristic of living cells that they are internally negative with respect to their extracellular environment by some 50 to 90 mV.

For this reason, an injury to a primitive cell, or a modern one, would produce an instantaneous and steady flow of charge. Since cells are internally negative, one would expect the flow of current to *enter* the hole in the membrane. This flow of charge should be maintained for some time (as long as the cell is alive) by the very ionic pumps which served to produce the voltage across the membrane. In fact, this appears to be the case, and has been described by physiologists as early as the middle of the last century. For example DuBois-Reymond (1843), Lorente de Nó (1947a, b) and Cowan (1934) demonstrated that near an injury, nerves are externally electronegative with respect to the uninjured portions (thus, we can infer that current enters the injured area). Moreover, this difference in potential (on the order of 10s of millivolts) persists for days; even over a week (Lorente de Nó, 1947a).

The same principles can be applied to the organism as a whole. The "membrane" of the organism is its integument. Most all animals studied maintain large potential differences across their skins such that skins are usually internally positive with respect to the outside. These trancutaneous potential differences have been studied most extensively in amphibia (Ussing, 1964; Kirschner, 1973), however they occur in a variety of other animals, as well as in humans (Barker *et al.*, 1982). Since skins are internally positive, by convention then, we would expect a steady current of injury to be driven *out* of a hole made in the skin. Once again, such a flow of current can be inferred from the measurements of wound potentials made by DuBois-Reymond (1860) and Herlitzka (1910). They reported that small wounds made in human skin were externally positive with respect to the adjacent uninjured areas—therefore current appeared to leave these small abrasions. This report set an experimental precedent for later measurements of current found leaving the amputated stump of a childs finger tip some 120 years later (Illingworth and Barker, 1980).

These interesting observations, made by the fathers of modern physiology, and their suggestions that steady electrical changes may be directly involved with the cells, or tissues' overall response to injury, have been largely forgotten or ignored for the last half of this century.

Recently the most direct evidence suggesting a critical role for steady ionic current during the regenerative process in vertebrates comes from studies of amphibian regeneration. Very interesting circumstantial evidence can also be found in three other areas—the regeneration (healing) of vertebrate bone; the responses of nerve to injury; and the healing of skin wounds. I will focus here on the role of current in vertebrate injury and regeneration. In a future piece, Kenneth Robinson and I will turn to electrical controls found in developing cells.

III. The Role of Steady Current in Amphibian Regeneration

A. SURFACE POTENTIAL MEASUREMENTS OF AMPHIBIAN LIMBS

Gaetano Viale, Herlitzka's student, studied wound potentials in injured nerve and in the regenerating tadpole tail. His electrical measurements, reported in 1916, sparked the imagination of the famous Italian embryologist, Alberto Monroy. Using a galvanometer and contact electrodes, Monroy (1941) made careful measurements of potential differences on the surface of the body of intact and regenerating salamanders (tritons). He found that the distal portion of an intact extremity was always electropositive with respect to its base. After amputation, limb and tail stumps showed a sharp increase in this distally positive potential difference followed by a gradual decline in its intensity over the next 46 days of measurement (however in limbs, the magnitude of this potential was still elevated over the contralateral uninjured extremity). Monroy explored what the source of the current producing these potential differences might be. Denervation experiments, and removing the heart failed to decrease these measured voltages; thus, Monroy reasoned that the skin must have been involved with the production of the measured potentials—but he was uncertain.

This type of "surface potential measurement" has been undertaken in more modern times by a variety of investigators. However they have not always enlarged our understanding of the electrical changes occurring after limb amputations in urodele amphibians; in many cases they have just confused it.

Becker (1960) disagreed with Monroy, suggesting that the intact limb was distally electronegative with respect to its base. Furthermore, Becker found that the distally positive values measured soon after amputation completely reversed themselves to substantial negative values by about 1 week after amputation. This distal electronegativity was also measured by Rose and Rose (1974), but only with respect to a reference position approximately 2 or 3 mm more proximal to the stump's cut end. In fact distally *positive* values were obtained if one compared this position with a reference position still more proximal on the limb stump's exterior. Rose did not report measurements for the first week postamputation. More recently Lassalle (1979) has reported distally positive surface potentials soon after amputation, a reversal to distally negative values, and an eventual return to distally positive values by 1 month postamputation.

All of these investigators have tried to determine the source of these surface potentials. Becker (1960) has claimed that the nervous system is the generator, and has expanded this notion to a general theory that surface detected "bioelectric fields" are a primitive form of "data" or "control" system predating yet coexisting with the CNS of complex vertebrates (Becker, 1974). Though imaginative, these notions have no empirical support. For example no data are ever presented for denervation experiments that were supposed to have abolished the surface voltages measured in several species of salamander (Becker, 1961). Neither Rose and Rose (1974) nor Lassalle (1974b) have been able to confirm this claim, nor is it supported in Monroy's seminal experiments. Thus, there is no reliable evidence for the assumption that the nervous system is the electrogenic source of the externally measured voltage gradients on the bodies of salamanders.

The skin of amphibians, however, is known to posses a powerful battery. Inwardly positive transcutaneous potential differences of over 80 mV are known to occur across skin (from inside to outside) (Ussing, 1964; Kirschner, 1973). In amphibians these voltages are known to be particularly sensitive to the concentration of Na^+ in the pond water. Lasalle (1974b) found that ablation of the skin, osmotic shock, and altering the external Na^+ concentration drastically altered the voltages measured on the outside of the limb. Additional support for the skin's role as the source of these potentials came from the work of Fontas and Mambrini (1977a, b). They reported a rough correlation between the magnitude of the skin's transcutaneous potential, and the magnitude of external voltages measured over localized areas of the creatures body. In summary, the "translimb" potential type of measurement suggests a distally electropositive voltage gradient in response to amputation, with the skin of the stump serving as the source of these potential differences. Considering that the techniques for measuring these potentials has changed very little since 1939 when Monroy began his investigations, it is not surprising that little new information has been gained beyond Monroy's original efforts.

All of these investigators used some sort of contact electrodes, spaced apart at varying distances on the outside of the limb, to measure the difference in potential between them while the animal was in an aerial environment. Since amphibians are semiaquatic in habitat, this serves to maximize the resistance between the electrodes, achieving a measurable voltage for the investigator. It is certain that the resistence along the thin film of moisture on the limbs surface (during measurements made with the animal out of water) are radically different from the resistances encountered in the extended environment of the aquarium or pond. Since the magnitude of such voltages is absolutely dependent on the electrical resistance between these two electrodes, the absolute magnitude of these potentials are of no physiological significance. Further complications arise when one wishes to compare all of these measurements qualitatively. If the most distal electrode is in contact with whole skin, the measured voltage will be associated with an IR drop along the skins' surface, between the electrodes. If the most distal electrode is in contact with the cut surface itself, the measured voltage will in part be due to the potential that exists *across* the skin. This potential will be partially sampled because the distal electrode will be at subdermal potential and referenced to the skins exterior.

However, the value of surface potential measurements is most obscured for this simple reason. It is doubtful whether any voltage gradient measured on the *outside* of a limb is even pertinent to the cellular changes occurring *within* the stump during the regeneration process. Salamanders can regenerate in bathing media of varying conductance (Huang *et al.*, 1979)—this fact strongly suggests that the magnitude of external fields is simply unimportant to the regeneration process. It is the internal voltage gradients that are pertinent (see Section III,G).

Another approach was needed, and for this reason, Joseph Vanable, Jr., Lionel Jaffe, and I began investigations designed to more precisely determine the electrical character of regenerating limbs, to determine a way to alter these changes in a predictable way, and in doing so, to test the relevance of such endogenous electricity to the regenerative process. The instrument that we used to measure this endogenous current was the ultrasensative vibrating probe. This device had just been developed at Purdue University by Lionel Jaffe and Richard Nucittelli (1974) about 1 year prior to our studies on regenerating amphibians.

B. The Vibrating Probe System

The vibrating probe is designed to measure the minute ionic currents that emerge from a variety of developing single cells, immersed in a natural medium. The electrode itself is a platinum black ball, about 20 μm in diameter. This ball is vibrated (at about 400 Hz) between two positions (typically 30 μm apart) by means of a piezoelectric bender element to which a small voltage is applied. When the electrode is vibrated near a biological current source (or a calibration source) immersed in a natural medium, it measures the minute voltage differences in the plane of its vibration between the extremes of its 30 μm excursion. This is accomplished by using a phase frequency lock-in amplifier to both set the frequency of vibration, and to amplify the signal. These two modalities are then simply tuned together. Since we directly resolve this minute voltage difference and we can easily measure the resistivity of the medium with a conductance bridge, than we calculate the current density entering or leaving the source using an analog of Ohm's law for extended media: $I = V/\rho\Delta$ where I is the current density in amps/cm^2; V, the voltage-gradient in volts/cm; ρ, the resistivity of the medium in ohms cm; and Δ is the amplitude of vibration. A picoampere source is used to supply a known current by which we calibrate the vibrating probe system.

In summary then, the vibrating probe system allows determinations of current density entering or leaving a single cell (or whole tissue) immersed in a natural medium (with a spatial resolution of 20 μm). The determinations of current density and potential difference are on the order of 100- to 1000-fold more sensitive than what can be achieved by conventional (static) electrodes. On a more macroscale we used this device to map the pattern and determine the magnitude of currents driven through regenerating amphibian limbs.

C. STEADY CURRENT IS DRIVEN OUT OF REGENERATING AMPHIBIAN LIMBS

When the probe was used to scan the surface of intact limbs, we found that small currents (on the order of 1 μA/cm^2) entered the limb at most all locations (Fig. 1A). After amputation this pattern and intensity changed dramatically: substantial current densities (some 100-fold larger) were measured leaving the end of the stump, usually localized in the dorsal postaxial region of the cut face, and completed the circuit through the medium by entering the intact, undamaged portion of the stump (and possibly the body of the animal as well) (Fig. 1) (Borgens et al., 1977a). This polarity of current flow about the limb stump would be associated with a distally positive, external voltage gradient. Thus, the polarity of the majority of translimb potential measurements discussed earlier are in agreement with these measurements. The magnitude of the outcurrent was variable from day to day, however it characteristically declined to less than 5% of the original values by about the tenth to fourteenth day postamputation. We then observed transient shifts in its direction. This period appeared to correlate with the first overt appearance of the blastema. All of the above measurements were performed on Notophthalmus viridescens, the red-spotted newt, however qualitatively similar measurements have been made on Ambystoma tigrinum (Borgens et al., 1979c), Ambystoma mexicanum (Vanable et al., 1980), and four different species of plethodontid salamanders: Desomagnathus quadrimaculatus, Batrachoseps attenuatus, Aneides lugubris, and Ensatina eschscholtzi xanthoptica (Borgens, unpublished measurements).

We next asked what the source of this endogenous stump current might be. The notion that the nervous system was the source of these stump currents made little sense. As discussed current is known to enter the ends of severed nerves (see

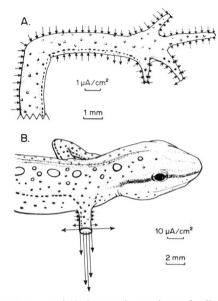

FIG. 1. The current pattern around (A) intact red spotted newt forelimb and (B) forelimb stump, 20 hours after amputation. Each arrow indicates the direction and magnitude of current density, as measured 340 μm from the surface. Note the 10-fold difference in current density scales in (A) and (B), allowing visualization of the small current densities of the undamaged forelimb. From Borgens *et al.* (1979b).

Section IV,A). Current entering the cut ends of nerve trunks at the stump's end would be opposite in direction to the stump currents which leave the cut face. Thus, severed nerves would be expected to reduce the net outcurrent, not serve as its source. In fact, a stump with few nerves would be expected to have larger currents leaving the cut end. We tested this experimentally. Functionally denervated newt limbs do produce more current than the contralateral control (innervated stump) (Borgens *et al.*, 1977a). Certainly then, nerves are not the source of the currents leaving amputated limbs.

We next tested the skin as the possible source. As mentioned, amphibian skin maintains a large Na^+-dependent transcutaneous potential difference across itself. Isolated skin preparations have been measured to drive substantial current as well: on the order of 50 μA/cm^2 (Ussing, 1965). Our probe measurements were always made in an artificial pond water (APW) of known ionic composition. If we replaced Na^+ with choline or simply removed Na^+ from the media, the stump currents were rapidly reduced by more than 90% of their original values. If we raised the Na^+ concentration in this medium about 5-fold, then the current density, on the average, increased about 5-fold. Skin physiologists also use several chemicals which are known to specifically block Na^+ channels at the outside (mucosal) face of the skin, preventing access of Na^+ to those cell layers which are actively producing the transcutaneous potential difference. We used two of these compounds: the methyl ester of lysine (MEL), and amiloride (Kirschner, 1973; Benos et al., 1976). When either of these is dissolved in the APW in which measurements are being made, the stump currents are dramatically lowered or abolished (Borgens et al., 1977a). All of these effects, including the modulation of the Na^+ concentration of the medium, are reversible after a return to the standard APW and a period of adjustment.

We concluded then, that the intact skin of the limb stump, and possibly of the body itself, is the "battery" which drives charge out the end of the stump. Additionally, this interpretation is consistent with the observations of Lassalle, Fontas, and Mambrini, and demonstrate that Alberto Monroy was correct in his intuitions. Most importantly, we had learned several ways to test the relevance of stump currents to the process of limb regeneration. If we chronically inhibit the production of these skin-driven currents, or substantially lower them, will this have any effect on limb regeneration?

D. The Effects of Lowering Stump Currents

We performed a variety of experiments in which the level of current leaving the stump was greatly reduced. We topically applied amiloride to only the limb stumps of large tiger salamanders and we maintained newts and tiger salamanders in Na^+-deficient media (Borgens et al., 1979c). We maintained newts in pond water containing MEL (unpublished research). In many cases these treatments either dramatically retarded limb regeneration, completely inhibited it, or caused it to be abnormal in character.

When tiger salamander stumps were swabbed with a 0.5 mM solution of amiloride in pond water every other day for 9 to 11 weeks, 36% of the experimental population (33 animals) was completely inhibited from regenerating, 15% regenerated in a grossly abnormal fashion, and 49% regenerated as normally as did the control group. Of the control group (34 animals) 100% regenerated fine looking forelimbs (Fig. 2) (Borgens et al., 1979c).

When tiger salamanders or newts were kept partially immersed in a Na^+-deficient pond water, we observed a most curious response. All of these animals showed absolutely no sign of regeneration for many weeks following amputation—at this period the controls had produced late cone stage regenerates, or even finger anlagen. About this time the experimental population did begin to regenerate, but at a greatly accelerated rate. They overtook (an in most cases passed) the control population in the rate of regeneration and developmental stage. It was as if some factor(s) were accumulating during the many weeks of inhibition. Furthermore, histological examination of inhibited newt limbs prior to the point where the entire population would be expected to escape inhibition was,

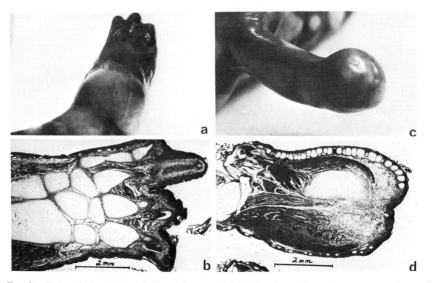

Fig. 2. External view (a), and photomicrograph (b), of a tiger salamander regenerate (control group) 11 weeks after amputation through the midforelimb. Note the well-delineated radius, ulna, carpals, and metacarpals. The rest of the digits were in other sections. External view (c), and photomicrograph (d), of a nonregenerating, amiloride-treated limb, 11 weeks after a similar amputation as shown in (a) and (b). Note the extensive callus formation, and the absence of forelimb, wrist, and hand elements. From Borgens *et al.* (1979c).

itself, curious. The stump end was covered by whole skin, and the severed bone was capped with callus and a typical connective tissue pad (see Schotté and Wilber, 1958; Rose, 1944). This is typical of nonregenerating amphibian limbs (and incidentally mammals as well). By all rights, these limbs should not have regenerated at all. One more fact is worth mentioning: When we spot checked "low Na^+" newts for stump currents, we found that after a few days they began to once again drive current out the cut end. Moreover, this "adaptation current" was Na^+ independent: It was not affected by a fresh change to Na^+-free media, nor was it affected by amiloride. We think it probable that the skin adapts to the low Na^+ environment, begins to drive Na^+-independent currents, and in this way initiates the process of limb regeneration, after a variable delay period (Borgens *et al.*, 1979c).

Such adaptation currents may also explain why about one-half of the amiloride-treated tiger salamanders regenerated normally; however, we did not test these animals for the presence of adaptation currents. Animals treated with amiloride may also have shed the amiloride-treated skin just after an application was made. The freshly exposed skin after a molt is known to be strangely resistant to the effects of amiloride (Nielsen and Tomilson, 1970).

Can these results be explained by any other means than a reduction of skin-generated current? One might suggest that the net effects of the Na^+ dilute media was to deprive the animals of Na^+, producing poor health and poor regeneration. First, it is known that amphibians can survive prolonged immersion in Na^+ dilute media without suffering a loss of total body Na^+ (McAfee, 1972). However, we did determine the blood Na^+ level by flame photometry for tiger salamanders immersed in Na^+ dilute APW and their controls. There was no significant difference between these groups. The inhibitory effects of amiloride were probably not due to general toxicity, since so little of the animals total skin was treated (less than 1% of the total body surface). New experiments by Joseph Vanable and Steven Datena (unpublished) strongly support this interpretation. Newts will regenerate normally in the presence of amiloride *if* its application is delayed for about a week and a half. Not only does this suggest that amiloride is not toxic, but that the current-dependent steps in limb regeneration are early ones. Besides inhibiting the production of skin-generated current by chemical means, one could also negate—even reverse—the flow of stump current by electrical means. Vanable and Hearson have preliminary data suggesting that pulling current through axolotl stumps in the reverse direction (distal–proximal), using a surgically implanted electrode connected to an external voltage source inhibits limb regeneration (see Vanable *et al.*, 1980).

D. THE STUMP CURRENT IN FROGS

I have discussed the nature of the evidence indicating a possible role of skin-driven current in amphibian limb regeneration. The classical work describing amphibian skin's potent battery was performed on frog skin; therefore, since anuran skin possess an inwardly positive potential difference, we would expect such skin to drive current out the cut end of the stump in the same manner as urodeles. However adult frogs (especially the genus *Rana*) do not regenerate their limbs. One might reasonably ask why stump currents may help initiate limb regeneration in urodeles but not in anurans. The answer to this question may lie in a difference in anatomy between the limbs of frogs and salamanders. Frog limbs possess a very loose skin that overlays large subdermal lymph spaces. Urodeles do not; their skin adheres tightly to the underlying musculature. These subdermal lymph spaces might provide a shunt path of low electrical resistance for the skin generated current [body fluid resistivities are approximately 100 Ω cm, while the resistivity of most soft tissues is approximately 10-fold greater (Schwan, 1963)]. In other words, current generated by the skin might be driven through these lymph portals, and relatively little current would traverse the core tissues of the stump from which the blastema arises. We tested this hypothesis directly with the vibrating probe system.

Immediately after amputation, before any epithelialization of the stump end occurs, we found the pattern of current density leaving a frog's limb stump to be substantially different from that of a newt (Borgens *et al.*, 1979d). The peak densities of current were always found leaving the periphery of the frog's stump—adjacent to the subdermal lymph spaces. They were never highest in the central regions of the stump. In these areas the current density averaged 7 μA/cm^2, some 2- to 3-fold less than at the periphery of the same limb. In addition, these core densities in frogs were about 4-fold less than the peak current densities found leaving the central regions of comparably treated newt stumps (averaging about 30 μA/cm^2).[1] Moreover, the pattern of current leaving a newt stump is quite different than the pattern observed in frog stumps. In newts, current densities are *never* highest at the periphery as in frogs; they are always highest found leaving the central or postaxial region of the stump (Fig. 3a and c).

Thus, it is apparent that a substantial shunting of skin-generated current does occur through the subdermal lymph space in frogs, resulting in relatively low current densities (hence electrical fields) within the core tissues.

Also pertinent to these measurements are older observations made by Oscar Schotté and Margaret Harland (1943). They carefully studied the regeneration of *Rana clamitans* tadpole limbs as they are undergoing metamorphosis. They concluded that there was no direct relationship between the structural differentiation of the limb and the progressive loss of regenerative ability, which occurs in a proximal–distal direction in tadpole limbs. However, as noticed by Rose (1944), a careful reading of this paper demonstrates that during metamorphosis the subdermal lymph spaces make their appearance at the approximate level of the limb where regenerative ability was observed to be lost. As summarized by Schotté and Harland: "It is possible that the skin plays a role in fostering or in preventing the formation of a blastema according to the levels of amputation, the increasing toughness of the skin from distal to proximal being quite conspicuous. To the toughness of the skin should be added the complicating influence of the lymphatic sacs, which can be seen to become increasingly more prominent from the metatarsals to the thigh" (Schotté and Harland, 1943). Thus, it is possible that the loss of regenerative ability in tadpoles may in part be due to the shunting of endogenous stump currents through the developing lymph spaces. How might one test if this shunting, which certainly occurs in *adult* frogs, is relevant to their lack of regenerative ability? If the electric field within an adult frog stump is

[1]This value of 30 μA/cm^2 was itself low in newts when compared to the peak current densities measured during the next 30 hours—which usually range from 60 to 100 μA/cm^2 (Borgens *et al.*, 1977a). The low values measured in stumps of all amphibians *immediately* after amputation are probably due to the effects of intense injury current *entering* damaged or dying cells. This would subtract from the net current driven out the stumps end by the skin. The substantial increase in the stump outcurrents probably relates to the rapid decrease in this injury current (in nerves this steep decline occurs within the first 20 hours after axotomy) (see the following discussion on nerve injury current, Section IV,A).

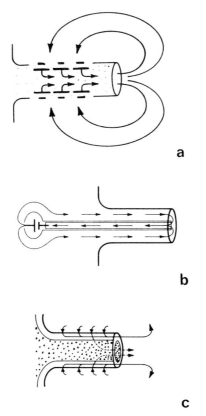

a

b

c

Fɪɢ. 3.   A comparison of circuits driving endogenous currents through a urodele stump (a); artificially imposed currents through an adult frog stump (b); and endogenous current flow through an adult frog stump (c). In (a) and (c) the skin battery drives charge from outside to inside. The resulting voltage drives positive change toward and out of the leak at the stump's end. The circuit is completed by a return of charge to the skin's exterior via the medium. Note that in (a) current traverses the salamander stump's core tissues, but in (c) much of the skin-driven current is shunted through the frog's subdermal lymph space. In (b) the insulated wick cathode pulls current out through the core tissues of the frog's limb stump—in the same direction as the endogenous current flow in salamanders (a) and of a greater density than found in the core tissues of untreated adult frogs (c). From Borgens *et al.* (1977a,b, 1979b,d).

insufficient to initiate limb regeneration, one approach might be to artificially enhance these fields by means of an implanted battery and electrodes.

F. Eʟᴇᴄᴛʀɪᴄᴀʟʟʏ Iɴɪᴛɪᴀᴛᴇᴅ Lɪᴍʙ Rᴇɢᴇɴᴇʀᴀᴛɪᴏɴ ɪɴ Fʀᴏɢꜱ

S. D. Smith published two studies (1967, 1974) which suggested that if a distally negative electric field is imposed within the terminal portion of an adult frog stump, then limb regeneration can be initiated. In the first study, Smith implanted small bimetallic strips into the stump's end. An implant such as this electrolytically produces a small current between the ends of the bimetallic strip. Smith found that such implants did induce a measure of regeneration in experimental frog stumps. He later elaborated on this idea and implanted a small suitably insulated hearing aid battery beneath the skin on the backs of frogs, routed the negative electrode through the limb stump, and left the uninsulated portion of the negative electrode (cathode) in contact with the terminal stump tissues. A cathode placed this way would pull current through the limbs in a proximal to distal direction (refer to Fig. 3b). This procedure resulted in more striking regenerative responses. Though these studies were ingenious and innovative, they were ambiguous. Because Smith delivered current to the stump tissues with metal electrodes he could not state unequivocally that the results were due to the effects of electric current. The results could have been produced by the electrode products which would have contaminated the stump tissues. This contamination is inevitable any time one delivers current through metal into a conductive solution such as body fluids. Also, Smith used no sham-treated controls; no inoperative batteries and electrodes were implanted to control for the effects of the chronic irritation produced by the physical presence of stainless-steel electrodes hooked into the stump tissues. It has been long appreciated that chronically irritating the ends of frog stumps (with pin pricks or other noxious stimuli) can induce a measure of regeneration (reviewed by Polezhaev, 1972). Smith also made the claim that the implanted battery system, in one case, initiated the complete regeneration of a limb. Smith (1974) describes this limb as "absolutely indistinguishable from a normal one." However, the only evidence presented to support this astonishing claim was a photomicrograph of normal wrist and hand histology. Though this regenerated limb was described as taking 1 year to develop, no photographic record of its developmental history has ever been published. Thus this claim has to be regarded as unsupported, owing to the absence of proper documentation.

For all of the above reasons, we decided to repeat Smith's experiment, except we delivered current to the stump tissues with wick electrodes (Silastic medical grade tubing filled with cotton string and amphibian ringers) (Fig. 4a). Current would be delivered to the stump by a conductive solution similar to body fluids, and thus no electrode products could contaminate the stumps end. We could be sure, then, that any result would be mediated by electric current, and not electrode products. Second, we included a series of sham-treated animals to control for any mechanical effects of the indwelling electrodes. An additional series of animals were treated with distally *positive* current (Borgens et al., 1977b). In 100% of the cathodally treated frog stumps, 200 nA of current (applied for about 3 weeks) initiated varying degrees of limb regeneration. The histology of these regenerates demonstrated that multitissue regeneration had indeed occurred (Fig. 5). Large amounts of new muscle, nerve, new bone, and separate cartilagenous

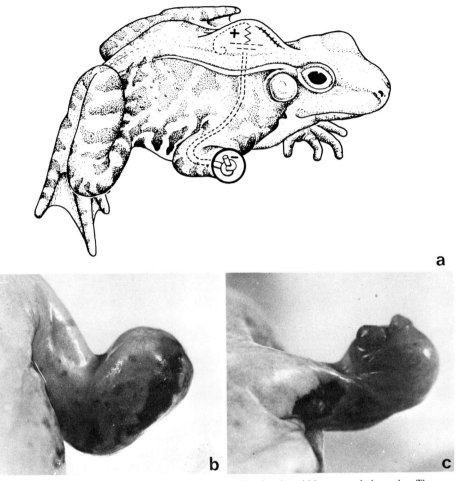

FIG. 4.   (a) Grassfrog (*Rana pipiens*) implanted with a hearing aid battery and electrodes. These units were inserted at the time of amputation and removed about 4 weeks later. In cathodally stimulated animals, the uninsulated portion of the wick electrode was in contact with tissues at the stump's surface. The positive electrode was coiled beneath the unit in contact with the underlying tissue. External views of (b) sham-treated limb, 3 months postamputation; and (c) cathodally stimulated limb, 6 months and 3 weeks postamputation. Sham-treated limbs healed over smoothly with pigmented skin; their appearance did not change for the next 3 to 7 months of observation. The atypical regenerates initiated by cathodal stimulation continued to change in form for as much as 11 months postamputation. From Borgens *et al*. (1977b).

structures (which resembled the anlagen of hand or wrist bones) were observed. Occasionally, even connective tissue structures resembling tendons or ligaments were found in these atypical regenerates. In the experimentals, it is probable that some of the artificially imposed current was shunted through the lymph spaces in

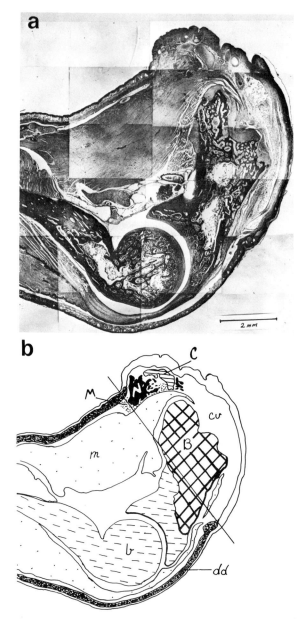

Fig. 5. (a) Low power photomicrograph of a cathodally stimulated limb stump. (b) Tracing of photomicrograph. M, regenerated muscle; m, original muscle; N, regenerated nerve; c, catilagenous island; B, regenerated bone; b, original bone; dd, dense dermis; cv, loose connective tissue and vascular tissue. The line indicates the approximate level of amputation. The grassfrog was sacrificed 9 months, 1 week postamputation. From Borgens *et al.* (1977b).

a manner similar to the natural current. However, the insulated portion of the electrode was pulled tightly against the musculature of the stump end; the circuit must have been completed through this area. Therefore, some current was pulled through the core tissues. Of the sham-treated controls, 100% healed over in a manner typical of adult grass frogs (Schotté and Wilber, 1958; Rose, 1944). The end of the stump was covered with skin; beneath we observed a large amount of bone callus with a connective tissue pad capping the end of the severed radio-ulna (Fig. 6).

In the series where the anode was located at the stump's end, current would be driven in an opposite direction within the stump when compared to either the endogenous current in salamanders or the cathodal current, which initiated regeneration in the experimental series. In all of these anodally treated frog stumps, gross degeneration of the limbs was evident months after the electrodes were removed. No evidence of regeneration was ever observed. The limbs were atrophied, with much tissue destruction evident histologically. However, with this arrangement of electrodes, we could not completely rule out the possibility that the *degeneration* was due to metal ion contamination (Borgens *et al.*, 1977b) (Fig. 7).

How well do the artificially imposed, distally negative fields in frogs mimic the natural fields in urodele core stump tissues? These electrical applications are similar in (1) polarity (Fig. 3, compare a and b), (2) duration (weeks), (3) position [the position of the electrode at the stumps end most effective in initiating regeneration in frogs is postaxial (Smith, 1974) This is the region of peak endogenous currents in salamanders (Borgens *et al.*, 1977a)], and (4) field strength. The contact area beneath the exposed portions of the wick electrode at the stump's end was approximately 1 mm^2, an area of about 0.01 cm^2. Thus the 0.2 μA of total current that was pulled through this area resulted in a current density of about 20 $\mu A/cm^2$. This level of current is comparable to the natural levels measured in regenerating salamanders.

In summary, it appears that if one artificially enhances the density of current traversing the stump's core tissues in nonregenerating adult frogs, limb regeneration can be initiated. Moreover, this suggests that in frogs, the natural level of current traversing this area is indeed insufficient (due to lymph space shunting) to initiate limb regeneration.

G. Are the Stump Currents Relevant to Limb Regeneration?

Recently Lassalle (1979, 1980) has argued that the surface potentials measured along regenerating amphibian limbs have nothing to do with the process of limb regeneration. In principle I would agree with Lassalle; the surface potential measurements are very unreliable indicators—and, as previously discussed, are not even the pertinent voltage gradients to be concerned with. The pertinent

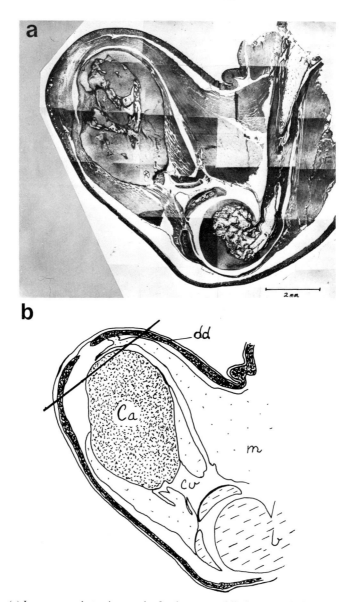

FIG. 6.   (a) Low power photomicrograph of a sham-treated limb stump. (b) Tracing of photomicrograph. Ca, callus; b, m, dd, and cv, as in Fig. 5. This grass frog was sacrificed 6 months, 3 weeks postamputation. From Borgens *et al.* (1977b).

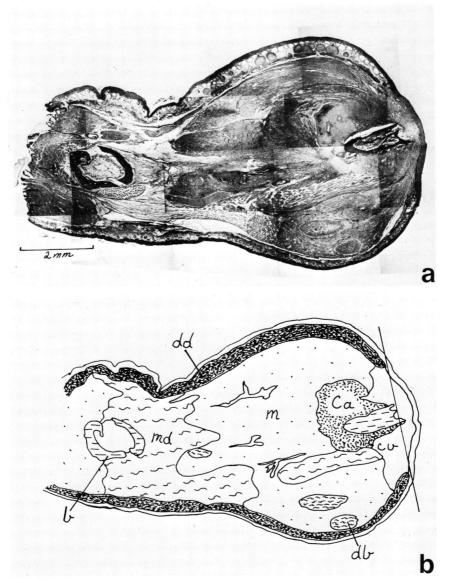

FIG. 7. (a) Low power photomicrograph of an anode-stimulated limb stump. (b) Tracing of photomicrograph. md, areas of muscle degeneration; db, pockets of cellular debris; Ca, b, dd, m, and cv, as in Figs. 5 and 6. This grassfrog was sacrificed 4 months postamputation. A study of other sections showed that the distal remnant of the radio-ulna was not connected to more proximal fragments. From Borgens *et al.* (1977b).

fields are *within* the limb, and these have yet to be directly measured.[2] However, Lassalle speaks to three other issues which he feels suggest that, overall, electrical factors are not important to limb regeneration. We should consider these opinions in some detail: (1) Lassalle says that healing skin wounds show the same pattern of surface potentials as regenerating limbs; thus, the potentials cannot be relevant to both processes. (2) Stump surface potentials disappear or invert during shedding of the skin, therefore, the potentials cannot help initiate limb regeneration. (3) Lassalle claims not to have observed surface potentials at all in larval *Pleurodeles* and axolotls; they appear only after metamorphosis in *Pleurodeles* or artificially induced metamorphosis in the axolotl. He goes on to cite the work of others that suggest larval anurans (good regenerators) have no transcutaneous potentials across their skins.

1. With regard to Lassalle's first criticism: skin wounds in most all animals studied to drive steady current out of damaged areas. This does result in the wound being electropositive with respect to adjacent uninjured areas. However, what Lassalle appears to misunderstand is the limb circuit (Fig. 3a). He treats the skin, the epidermis, and the limb stump itself as separate entities, forgetting that not only will the skin battery drive charge out small holes in itself—but also charge will be driven out a large hole as in the case with an amputation. In the latter case, this flow of charge will produce a bulk electric field within the stump. Small holes in the skin and epidermis will involve rather localized charge movement—and localized reparative processes. The role of ionic current in these two events need not be mutually exclusive as Lassalle would lead us to believe. Quite to the contrary, some evidence suggests this charge movement is relevant to wound repair as well as limb regeneration (see Section V).

2. During the molting period, a variety of events occur which would certainly affect the external surface resistances between the two contact electrodes Lassalle and others have used to measure translimb potentials. There are changes in the skins permeability to water, and a net loss of Na^+ through the skin that are concomitant with the molt (reviewed by Larsen, 1976). In fact various investigators have used the increase in electric conductivity of the bathing media caused by the molt as an index of these two processes (Jorgensen, 1949; see also Larsen, 1976). This fact alone helps to explain Lassalle's observations. An increase in conductance of the surface moisture of a salamander's limb would be synonymous with a decrease in its resistivity (they are inversely proportional). A large drop in surface resistivity would be sufficient to significantly lower the

---

[2]However, one can estimate the approximate magnitude of these fields. If 100 μA/cm^2 of current traverses the stump tissues with a resistivity of approximately 1000 Ω cm (Schwan, 1963) then the resultant field (within the stump) would be on the order of 10 mV/mm.

voltages measured between the contact electrodes on the outside of the limb. Add to this other complications caused by the molt such as glandular secretions, the formation (and separation) of a double skin layer, and the production of a molting "slime" immediately before shedding (Larsen, 1976)—and it is no surprise that Lassalle measured large decreases and transient inversions during the molt period. However, these molt complications only affect the measurement of *surface* potentials which still must be regarded as not especially pertinent to regenerative changes occurring within the stump. One might ask, however, if there are alterations in the natural voltage produced *across* the skin during a molt; this would affect the magnitude of current driven out of the stump's end, and the pertinent fields produced within it. Over a period of about 3 to 5 hours during slough formation there is a slow decline, and then increase in, the active Na^+ flux across the skin (measured as $\mu A/cm^2$ of active Na^+ current in voltage clamped skin. (Larsen, 1976).

Such a short decrease of about 40% in skin battery output is well within the normal range of variation measured in stump currents in regenerating salamanders (Borgens *et al.*, 1977a). Since this is a very transient phenomena and the periods of time between molts can be considerably variable (from a few days to a few weeks) there is no reason to suppose the transient variations in stump currents during a molt would affect current-dependent steps during the process of regeneration.

3. Lassalle's last criticism stems from his inability to measure surface potentials in the axolotl and in *Pleurodeles* larvae and his suggestion that skin potentials should be lacking in tadpoles. First, stump currents have been measured in axolotls (Vanable *et al.*, 1980), comparable to other salamanders. The peak current densities do decline more rapidly than those measured in newts, but axolotls regenerate more quickly. All that is necessary for an amphibian to drive such stump currents is an inwardly positive transcutaneous potential difference. This property of the skin is very pervasive among amphibia (Kirschner, 1973; Bentley, 1975). Lassalle's inability to measure voltages on the outside of some urodele stumps is probably indicative of the capricious nature of surface potential measurements themselves, and their inability to resolve smaller voltage gradients. Concerning anuran larvae: there is a report in the literature suggesting they do not possess an inwardly positive voltage across their skin (Taylor and Barker, 1965). These measurements were largely performed on *excised* skins that were clamped in what were apparently "Ussing chambers," and so damage to the delicate larval skin cannot be ruled out. When care was taken to reduce the effects of such "edge damage," transepitheleal potentials could be measured across the skin of larval anurans (Cox and Alvarado, 1979). Recently McCaig and Robinson (1980) report that in *Xenopus,* a trancutaneous potential difference is present during most of development. It is present as early as neural tube

formation. These measurements of skin potential in intact larvae were made using microelectrodes, and are not complicated by possible damage to the delicate skins.

In larval amphibians as in fish (Kirschner, 1973) inward charge movement can also be mediated by the branchiae. Apparently the gills of larval *Xenopus* also pump charge inward through the branchiae as well as the skin, and this flux seems to be a major component of a net current measured *leaving developing limb buds* as well (K. R. Robinson, personal communication). This is an interesting finding (measurements which I have recently confirmed studying the developing hindlimbs of axolotls). Net current leaves the limb placode, increases in intensity, and localizes around the bud as the bud emerges. Thus there is a parallel between current leaving developing limbs as well as in regenerating ones. Moreover, it makes no difference to limb tissues what the major source of this outcurrent is (the skin or the gill). (It would be interesting to see if inhibiting this current in developing limb placodes would inhibit the development of the limb.)

Overall it seems that Lassalle has performed no experiment, nor has he put forth any argument that detracts from the thesis that the internal fields associated with skin generated current are relevant to amphibian limb regeneration. On the other hand the evidence supporting its critical role is this: (1) The appearance and decline of endogenous stump currents in newts and axolotls roughly correlate with the first signs of limb regeneration. (2) Any one of four different methods used to radically reduce, inhibit, or reverse the flow of current through the stumps of a variety of salamanders also significantly retards the limb regeneration process, causes it to be grossly abnormal, or completely inhibits it. (3) Adult frogs that do not regenerate their limbs have stump currents that are shunted around (but not through) the core tissues of the stump from where the blastema arises. This shunting is caused by the subdermal lymph spaces present in adult frogs but not present in their larvae, or in salamanders or newts. Moreover, the appearance of these lymph spaces during larval metamorphosis in frogs coincides with that position of the limb where regenerative ability is lost. (4) If the deficient fields within an adult frog's core stump tissues are enhanced by artificial means, regeneration is initiated. If they are reversed, the limb fails to regenerate.

In summary then, the available experimental evidence supports the notion that stump currents, like nervous tissue (Singer, 1965), may be a necessary component of limb regeneration.

H. Stump Currents and the Regeneration Literature

Since the early part of this century, many experiments have been performed which either inhibit the regeneration of urodele limbs, induce supernumerary limbs, or initiate limb regeneration in postmetamorphic frogs. It is interesting to review some of these experiments in light of our new understanding of the role of skin-generated stump currents.

Many students of regeneration have been interested in the polarized nature of regenerating limbs. Segments of limbs were removed, turned 180°, and regrafted to the stump (Graper, 1922). Limbs were also notched, producing an open expanse of epithelium in various locations on the limb (Della Valle, 1913; Eiland, 1975), intact limbs were ligated (Kasanzeff, 1930), or segments notched and ligated (Della Valle, 1913) to produce new surfaces from which regeneration could occur (reviewed by Rose, 1970). In these experiments, mutiple surfaces on a limb covered only with a wound epithelium usually produced multiple regenerates. The emphasis here was that no matter what the orientation of a limb segment, or surgically exposed area, distal structures always formed. However, it is of interest to ask *why* multiple structures formed at all. I have suggested that to initiate limb regeneration, one may need an outward flow of skin generated current, i.e., an electrical "leak" through a low resistance epithelium. What all of these experiments have in common is that exposed limb tissue, covered only by a wound epithelium, was either purposely (or inadvertently) produced by the procedure. For example, by reversing a limb segment end for end (180°), the result would be that the attached areas would be of different diameters, and several patches of stump tissue covered by an epithelium would be produced. Current would be expected to be driven out of these skinless areas, at places where regenerates may eventually be produced.

On the other hand, a graft of *whole* skin across the stump face is sufficient to suppress limb regeneration in urodeles. Here my suggestion would be that the relatively high electrical resistance of skin (when compared to a wound epithelium) would be expected to greatly reduce the outflow of stump current, and to inhibit regrowth of the limb. Mescher (1976) has shown that wherever such a skin flap is incomplete (by accident or design) at the periphery of the stump face, a regenerate will form here, growing out at an angle to the long axis of the stump. Current would likewise be expected to leave this incompletely grafted area.

The electrical properties of normal amphibian skin may also help to explain several interesting observations made by Charles Thornton. He removed a cuff of forelimb skin from *Ambystoma talpodeum,* and replaced it with an autograft of head skin. Once this had healed properly, the limb was amputated through the grafted area. No limb regeneration was observed (Thornton, 1962). It is known that the transcutaneous potential difference does vary in its magnitude in different portions of the body in frogs (Fontas and Mambrini, 1977b). Could head or belly skin in certain urodeles possess a less powerful battery, and fail to produce a sufficient field within the stump to initiate regeneration? An alternate explanation may be that the wound epithelium covering the stump face which was derived from head skin may be electrically less leaky than the normal wound epithelium derived from limb skin.

The thesis that a skin generated outwardly flowing stump current helps initiate limb regeneration also provides an explanation for the many (apparently unrelated) methods that have been successful in initiating limb regeneration in adult frogs.

Polezhaev (1972) reviewed many early reports that a variety of noxious stimuli to the stump end in postmetamorphic frogs will initiate regeneration. These procedures ranged from simple pin pricking, to the topical applications of caustic agents such as iodine and nitric acid. He suggested that these treatments were successful because they prolonged a phase of dedifferentiation of the terminal stump tissues, and allowed regenerative processes to commence, instead of the precocious scar tissue formation that normally occurs. These treatments may have kept the stump covering (be it whole skin or an epithelium) traumatized and electically leaky. It is probable that current would continue to flow through such chronically disturbed areas as in any wounded skin. This would at least increase the duration of current flow through the core tissues in frogs, perhaps compensating for weak densities traversing this area. Hypertonic salt treatments as reported by Rose (1942, 1945) may have had the same effect as well. However it is important to point out that high concentrations of Na^+ bathing the intact limb skin will also greatly increase the level of *current* driven out of the stump by the skin (Borgens *et al.*, 1977a). The transcutaneous potential in frogs (as well as most fresh water animals) is exquisitely sensitive to external Na^+ concentration, demonstrating typical saturation kinetics. That is, the active Na^+ current is dependent on the Na^+ concentration when it is dilute in the bathing media, but it has an upper limit. The K_m (the external concentration of Na^+ at which the Na^+ current is half maximal) for *Rana pipiens* is only 5 mM Na^+ (Brown, 1962; Kirschner, 1970). Thus high concentrations of Na^+ in the bathing media will boost the output of the skins battery to saturation levels. However, Rose's most complete regeneration did not occur in response to saturated salt solutions; 52 mM (0.3%) NaCl produced the best regeneration that Rose ever reported (Rose, 1944). In fact, quite remarkable regeneration of amputated fingers occurred in frogs in response to these treatments in which hypertonic salts were not used at all. It is difficult to see how an application of 52 mM NaCl could be construed as an irritant (this is less than the concentration of Na^+ in Ringers). However, again please note that this Na^+ concentration is 20- to 50-fold greater than the usual concentration of Na^+ in pond water (between 1 and 3 mM), the usual bathing media the skin is exposed to. This interpretation of Rose's results is strengthened by the report of Polezhaev (1972) that applications of 0.8% $NaHCO_3$ was sufficient to initiate limb regeneration in adult frogs. This is 95 mM in Na^+, about that of Ringers; once again, hardly to be considered as caustic.

The most remarkable degree of induced limb regeneration in adult frogs was reported by Schotté and Wilber (1958). They implanted supernumerary adrenal glands into the throat tissues of adult *Rana clamitans*. Schotté argued that the extra level of systemic glucocorticoids would inhibit the rapid formation of the connective tissue scar that always forms after amputation in frogs. This may well be true, however, glucocorticoids and mineral corticoids produced by the adrenals are also known to boost the production of the potential difference across the skin (Bishop *et al.*, 1961; Myers *et al.*, 1961) and as a result, the level of current driven out the stump's end.

One of the well-established prerequisites for limb regeneration in amphibians is an adequate nerve supply to the extremity (Singer, 1965). If one reduces the numbers of nerve trunks servicing a salamander's limb by selective denervation, one can reduce, even abolish, the regenerative process (reviewed by Singer, 1952). One can quantitate this by comparing the cross-sectional area of the stump to the cross-sectional area of the severed nerves contained within it. Moreover, those vertebrates that show a weak regenerative response [adult frogs (Singer *et al.*, 1967), lizards (Zika and Singer, 1965), and mice (Rzehak and Singer, 1966)] are also deficient in the "axoplasmic ratio." Singer (1954) showed that if one surgically reroutes the sciatic nerve of the hindlimb to the forelimb stump in postmetamorphic frogs (increasing the nerve supply) regeneration can be initiated. When we induced a measure of limb regeneration in adult frogs by an imposed electric current, we observed a striking hypertrophy of nerve within the newly regenerated structure—up to 20% of the new tissue, in fact, was nerve (Borgens *et al.*, 1977b). We have suggested that a possible mechanism of effect is that the imposed current first acted on nerve tissue within the stump to increase its amount (Borgens *et al.*, 1977b, 1979b) and rate of growth (Borgens *et al.*, 1979a). In other words, it is possible that we achieved an electrical hyperinnervation of frog stump tissue, the effect being the same as Singer's surgical hyperinnervation. This notion also suggests that an analogous mechanism may operate during the normal regeneration of salamander limbs. That is, skin-generated electrical fields within the stump serve to ensure a proper innervation sufficient to support limb regeneration. This notion also directs our attention to a basic question: What are the target(s) of the fields within the limb?

I. CELLULAR TARGETS OF THE FIELDS

If we wish to speculate which tissues within the amphibian stump are most responsive to either the natural or artificially applied electric fields, first we should estimate the magnitude of these voltage gradients. As mentioned previously, the current densities measured traversing salamander stumps, and those calculated to occur beneath the indwelling electrode in *Rana* stumps are quite comparable (see Section III,E). Thus, the electrical field within a salamander stump, and the fields produced in a more local area beneath the cathode in frog stumps should be comparable as well. In frogs we should expect current densities on the order of 20 μA/cm^2 traversing this area. Such densities should produce a field, given tissue resistances of 1000 to 5000 Ω cm, on the order of 2 to 10 mV/mm. This is a weak field, yet still within the low range of applied fields known to affect developing cells in culture (Hinkle *et al.*, 1981; Jaffe and Poo, 1979; Jaffe and Nuccitelli, 1977). If we consider the *geometry* of the various cells within the stump this will give us more insight to their ability to respond to natural or applied weak electric fields.

The geometry of elongate cells such as nerve and muscle suggests they would be more plausible target cells than small ($\cong$ 30 μm) isodiametric cells such as fibroblasts, osteoblasts, and the like. Moreover nerves are especially attractive as possible targets for several reasons. Their great length, parallel to the long axis of the electrical field, would perhaps allow them to respond to a very shallow voltage gradient. Also, the large surface area of their dilated growth cone (hence large membrane conductance) and the low cytoplasimic conductance of the axon make them especially penetrable by naturally or artificially imposed fields (see this discussion in Jaffe and Nuccitelli 1977; Hinkle *et al.*, 1981). Lastly, they are well known to be an important prerequisite to the process of limb regeneration in urodeles (Singer, 1965). Moreover, there is a growing experimental literature demonstrating that the proliferation and even the direction of growth of neurities in culture can be controlled by weak applied fields. Suffice it here to suggest that a mechanism in which current affects limb regeneration could be that it first affects the direction (Hinkle *et al.*, 1981), rate (Borgens *et al.*, 1979a), and amount of growth (Borgens *et al.*, 1977b) of nerves within the stump—and that these nerves then exert their very special effect on limb regeneration.

The bulk of other mesodermal cells within the stump (with diameters of approximately 30 μm) would be less likely candidates for direct targets of the limb fields. Weak fields on the order of 10 mV/mm would be expected to produce a voltage drop of about 0.3 mV across such a cell. This is significantly less than the voltages known to influence their development, or movement (Hinkle *et al.*, 1981; Jaffe and Nuccitelli, 1977; Jaffe, 1979).

Besides elongate cells, cells or tissues that would have an especially high resistivity would be another likely candidate. For example, the wound epithelium that covers the stumps cut face would be such a tissue—and one known to be important to limb regeneration (Thornton, 1968). Initially this epithelium is only a few cell layers thick, but this proliferates to form a special structure, the Apical Cap Epithelium (morphologically resembling the Apical Ectodermal Ridge in developing limbs). In experiments where current is artificially pulled through frog stumps, a total voltage of about 1 V would be imposed across as little as a 10 cell thick wound epithelium. This would result in a voltage drop as much as a 100 mV per cell. Since many developing cells can be influenced by fields as small as 10 mV per cell or less (Jaffe, 1979), such an epithelium could be influenced by imposed fields. Additionally, we have observed that the current leaving sala-mander stumps declines with time after amputation. As the wound epithelium stratifies to an apical cap, the total resistance to the flow of skin generated current might be expected to increase as well—this may explain the steady decline in current density measured leaving a salamanders limb (Borgens *et al.*, 1977a). It should also be considered that the amount of intercellular space and the character of the electrical connections between individual epithelial cells (tight junctions) may change with time. Soon after amputation, the amount of intercellular space is enlarged in the wound epithelium, and the cells themselves are more loosely arranged when compared to the epithelial layers in normal limb skin (reviewed by Singer and Salpeter, 1961). This suggests that the wound epidermis would be expected to be electrically leaky during this early period. However, it should be kept in mind that the electrical milieu of an amputated stump is indeed complex, especially near the stump's end. As we shall see, truncated nerve and bony tissue produce their own characteristic electrical behavior which would be modified by the skin-driven current traversing the stump.

IV. Injury, Ionic Current, and Nerve Regeneration

The fathers of modern physiology and neurophysiology all knew that injury to a nerve produced a steady current that entered the lesioned area. They actually measured distally negative voltage gradients along the outsides of severed nerve trunks, and correctly inferred that current should flow into the damaged area. These measurements were largely carried out to understand the nature of the action potential; however Lorenti de Nó suggested that this flow of current into the injured area was probably involved with the regenerative process as well (see Borgens *et al.*, 1980). This old, half forgotten literature was displaced (or re-placed?) by the rapid increase in knowledge of nerve-muscle physiology brought about by the introduction of the microelectrode after the Second World War. The new emphasis was on the accurate recording and characterization of the electrical events (the action potential), occurring in milliseconds. Steady electrical events, lasting days or weeks, received little or no attention from electrophysiologists.

Recently we have begun reinvestigating the role of "injury currents" in nerve regeneration. The giant reticulospinal neurons of the larval lamprey was chosen as the model system. These giant axons are known to eventually regenerate across a complete transection of the spinal cord, coinciding with a return of behavioral function (Rovainen, 1974; Wood and Cohen, 1979, 1981). The cell bodies of these neurons are large (ca. 100 μm in diameter) and occupy charac-teristic positions within the lamprey brain (Fig. 8a). They can be easily visualized with a low power stereomicroscope (Rovainen, 1967). The axons (ca. 40 μm in diameter) descend the spinal cord in well-characterized tracts, and can

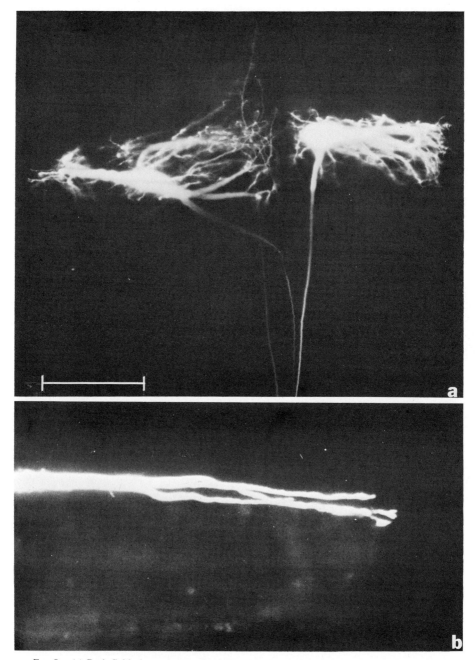

FIG. 8.   (a) Dark-field photomicrograph of two reticulospinal neurons within the brain of a lamprey larva. These cells were filled with a fluorescent dye (Lucifer yellow) by iontophoreisis,

also be easily visualized and impaled with a microelectrode (Wood and Cohen, 1979) (Fig. 8b). Once more, the entire, intact CNS can be removed to a simple organ culture and is viable for periods of about 1 week (see Borgens et al., 1980). Using the vibrating probe system, we characterized the particularly intense currents of injury that entered the cut face of completely severed lamprey spinal cords in organ culture. All of the observations I discuss below were made on that segment of the spinal cord in continuity with the brain (the proximal segment).

A. Intense Ionic Currents Enter the Severed Spinal Cord of Lamprey Larvae

Immediately after transection, current densities of about 0.5 to 0.8 mA/cm^2 were measured to enter the cut face of the spinal cord. Within an hour these currents decline to about a quarter of their initial values, and by 2 days posttransection they decline further to a stable level of about 4 μA/cm^2 (Figs. 9 and 10). We measured these incurrents for the next 4 days until the experiments were concluded (at a time where the CNS preparations were still viable and electrophysiologically active) (Borgens et al., 1980). Thus, large currents enter the area of injury for at least 6 days following transection. Since the vibrating probe resolves current flowing into the cut spinal cord *from* the medium, we attempted to find the area where intense currents were leaving this preparation, but were unable to do so. Small alternating regions of *outcurrents* and *in*currents (about 1 to 2 mm long) were found along the lateral margins of the spinal cord all the way to the brain. However these appeared to balance each other, and do not provide an excess of charge flowing out of the spinal cord to account for the intense currents entering the cut face. For technical reasons we were unable to investigate the venteral surface of the cord, or the brain itself. So at this writing, the place(s) where current emerges from the CNS preparation is still a mystery. We carefully scanned most of the surface of the cut face and were able to see a difference in the profile of current entering the transection. Incurrents were always greatest in magnitude directly adjacent to the end of a single severed giant axon or a group of them. For example, the pair of Mauthner cell axons can be easily identified descending the spinal cord. The left Mauthner cell projects its

---

visualized, and photographed using transmission fluorescent optics. The cell on the left is a Mauthner cell, whose axon crosses the midline of the brain, and projects down the contralateral margin of the spinal cord (which is not in the photographic field). The cell on the right is a Müller cell of the Bulbar group, whose axon projects ipsilaterally down the spinal cord. (The third axon visable is from another dye-injected cell body more rostral to these, and out of the photographic field.) Scale bar: 0.25 mm. (b) Dark-field photomicrograph of the terminal end of a regenerating Mauthner cell axon. Within the spinal cord, lamprey reticulospinal axons are of large size; the axon giving rise to these 3 neurites is approximately 45 μm in diameter.

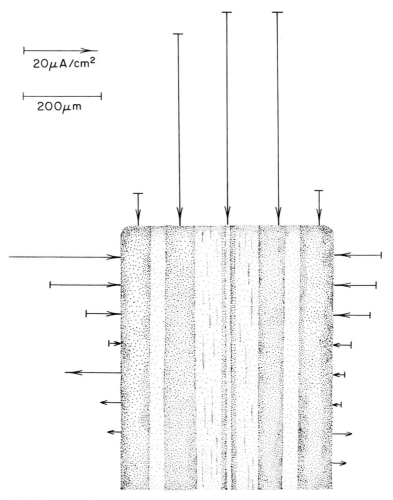

FIG. 9.   The pattern of current density entering and leaving the cut end of the proximal segment of
a lamprey spinal cord 4 hours posttransection. The arrow lengths are proportional to the current
density. The giant reticulospinal axons are shown as light bands parallel to the long axis of the cord.
From Borgens *et al.* (1980).

axon down the extreme right lateral margin of the spinal cord. The right Mauthner cell projects its axon down the left side. The other giant axons (originating
from Müller cells) are centrally located, clustered about the central canal (Wood
and Cohen, 1981). This enabled us to scan the pattern of current entering the cut
end of an individual giant axon (from a Mauthner cell), uncomplicated by the
close proximity of other giant axons. We always observed a sharp peak in current

density over its center. Additionally, since most of the surface exposed by the transection consists of cut axons rather than glial cells, it is reasonable to assume that most of the incurrent actually enters the cut ends of the axons themselves.

Since there is an apparent longitudinal flow of charge within (at least) the terminal portion of the severed cord, we performed some preliminary tests to determine what ions may be carrying this charge. By substituting or replacing ions in the defined medium in which measurements were being made, we determined that about 50% of the total incurrent was carried by Na^+ ions, and that much of the balance was probably carried by Ca^{2+}. The general features of this injury current and its ionic composition are supported by recent investigations of the membrane properties of regenerating axonal tips. Meiri *et al.,* (1981) have investigated changes in membrane conductance in the tips of regenerating cockroach giant axons, using conventional microelectrode technique. Immediately after these axons were severed, there was a decrease in the membrane potential and input resistance near the cut end. Two hours after transection, the resting potential repolarized slightly, but showed no real recovery to normal values for the next 2 days. After this time the membrane potential gradually recovered, taking about 10 days to reach normal values of about -80 mV. The input resistance recovered in a roughly similar fashion, commencing 2 hours after axotomy. The steep decline in the initially intense injury currents measured in lamprey spinal cord also takes about 1 to 2 hours, when it reaches substantially smaller and more stable values. One would expect that the intially intense flow of

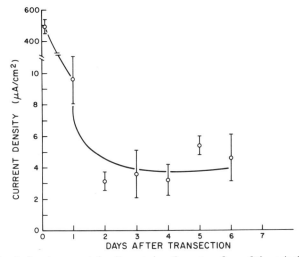

Fig. 10. The decline in current density entering the cut surface of the spinal cord (proximal segment) during the first week posttransection. Note the relatively stable level of current entering the cord's cut surface after the first day posttransection. From Borgens *et al.* (1980).

positive charge into the damaged axon would (1) maintain a depolarization of the membrane at the injured tip, and (2) occur in conjunction with a dramatic drop in input resistance. Moreover, these investigators determined the change in conductance for various ions with time after axotomy. Initially, the membrane at the cut end of a giant axon is electrically leaky, and is ionically nonselective. However, a recovery of ionic selectivity was observed to begin by about 4 hours after axotomy, with the greatest conductance being for Na^+ and Ca^{2+}.

B. WILL AN APPLIED ELECTRIC FIELD ENHANCE NERVE REGENERATION?

I have discussed the possibility that endogenous currents of injury may be a requirement for axonal regeneration. If this is so, can one artifically enhance the regeneration of nerves by impossing an electrical field across them? This apparently has been achieved in experiments where limb regeneration in adult frogs was initiated by implanted batteries. Cathodal stimulation of limb stumps in *Rana* resulted in an extraordinary amount of nerve (as determined by histology) within the regenerate structure (Borgens *et al.*, 1977b). When similar experiments were performed on large adult *Xenopus* (Borgens *et al.*, 1979a), gross amounts of nerve was found throughout the cartilagenous core of the regenerating structures. In both of these experiments nothing comparable was observed in sham-treated animals. One can suggest that artificially applied fields induced this hypertrophy of nerve (refer to Section III,H).

Moreover, there have been a variety of reports dating back to the early 1900s that growing nerves in culture can be affected by applied electric fields. However in large part these experiments are difficult to interpret for various reasons (reviewed by Jaffe and Poo, 1979; and Hinkle *et al.*, 1981). The best of this group was the 1946 paper by Marsh and Beams reporting that applied fields could affect the amount and the direction of growth of neurites emerging from cultured chick dorsal root ganglia (DRG). A modern reinvestigation of this phenomenon using explanted chick DRG (Jaffe and Poo, 1979) has largely confirmed that the rate, and proliferation of neurites will be enhanced toward the cathode in the applied electric field. No evidence of a "turning" or deviation of neurites toward the cathode was seen however, as Marsh and Beams originally proposed. Recently the galvanotropic responses of single differentiating *Xenopus* neuroblasts *in vitro* has provided a particularly elegant demonstration that indeed, elongating neurites will also turn in the applied field and actually grow toward the cathode (Robinson and McCaig, 1981; Hinkle *et al.*, 1981) (Fig. 11). In some cases, this will entail a nearly complete reversal in the neurities direction of growth. In the past, results such as these were controversial. Paul Weiss (1934) suggested that the galvanotropic responses of nerves in culture could be attributed to an electrically induced reordering of the substrate on which they were grown. As Lionel Jaffe has pointed out, modern investigations of electrically induced birefringence in a variety of elongated macromolecules suggests that such a reordering could never have taken place. To polarize or orient molecules such as collagen requires fields ranging from 10 to several hundred volts per mm. Cultured chick DRG responds to fields less than 100 mV/mm (see Jaffe and Nuccitelli, 1977). Moreover, in the experiments of Hinkle, McCaig, and Robinson, *Xenopus* neuroblasts were cultured on *tissue culture plastic,* rather than a collagen substrate.

Thus, there is a body of evidence suggesting that growing nerves (severed peripheral nerves in anuran limb stumps, neurites elongating in culture) are particularly responsive to applied electric fields. One might ask if the regeneration of CNS nerves can be enhanced by electrical means. Ernesto Roederer, Melvin Cohen, and I have recently completed experiments designed to test whether applied fields can enhance the regeneration of identifiable reticulospinal neurons within the completely severed spinal cord of larval lampreys.

We imposed a weak electric field across the completely severed spinal cord of eel larvae for the first 5 to 6 days posttransection (Fig. 12), and assayed for morphological and physiological effects of this treatment at about 55 days posttransection. Under normal conditions, there is only modest regeneration of these identifiable giant axons (from Müller and Mauthner cells) across the lesion by about 100 days posttransection (Rovainen, 1974; Wood and Cohen, 1979, 1981). Thus, our sham-treated control animals provided us an unambiguous baseline to compare the effects of our treatments. About 10 μA of total current was applied across the severed cord, using an external voltage source and indwelling ''wick'' electrodes (flexible salt bridges), with the cathode caudal, and the anode rostral to the lesion. This polarity of current flow (proximo-distal) and its associated electric field (distally negative) would be similar to that produced naturally in the extracellular space by the endogenous current of injury entering the cut face of the proximal segment of the spinal cord. We assayed the responses of this treatment morphologically (by the iontophoresis of fluorescent dye into indentifiable axons) and electrophysiologically (using extracellular stimulation of the spinal cord and simultaneous extracellular and intracellular recording of evoked action potentials). Clearly, distally negative applied fields enhanced the regeneration of these giant axons into and across the transection: (1) In electrically treated cords the greatest proportion of dye-injected axons was found within the area of the lesion or to have crossed it by about 58 days posttransection. Most of a comparable population of identifiable axons did not even reach the lesion in the sham-treated control group. (2) The morphology of most of the terminal ends of giant axons found in experimental cords was indicative of actively growing regenerating fibers (Fig. 13a). Most of the ends of control fibers were relatively undifferentiated morphologically with the appearance of axons that were in a less active state of growth, perhaps even in stasis (Fig. 13b). (3) In most sham-treated control cords, action potentials *did not* propagate across the lesion when the extracellular stimulus was applied to the cord (just behind the brain) and the

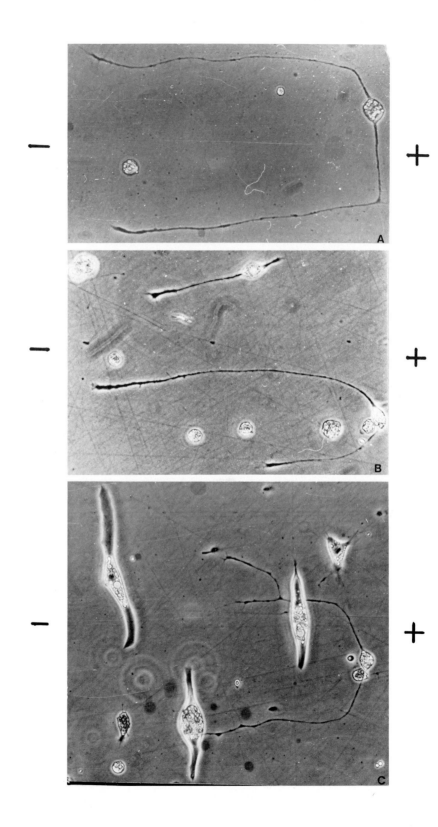

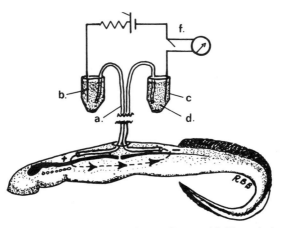

Fig. 12. Electrical manipulation of lamprey larvae. Larvae, 10–15 cm in length, were anesthetized and pined out in lamprey Ringers. An incision, 5 to 7 mm long, was made in the dorsal midline so that the spinal cord could be exposed and completely severed. This was performed at a point about midway between the head and the end of the tail. The end of one wick electrode (a) was routed through the small incision on the back of the animal, beneath the skin. The uninsulated open tip of this electrode was left in a position about 2 cm anterior to the spinal cord transection. Similarly, another wick electrode was placed about the same distance posterior to the transection. The animal was placed in a plastic container filled with aerated well water. The loose ends of the wick electrodes were immersed in two small plastic reservoirs (c) of lamprey Ringers (d), and connected in series with an external voltage source outside the animal's container via Ag–AgCl electrodes (b). The voltage source (e) was a 30-V battery (Eveready No. 413). A total of about 10 μA was driven through this simple circuit. The arrows indicate the direction of current flow through the animal. The circuit could be checked by inserting an electrometer at position f. Sham controls were treated in an identical manner except that the ends of the indwelling electrodes were not connected by any external power supply. From Borgens et al. (1981).

Fig. 11. The effects of an applied electric field on disaggregated and differentiating *Xenopus* neurons in culture. In all three photomicrographs, the cathode was immersed in the culture medium to the left of the photographic field; the anode to the right. Both electrodes were several cm distant from the cells presented here. In all cases, cells were cultured on nonpolarizable tissue culture plastic, and were experiencing fields of approximately 100 mV/mm. In (A), (B), and (C) note the exaggerated deviation of the elongating neurites toward the cathode. Note also in (B), upper neuron, that the neurite emerging toward the cathode has grown out much further than the neurite emerging toward the anode. In (C) the typical arrangement of both neuroblasts and myoblasts within the electrical field is shown. Neurites grow roughly parallel to the axis of the electrical field, and are directed toward the cathode, while differentiating myoblasts develop their bipolar axis of symmetry perpendicular to the direction of the electric field. This orientation, observed here in culture in response to an applied electric field, is similar to what is observed between developing neurites and myoblasts in the *Xenopus* embryo (see Hinkle et al., 1981). Photos courtesy of Kenneth R. Robinson.

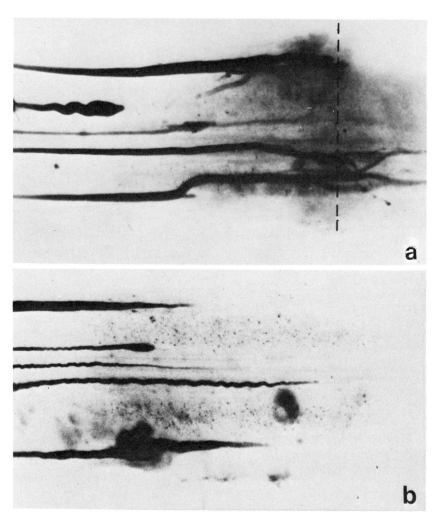

FIG. 13. Photomicrographs of spinal cord wholemounts. (Reverse positive prints made from dark-field photomicrographs of Lucifer yellow injected preparations.) In some of these photomicrographs, the lateral margins of the spinal cord are not visible. The width of most cords ranged from 0.8 to 1 mm. (a) A good example of axon regeneration in an electrically treated spinal cord 59 days posttransection. The Mauthner cell axon (at lower margin) gave rise to a large process which, like the Müller cell axon medial to it, coarsed throughout the lesion and traversed it. The contralateral Mauthner cell axon (upper margin) had many large and small diameter processes within the lesion, but these were not observed to cross it. One small unidentified axon (centrally located and faintly filled) traversed the lesion, while another giant axon ended in a giant swollen tip proximal to the lesion. (b) A typical sham-treated spinal cord 50 days posttransection. The lesion is not within the photographic field and was about 3 mm further distal (to the right-hand margin of this figure). Two centrally located Müller cell axons taper to fine sharp points, while another ends in a swollen tip. One Mauthner axon (at the upper margin) ends in a tapered point; the contralateral Mauthner axon ends in a swollen tip which gives rise to a tapered process. From Borgens *et al.* (1981).

extracellular recording electrodes were just caudal to the lesion—or if the direction of stimulation-recording was reversed. In most experimentally treated spinal cords action potentials were propagated in both directions across the lesion. (4) We used simultaneous intracellular and extracellular recording followed by a dye injection from the intracellular recording electrode to morphologically characterize those axons responsible for the extracellular spinal cord records. We observed that individual axons which did not propagate an action potential across the transection in both directions, indeed, did not cross the lesion. Rarely did such an axon even terminate within the area of the lesion. Conversely, all axons responsible for an action potential that was recorded to cross the lesion in both directions were themselves found within the lesion, or to have physically crossed it. In no case where an action potential propagated across the lesion did we observe the identified fiber to end proximal to the lesion itself.

Therefore, we have been able to build a strong case for the electrically enhanced regeneration of identified neurons in the severed spinal cord of this primitive vertebrate.

C. What Are the Physiological Effects of the Injury Current?

The immediate consequence of axotomy is to expose the axoplasm to the external milieu. Thus, ions and molecules are free to move down their concentration gradients because the electrochemical gradient generated by the membrane, now gone, cannot be maintained. Ca^{2+} and Na^{2+}, being in high concentration in the extracellular medium but very low in axoplasm, will enter the axon. K^+ will tend to be lost to the extracellular medium, and the axoplasm itself will (for a short while) be extruded from the cell (Lubińska, 1956; Zelená, 1969). Within a variable period of minutes to hours (depending on the nerve cell), apposition of old membrane, and synthesis of new, will once again seal off the end of the nerve from the external world. This will end the extrusion of axoplasm, and perhaps stop the free exchange of some molecules, certainly those of any size—but it will hardly affect the free exchange of ions because this new membrane is nonselective and electrically leaky. Our measurements (Borgens et al., 1980) and those of Meiri et al. (1981) show the membrane eventually recovers its selective character—selective to Na^{2+} and Ca^{2+} in particular. This Ca^{2+} conductance should lead to a great increase in free Ca^{2+} in the terminal portions of the axon. Moreover, small increases in internal Na^+ are known to cause a release of internally stored free Ca^{2+} (from the mitochondria) (Carafoli and Crompton, 1976), accentuating the steep Ca^{2+} gradient which should form after axotomy. Can we correlate these ionic changes with the great alterations in axonal morphology usually observed after a transection?

High internal Ca^{2+} is certain to cause an immediate dissolution of contractile and tubular elements in the axoplasm as Schlaepfer (1974) has suggested. Most instructive are two of his observations: If the Ca^{2+} ionophore A23187 is applied to peripheral nerves in culture, a degeneration of neurotubules and neurofilaments is initiated. This breakdown of the axoplasm was identical to what was observed in the degenerating distal segment of the nerve (that portion separated from the cell body) (Schlaepfer, 1977). Second, Wallerian degeneration of a distal nerve segment can be inhibited for some time by simply culturing them in Ca^{2+}-free media (Schlaepfer and Hasler, 1979; Schlaepfer and Bunge, 1973). Axotomy sometimes produces a clear, amorphic "pellet" at the axon terminus (Zelená, 1969; Zelená et al., 1968; Watson, 1976). This area of almost complete degradation of axoplasm is probably related to the Ca^{2+} component of the injury current (see Discussion in Borgens et al., 1980; and Watson, 1976). As Hamutal Meiri has pointed out to me, the collapse of the axonal cytoskeleton commensurate with changes in the fluidity of axoplasm brought about by increases in axoplasmic Ca^{2+} may additionally lead to the rapid apposition of wounded terminal membranes and a sealing of the cut end.

Such an increase in Ca^{2+} in the tips of nerves may also be involved, not only with degenerative changes, but with their regeneration and growth as well. The changes in membrane properties in giant cockroach axons that I have discussed (Meire et al., 1981) were, in fact, measured in *regenerating* axons (as confirmed by intracellular injections of cobalt). There were alterations in the physiology of the terminal ends of these nerves which also may relate to this increased Ca^{2+} conductance. For example, a shift from Na^+-dependent AP propogation, to Ca^{2+}-dependent AP propagation at the tip of the severed nerve was observed. Such Ca^{2+}-dependent AP propogation is observed during the normal development and differentiation of a variety of nerves (Spitzer, 1979), suggesting that these membrane properties are a general feature of regenerating and differentiating neurons. Moreover, terminal accumulations in cytoplasmic calcium may be a more general phenomenon in many types of elongating or apically growing cells, both in animals and plants: it is observed in the tips of elongating pollen tubes (Jáffe et al., 1975), during the establishment of polarity and germination of eggs of the brown alga *Fucus* (Nuccitelli, 1978; reviewed by Jaffe, 1979), during the acrosome reaction in sea urchin sperm (Schackmann et al., 1978; see Jaffe, 1980), and during pseudopod formation in amoebae (Nuccitelli et al., 1977). It is probably oversimplified to ascribe all of such cellular events soley to variations in cytoplasmic Ca^{2+}. However, it is certainly justified to underscore the *ionic* dependence of these events (such as cytoskeletal assembly and disassembly) as possibly being mediated by developmentally important ionic currents (see Tilney, 1979).

While there are certain to be ionic effects of this flow of charge into wounded nerves, there is likely to be voltage-mediated effects as well. The intense current flowing in and around nerves will produce an electrical field in the extracellular spaces, as well as form a voltage gradient within the axon itself. For example, we have calculated (Borgens *et al.*, 1980) that in the severed lamprey spinal cord, voltage gradients of 1–10 mV/mm could theoretically be produced by the current entering the cut axonal tip. Within hours after transection (and during the regeneration or early sprouting of nerves), there is an enormous accumulation of various cellular organelles at the growing tip. Mitochondria, smooth ER, and various vessicles increase in number on both sides of the transection. At least in the case of mitochondria, their increase apparently is related to a transport phenomenon, and *not* an increased synthesis at the tip (Zelena *et al.*, 1968; Zelena, 1969). We have suggested this accumulation may, in part, be a result of axoplasmic electrophoresis (Borgens *et al.*, 1980).

The voltage gradient *within* the axon will be distally positive, and thus negatively charged components would be expected to accumulate at the cut end, while positively charged components would tend to move toward the cell body. The movement of two intracellular dyes apparently behaves this way in cut nerves. We have injected Lucifer yellow (overall negatively charged) into giant axons. Once a weak dose of the dye has homogeneously distributed itself in the axon, the axon can be cut, and the preparations fixed within 5 minutes or so. What is usually observed is an intense accumulation of the dye at the terminus. Intracellular injections of positively charged cobalt (Meiri *et al.*, 1981), apparently behaves in an opposite fashion. In severed cockroach giant axons, injected cobalt does not stain the tips within the first 4 hours [a period when injury currents are most intense (Borgens *et al.*, 1980)]; only after a day posttransection can intracellular staining with cobalt permit a visualization of the tip. Moreover, there exists experimental confirmation that cytoplasmic electrophoresis can indeed occur. In the nurse cell–oocyte complex of *Cecropia*, a steady transcellular current maintains the oocyte cytoplasm some 10 mV more electropositive than the nurse cell(s) (Woodruff and Telfer, 1973, 1974; Jaffe, 1979). Fluorescently decorated molecules were injected into the nurse cell or oocyte, and their movement within this complex depended on their overall charge. An injected protein with an overall negative charge (rabbit serum globulin) moved across the intercellular bridge(s) from nurse cell to oocyte, but was unable to move in the opposite direction. Injected proteins with an overall positive charge (lysozyme) moved in the reverse direction, from oocyte to nurse cell (which is endogenously electronegative). However, if fluorescein-labeled lysozyme was treated in such a way to alter its charge (by methylcarboxylation of its amino groups), it behaved identically to the rabbit serum globulin. Injected proteins that were approximately neutral (hemoglobin and myoglobin) tended to move in either direction (Woodruff and Telfer, 1980). Thus it is conceivable that steady current flow into the ends of severed nerves may produce voltage gradients sufficient to move various components (molecules, even organelles) about within the axon itself. Voltage gradients in the extracellular space produced by the external flow of charge would be expected to move various components protruding from the axolemma as well. This will in part depend on the geometry of the spaces between injured nerves. The resistivity of interstitial fluid does not vary greatly among various animal groups (roughly 100 Ω cm). Thus for any given density of current flowing through the extracellular spaces between injured nerves, the intensity of the electrical field would be inversely proportional to the cross-sectional area of space the current is flowing through. Therefore large spaces between nerves (or other cells) would favor weak extracellular voltage gradients, and vice versa. However, it is possible that even very weak fields may electrophorese various macromolecules in the plane of the membrane (Jaffe, 1977). External fields when applied to embryonic muscle in culture will influence the distribution of fluorescently decorated membrane receptors (fluorescein labeled Con A or acetylcholine). Movement can be observed in fields as weak as 10 mV/mm (Poo and Robinson, 1977; Poo et al., 1978; Orida and Poo, 1978).

It is reasonable to argue that by enhancing the natural fields produced by nerve injury with an artificially imposed one, this may enhance the natural physiological responses of nerves to damage. The external fields produced by severing an axon are distally negative. Since all neurons tested grow or proliferate toward the negative pole of an applied field, the mechanisms of action in both cases may be the same. However, this is true only if the important developmental responses are mediated by the *external* electric field. Imposed currents penetrating the axoplasm driven by a distally negative applied field would be opposite in direction to the natural currents of injury. This could produce a rectification effect within the nerve which might dampen the early destructive effects of the injury current. In both cases, however, a more rigorous investigation of mechanisms is warranted.

Finally, there are a great many morphological and physiological changes that occur in neurons in response to damage (reviewed by Grafstein and McQuarrie, 1978; Cragg, 1970). As Veraa and Grafstein (1981) point out, the unknown ''signal'' initiating these events is one of the questions whose answer may provide important information about the mechanisms of axonal regeneration. I would suggest that a transcellular flow of ionic current induced by damage may just possibly be the early signal initiating regenerative responses in neurons.

V. Steady Current, Mammalian Regeneration, and Healing

A. MAMMALIAN REGENERATION

Complex vertebrates such as mammals usually do not regenerate body parts. However there are a few examples of mammalian regeneration; among them are the cyclical regeneration of antlers in deer (Goss, 1969), and the regeneration of holes punched in rabbit ears (Goss and Grimes, 1972, 1975). Even though neonatal opossum have been described as possessing a weak regenerative response (Mizelle, 1968), it now appears that this often cited example of mammalian regeneration may be in question (Fleming and Tassava, 1981). In order to restore the capacity to regenerate limbs, electric current has been imposed within the limb stump of rats (Becker, 1972; Becker and Spadaro, 1972; Libbin *et al.*, 1979). These experiments have resulted in claims of "partial limb regeneration" which have appeared both in the literature and in the popular press. However, a close inspection of the documentation reveals that nothing even approaching limbs has begun to appear. The major response to amputation has been the formation of curious epiphyseal plate-like structures, which in themselves are quite interesting. However, these calcifications are known to form within unamputated rat limbs in response to mild damage to the limb musculature or to the Achilles tendon. As Carlson has pointed out: "The formation of bony nodules of this sort or the reappearance of new epiphyseal plates on amputated long bones of young mammals is sometimes considered to be proof that an epimorphic regenerative process has occurred . . . these nodules form in the absence of the developmental events that are characteristic of epimorphic regeneration" (Carlson, 1978).

Moreover, a close inspection of Becker's (1972) and Becker and Spadaro's data (1972) shows that there is no difference in the "growth" responses elicited by implanted current generating devices producing total current on the order of 5 × 10^{-9} A, and implanted devices delivering 50- to 5000-fold less (which for all biological purposes is electrically quiescent). In the preliminary report these "low current devices" were in fact described as "control series" applications (Becker, 1972). This underscores that the responses in these seminal reports were probably innate and not even related to the application of an electric current.

Libbin *et al.* (1979) have attempted to duplicate these findings and are straightforward about the similarity between experimental and control treatments. They write "Indeed, there exists such a degree of interrelation between a) purely electrical responses, b) those responses attributable to mechanical factors deriving from the physical presence of the device within the limb terminus, and c) those responses initiated by loss or injury alone (innate responses) that an analysis of their unitary contributions cannot readily be made." However, these investigators still seem to believe that the cartilagenous nodules they observe in controls and experimentals still constitute evidence of "partial limb regneration." They further suggest the control group applications are not really control groups after all, but in reality are just another experimental group. If we do not consider these epiphyseal plate appearing structures, what we are left with is a confusing mixture of fracture callus, what appear to be bone chips (referred to by one investigator as anlagen of a supernumary radius and ulna), and no reliable indication of the exact plane of amputation. Lastly, these growths form within 3 days (at the earliest) to 7 days postamputation, a suspiciously short period of time to partially regenerate a limb. In essence, these studies essentially describe the bit of muscle regeneration, and proliferation of partially calcifying cartilage known to occur naturally in damaged muscle and bone. It is my opinion that these responses cannot be considered as electrically stimulated limb regeneration.

It is becoming clear, however, that mammals may have the ability to replace more of their extremities than is commonly believed. Scharf (1961, 1963) described the emergence of some very interesting outgrowths in response to the amputation of digits in young rats. Schotté and Smith (1961), however, vigorously argued that this growth response should not be referred to as "regeneration" (as the word is applied to amphibian regeneration), since there was no evidence of morphogenesis.

I have recently reinvestigated the capacity of rodents (weanling mice) to replace the tips of their toes (Borgens, 1981, in preparation). In some cases, the replacement of the foretoe tip, after amputation through the distal phalange is quite striking from external observation. As Schotté suggested, however, the histogenesis of this structure is different from the epimorphic regeneration of the urodele limb since there was no evidence of morphogenesis. The tissues that are replaced in the finger tips only arise from the same preexisting structure. For example, if no nail bed rudiment is left in the toe stump, then no nail reforms in the restored fingertip. However, what does reform can be quite organized. The distal phalange elongates to nearly normal proportion, the subcutaneous material is all present, the terminal fat pad, and very striking nails reform from only a tiny rudiment of the nail bed germinal epithelium left in the stump. There is no internal evidence of the typical dense fibrocellular scar (Schotté and Smith, 1959, 1961) and no external indications of the original plane of amputation. Amputations more proximal to the last interphalangeal joint almost always result in a blunt stump complete with callus formation and a fibrocellular scar. Though the process of limb regeneration in urodeles is different from the process of fingertip replacement in mammals, they do share several things in common. First, a functional replica of a multitissue structure has been reproduced in response to amputation. Second, in frog larvae undergoing metamorphosis and losing the ability to regenerate their limbs, a distal amputation will result in regeneration, but a more proximal amputation on the same limb will not (Schotté and Harland, 1943). Thus the regenerative ability is lost in a proximal–distal direction. Only very distal amputations in mice or childrens' digits result in regrowth.[3] Third, a flap of skin sewn over the stump in *all* cases inhibits the regrowth. And last, comparable densities of steady current (10 to 100 μA/cm^2) are driven out of the stump in children (Illingworth and Barker, 1980), mice (Borgens, unpublished measurements), and amphibians (Borgens *et al.*, 1977a). However, there is at present no direct evidence linking the flow of this charge to the process of fingertip regrowth.

It is plausible though, and tantalizing to speculate, that the reason why whole skin blocks these processes is that it blocks the current flowing out the stumps end (the electrical resistance of mammalian skin being on the order of 10 to 100 MΩ, a substantial resistance).

B. The Mammalian Skin Battery and Wound Healing

In the previous discussion, no mention was made of the battery which drives charge out fingertip stumps. This has not been directly tested, though it is almost certain that the source of the current is the intact skin of the stump (analogous to what has been described in amphibia). It is not well known to biologists that mammalian skin can act as a substantial battery in the same manner as amphibian skin. Even less well known is the fact that skin injuries in mammals will induce a steady flow of current out of the wound. The first to describe such a current of injury (in humans) was E. DuBois Reymond (1860). This was confirmed and extended by Herlitzka in 1910, who reported that about 1 μA was driven out of a small lesion made to the skin of his hand. Herlitzka's 1910 paper was apparently the last published measurement of mammalian wound currents until the recent report of current leaving childrens finger tip amputations (Illingworth and Barker, 1980).

In modern times the electrical properties of mammalian skin have largely been the domain of psychophysiologists who have reported measurements of transcutaneous voltages, but not current flow. Moreover, their emphasis has been on the relationship of skin potential to the physiology of sweating and emotional state (Edelberg, 1972, 1977; Stombaugh and Adams, 1971).

Only recently has there been a detailed investigation into the electrical properties of intact and wounded mammalian skin. Working on the cavy (guinea pig) and themselves, Barker *et al.* (1981) (Fig. 14) describe transcutaneous potential differences that range from 30 to 80 mV, internally positive with respect to the skins surface. The magnitude of the potential was quite depressed (less than 10

---

[3]This is apparently the reason no description of the replacement of mammalian digits (with the exception of children) has previously appeared in the literature. The older, and thorough, investigations of Schotté and Smith (1959, 1961), as well as recent observations of Neufeld (1980) have all dealt with amputations more proximal than the level of the last interphalangeal joint.

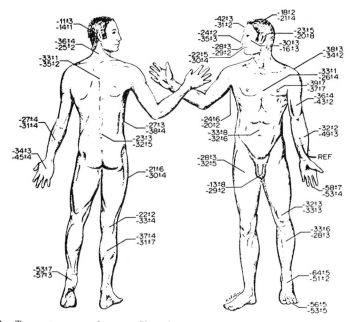

FIG. 14. Transcutaneous voltage profiles of man. Each set of data is the average (± SEM) transcutaneous voltage (in millivolts) measured on two separate investigators (upper and lower data points). The negative sign indicates the voltage *below* the subdermal reference (REF) level. (Transcutaneous potentials in amphibia, discussed in Sections II and III,C, are usually referenced to the skin's *exterior*. Thus, the *polarity* of voltages measured across the skin in humans and amphibians is the same.) (See Barker *et al.*, 1982.) Drawing courtesy of L. F. Jaffe.

mV) in hairy regions of the cavy; the higher values were measured in so-called glabrous (or gland-free) skin of the tarsal pads and the curious postotic bald spot, a centimeter-wide patch of skin caudal to the ear that is free of glands and follicles. In man, there was also a tendency for low transcutaneous voltages to be related to hairy regions of skin, but this was less marked than what was observed in the cavy. Additionally, transcutaneous voltages were substantially (and reversibly) depressed by applications of amiloride, indicating a Na^+-dependent skin battery similar to that found in amphibians. The ability of cavy and human skin to drive current through wounds was also investigated. About 10 μA/cm of wound perimeter was driven by the epidermal battery. This flow of current produces a lateral voltage gradient that is quite steep (200 mV/mm) near the wound but that falls off rapidly lateral to it (about a 3-fold decrease per quarter millimeter). These authors believe that these voltage gradients may play a role in the wound healing process. I will summarize here their main argument. First, drying the skin substantially slows the wound healing process; it also restricts the flow of current into the wound from its margins. Second, they point out that the electrical fields they measured occupy a position above the main intercellular seals in the living layers, that is, the outermost layers of the living epidermis. It is these cells which are the first to move into a mammalian epidermal wound. (Mammalian skin wounds are closed in this way by a migration of cells down into the open lesion.) Barker *et al.* (1981) also note that the range of cell mobilization after wounding roughly correlates with the rapid fall off of voltage gradients lateral to the wound. Cell mobilization begins within a few seconds to a few minutes after wounding, but only in an area about a tenth of a millimeter from the margin of the wound. Only a few tenths of a millimeter further, and the latency is on the order of 0.5 hour to an hour after injury. Finally I note one clinical report (Wheeler *et al.*, 1971) suggesting that artificially applied electric fields speed the healing of decubitus ulcers, a particularly intractable skin lesion. Thus it is suggested that the steady fields near an epidermal wound galvanotactically guide the epidermal migration which serve to close the lesion. As Barker *et al.* (1981) point out, although the signals and vectors which serve to direct this epidermal movement are entirely unknown their notions are testable.

Lastly, most of us that have studied transcutaneous voltages, injury current, and their role in wound healing and regeneration have wondered: what is the primary physiological role of the skin battery? In amphibians it has always been suggested that the Na^+-dependent transcutaneous potential subserves Na^+ up-stake from their dilute pond water habitats, i.e., the "skin battery" is primarily involved in osmoregulation. However, *dietary* Na^+ is sufficient to maintain salt balance in amphibians chronically maintained in a Na^+-free media (McAfee, 1972). Moreover, the inward Na^+ pump across the skin still operates in frogs that exist naturally in hypersaline environments (Gordon *et al.*, 1961). Clearly if the skin's role was primarily involved with regulating Na^+ uptake, one would expect a substantial decrease in the skin's uptake of Na^+ (mirrored by a decline in transcutaneous potential) under such conditions. This is not the case however. In *Rana cancrivora* and *R. tigrinum* (frogs that tolerate nearly brackish water conditions), the primary means of osmoregulation is by increasing plasma urea (much in the same way as elasmobranchs). Finally, it is a mystery what role a similar Na^+-dependent skin battery plays in mammals (see Barker *et al.*, 1981). In all animals, we would suggest, this transcutaneous voltage is primarily involved in wound healing and mediating regenerative events.

C. ENDOGENOUS ELECTRICITY IN BONE

Bone is a complex and dynamic living tissue. During the normal lifespan of an individual, bone not only grows (during the formative years) but it has the ability to repair itself in response to injury, or remodel itself continually in response to changes in mechanical stress or strain. This ability to adaptively remodel in response to stress has many times been referred to as Wolf's law, and is not only involved with changes in normal bone morphology—but in fracture repair as well. For example, when a fracture is misaligned, excess stress on the concave portion of the union will result in bone deposition, while bone will be resorbed in relatively stress free areas. The result: the bone will straighten with time during the healing process—and will support load as before the injury (see Weiss *et al.*, 1980).

It has become apparent over the last 25 years that the signals that mediate the remodeling, repair, and possibly even the growth of bone, may be electrical in nature.

Measurements of voltage gradients in intact, stressed, and unstressed bone, growing bone, and injured bone suggest three fundamental electrical properties:

1. Short lived potential differences (on the order of mVs for a few tenths of a second) can be induced by mechanical stress. Areas of compressive stress are electronegative with respect to the unloaded portion of a long bone (Fukada and Yasuda, 1957; Cochran *et al.*, 1968; Bassett, 1971). The generation of these potentials appears to reside in the organo-mineral composition of bone, and is apparently unrelated to whether its cellular components are alive or dead (Yasuda, 1974; Williams and Breger, 1975). Dry bone can act as a piezoelectric transducer (converting mechanical stress to an electrical signal). This property appears to reside not in its crystalline hydroxyapatite structure, but its collagen component (Bassett, 1971; Steinberg *et al.*, 1977; Korostoff, 1977). In living (hydrated) bone, stress-generated potentials may only in part be due to bone's piezoelectric properties (Williams *et al.*, 1979; Eriksson, 1974). During normal movement, bone is intermittently stressed and relaxed; this produces a movement of body fluids through the small channels in the bone texture. Electrical potential differences (streaming potentials) are generated under such conditions because the channel walls preferentially bind ions of one sign as fluid is forced through the bone from the compression side to the tension side. In fact, new evidence supports this mechanism as the most important source of electric signals in stressed bone under physiological conditions (W. S. Williams, personal communication).

2. Steady potential differences are also observed in living bone during growth. Areas of the long bone just below the epiphyseal plate were negative with respect to the epiphysis itself or to the cortical bone of the diaphysis. This area is associated with active cellular proliferation and remodeling during the elongation

of long bones. Moreover, these potentials dissappear if the bone is not viable (Friedenberg and Brighton, 1966; reviewed by Weiss *et al.*, 1980).

3. An injury to a bone also induces voltage gradients, negative in the area of the lesion with respect to undamaged portions. These steady potentials are also dependent on the viability of bone (Friedenberg and Brighton, 1966; see also Weiss *et al.*, 1980). As is the case with the ''growth potentials'' discussed above, we do not yet have a clear understanding of the source of these voltages.

Taken as a group, all of these measurements seem to suggest one central theme: areas of active growth, repair, or bone apposition during remodeling are electronegative with respect to less active areas. Thus, we can infer that current should enter actively growing areas of bone; however, no such measurements have been made. Moreover, there is little direct evidence that this electrical behavior is a cause or consequence of growth. Support for a critical role of this current flow comes from studies where artificially applied current initiates bone formation at the cathode, and bone resorption at the anode (Yasuda *et al.*, 1957; Yasuda, 1974; Bassett *et al.*, 1964; Friedenberg *et al.*, 1974). However, as pointed out by Jaffe and Nuccitelli (1977) it is unclear whether these effects are due to the imposed electrical fields or to the presence of electrode products. Brighton and Friedenberg (1974) suggest alterations in oxygen levels and pH in the immediate vicinity of the electrodes may indeed mediate these processes. So, at this writing, it is still unproved that direct current effects on bone remodeling (or enhanced fracture healing discussed below) are in fact electrically mediated. There is still much need in this rapidly growing area of interest for some critical experimentation into the mechanism of action of applied DC fields, and a determination of the tissue origins of steady electrical potentials in living bone.

D. ELECTRICALLY ENHANCED FRACTURE REPAIR

As with most of the phenomena that I have discussed in this text, studies into the therapeutic application of current to heal bone fractures actually began in the middle of the last century (see Spadaro, 1977). In modern times, such investigations apparently began with Friedenberg and Brighton and their co-workers (1971). Today, though novel, electrotherapy is gaining acceptance as a procedure for treating chronic bone fracture nonunions (see Fig. 15). Such intractible lesions are due to clinical conditions as varied as diabetes and cogential pseudoarthrosis (Patterson *et al.*, 1980), and electrical therapy is clearly a treatment of last resort. These nonunions apparently respond with a high degree of success to applications of a cathodal current or to the application of electromagnetic fields (Hinsenkamp *et al.*, 1978; Bassett *et al.*, 1977). (For a brief but excellent review

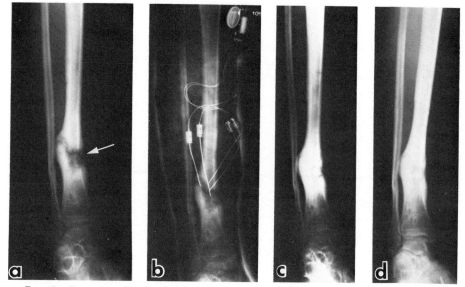

FIG. 15.   Treatment of a chronic bone fracture nonunion with steady electric current. (a) The arrow points to a nonunion of the right tibia in a 18-year-old girl with chronic osteomyelitus. This roentgenogram was made at the time of electrode insertion, 1 year after the original injury had remained unhealed. (b) Same as (a): note the four cathodes in place at the lesion site. (c) This roentgenogram was made at the time the electrodes (seen in b) were removed, after 12 weeks of constant direct current therapy. Note the early healing of the fracture. (d) This roentgenogram was made 12 weeks after electrode removal (in c): note the healed nonunion. Photos courtesy of C. T. Brighton.

of clinical applications, see Brighton, 1981. See also Brighton *et al.*, 1979; Bassett *et al.*, 1979; Weiss *et al.*, 1980.)

In studies where direct current has been used to enhance the healing of bone or cartilage in animals and humans, it is quite apparent that these responses are due to the implantation of cathodes. Anodal current initiates the destruction of hard tissue.[4] In most instances, the negative electrode (usually stainless steel or titanium in composition) is surgically inserted directly into the fracture gap. The anode is in contact with the adjacent soft tissues, or is externally in contact with the skin's surface. Steady current (on the order of tens of microamperes) is pulled into the lesion. However, one cannot estimate current densities, or the voltage gradients that the responding cells perceive, due to the great variation in geometry of the fractures and the varied numbers of indwelling electrodes that have been used.

[4]Support for this observation also comes from our studies where 0.2 μA of current was delivered to adult frog stumps via salt bridges (wick electrodes). Cathodal current initiated elongation of bone, anodal current, the near complete resorption of the forstump bones (Borgens *et al.*, 1977b).

Comparable bone lesions will also respond to the application of electromagnetic fields (Bassett *et al.*, 1974). An applied field of this type will induce current flow perpendicular to the long axis of the lines of magnetic force. This current flow must, in principle, be an alternating current, though several investigators claim that the electromagnetically induced current flow is DC in nature. The pulse parameters are set in such a way as to accentuate the duration of one polarity of the alternating peaks. Once again, we can not reliably estimate the effective current densities or internally imposed electrical fields. In fact, I know of no convincing electrophysiological measurements which provide details of the electrical environment imposed on the target cells by any of these procedures. Additionally, pulsed AC and DC stimulation appear to induce a measure of responses in bone (cataloged by Spadaro, 1977; see also Kenner *et al.*, 1975).

It is possible that all of these apparently unrelated methods are successful because they do modulate, in some way, the natural electrodynamics of hard tissue. It is reasonable to speculate, though, that the ratio between bone resorption and bone deposition in response to physical activity or injury may involve both its piezoelectric and direct current generating capabilities.

VI. Summary

1. In biological systems the steady unidirectional flow of ions constitutes an electric current. The flow of such current *into* an injury to a single cell, or *out of* the damaged region of an animals integument, may be one of the earliest evolutionary response—and signal—to wounding.

2. In tailed amphibians, skin driven current is pulled through the core tissues of a regenerating limb stump. In nonregenerating adult frogs most of this flow of current is shunted around but not through core tissues of the stump by the subdermal lymph spaces (present in adult frogs, but not in salamanders and newts). These lymph portals make their appearance in frog larval development coincident with the loss of the ability to regenerate their limbs.

3. If this flow of endogenous current is inhibited or reversed in regenerating amphibians (by a variety of means) then limb regeneration is inhibited, retarded, or caused to be abnormal.

4. A current, comparable to that traversing the core tissues in salamander stumps, can be pulled through the core tissues of an adult frog stump by artificial means. Limb regeneration is thereby initiated. If the polarity of this applied current is reversed, the limbs do not regenerate.

5. Mammalian skin possesses a "skin battery" similar to amphibians. Moreover, injured mammalian skin drives current out of the damaged area. This flow of current is associated with substantial voltage gradients adjacent to the wound, and may be implicated in playing a role in epidermal wound healing.

6. In the completely transected spinal cord of lamprey larvae, intense currents of injury enter the cut face of the cord. Peaks in current density can be measured adjacent to the bore of giant, identifiable reticulospinal axons contained within. Thus, current is driven into the severed ends of nerves.

7. If the electrical field associated with this injury current is enhanced by artificial means, one observes an enhancement of axonal regeneration into and across the lesion in completely severed lamprey spinal cords.

8. Bone not only produces injury potentials near a damaged area, but also produces small potentials in response to load. These electrical properties may help to explain observations of electrically enhanced fracture healing, and the observations that bone morphology can be modified by a variety of invasive and noninvasive electrical applications.

9. The mechanisms of action by which natural and applied electrical fields influence cells and tissues is still poorly understood, but probably involves the following: (a) the gross morphology of cells, the character of their cytoskeleton, and many cell-specific abilities may be under ionic control. Thus, any of these may be altered by steady ionic transcellular current; (b) the rate and direction of cell growth or the elongation of cell processes can be influenced by artificially applied electric fields; (c) various components within cytoplasm, and various components protruding from the cell membrane can be moved by natural or applied fields.

10. All of these observations are only a small portion of the emerging picture that steady electrical events are involved in a general way with the developmental physiology of plant and animal cells.

ACKNOWLEDGMENTS

I am grateful to Lionel Jaffe, Joseph Vanable, Jr., Kenneth Robinson, Hugh Wallace, David Redden, Wendell Williams, Carl Brighton, and Hamutal Meiri for their careful reading of this manuscript, and their suggestions.

REFERENCES

Ambrose, E. J. (1965). "Cell Electrophoresis." Little, Brown, Boston, Massachusetts.

Barker, A. T., Vanable, J. W., Jr., and Jaffe, L. F. (1982). *Am. J. Physiol.* in press.

Bassett, C. A. L. (1971). *In* "Biochemistry and Physiology of Bone" (G. H. Bourne, ed.), 2nd Ed., pp. 1–76. Academic Press, New York.

Bassett, C. A. L., Pawluk, R. J., and Becker, R. O. (1964). *Nature (London)* **204,** 652–654.

Bassett, C. A. L., Pawluk, R. J., and Pilla, A. A. (1974). *Ann. N.Y. Acad. Sci.* **238,** 242–261.

Bassett, C. A. L., Pilla, A., and Pawluk, R. J. (1977). *Clin. Orthop.* **124,** 128–143.

Bassett, C. A. L., Mitchell, S. N., Norton, L., Caulo, N., and Gaston, S. R. (1979). *In* "Electrical Properties of Bone and Cartilage. Experimental Effects and Clinical Applications" (C. T. Brighton, J. Black and S. R. Pollack, eds.), pp. 605–630. Grune & Stratton, New York.

Becker, R. O. (1960). *IRE Trans. Med. Electron. ME* **7**, 202–207.

Becker, R. O. (1961). *J. Bone Joint Surg.* **43A**, 643–656.

Becker, R. O. (1972). *Nature (London)* **235**, 109–111.

Becker, R. O. (1974). *Ann. N.Y. Acad. Sci.* **238**, 236–241.

Becker, R. O., and Spadaro, J. A. (1972). *Bull. N.Y. Acad. Med.* **48**, 627–641.

Benos, D. L., Simons, S. A., Mandel, L. J., and Cala, P. M. (1976). *J. Gen. Physiol.* **68**, 43–63.

Bentley, P. J. (1975). *Comp. Biochem. Physiol.* **50A**, 639–643.

Bishop, W. R., Mumbach, M. W., and Scheer, B. T. (1961). *Am. J. Physiol.* **200**, 451–453.

Borgens, R. B., Vanable, J. W., Jr., and Jaffe, L. F. (1977a). *Proc. Natl. Acad. Sci. U.S.A.* **74**, 4528–4532.

Borgens, R. B., Vanable, J. W., Jr., and Jaffe, L. F. (1977b). *J. Exp. Zool.* **200**, 403–416.

Borgens, R. B., Vanable, J. W., Jr., and Jaffe, L. F. (1979a). *J. Exp. Zool.* **207**, 217–225.

Borgens, R. B., Vanable, J. W., Jr., and Jaffe, L. F. (1979b). *BioScience* **29**, 468–474.

Borgens, R. B., Vanable, J. W., Jr., and Jaffe, L. F. (1979c). *J. Exp. Zool.* **209**, 377–386.

Borgens, R. B., Vanable, J. W., Jr., and Jaffe, L. F. (1979d). *J. Exp. Zool.* **209**, 49–55.

Borgens, R. B., Jaffe, L. F., and Cohen, M. J. (1980). *Proc. Natl. Acad. Sci. U.S.A.* **77**, 1209–1213.

Borgens, R. B., Roederer, E., and Cohen, M. J. (1981). *Science* **213**, 611–617.

Brighton, C. T. (1981). *J. Bone Joint Surg.* **63-A**, 847–851.

Brighton, C. T., and Friedenberg, Z. B. (1974). *Ann. N.Y. Acad. Sci.* **238**, 314–319.

Brighton, C. T., Friedenberg, Z. B., and Black, J. (1979). *In* "Electrical Properties of Bone and Cartilage. Experimental Effects and Clinical Applications" (C. T. Brighton, J. Black and S. R. Pollack, eds.), pp. 519–546. Grune & Stratton, New York.

Brown, A. C. (1962). *J. Cell. Comp. Physiol.* **60**, 263–270.

Carafoli, E., and Crompton, M. (1976). *In* "Symposia of the Society for Experimental Biology: Calcium in Biological Systems" (C. J. Duncan, ed.), Vol. 30, pp. 89–115. Cambridge Univ. Press, London and New York.

Carlson, B. M. (1978). *Am. Zool.* **18**, 869–882.

Cochran, G. V. B., Pawluk, R. J., and Bassett, C. A. L. (1968). *Clin. Orthop.* **58**, 249–270.

Cowan, S. L. (1934). *Proc. R. Soc. (London) Ser. B.* **115**, 216–260.

Cox, T. C., and Alvarado, R. H. (1979). *Am. J. Physiol.* **237**, R74–R79.

Cragg, B. G. (1970). *Brain Res.* **23**, 1–21.

Della Valle, P. (1913). *Bull. Soc. Natl. Napoli* **25**, 95–161.

Douglas, B. S. (1972). *Aust. Paediat. J.* **8**, 86–89.

DuBois-Reymond, E. (1843). *Ann. Phys. Chem. (Leipzig)* **58**, 1–30.

DuBois-Reymond, E. (1860). "Untersuchungen ueber tierische Elektrizitaet," Vol. II, p. 2. Reimer, Berlin.

Edelberg, R. (1972). *In* "Handbook of Psychophysiology" (N. S. Greenfield and R. A. Sternbach, eds.), pp. 367–418. Holt, New York.

Edelberg, R. (1977). *J. Invest. Dermatol.* **69**, 324–327.

Eiland, L. C. (1975). *J. Exp. Zool.* **194**, 359–371.

Eriksson, C. (1974). *Ann. N.Y. Acad. Sci.* **238**, 321–338.

Farrell, R., Disher, W. A., Nesland, R. S., Palmatier, T. H., and Truhler, T. D. (1977). *J. Am. Coll. Emerg. Phys.* **6**, 243–246.

Fleming, M. W., and Tassava, R. A. (1981). *J. Exp. Zool.* **215**, 143–149.

Fontas, B., and Mambrini, J. (1977a). *C.R. Acad. Sci. Paris Ser. D.* **284**, 2361–2364.

Fontas, B., and Mambrini, J. (1977b). *C.R. Acad. Sci. Paris Ser. D.* **285**, 229–232.

Friedenberg, Z. B., and Brighton, C. T. (1966). *J. Bone Joint Surg.* **48a**, 915–923.

Friedenberg, Z. B., Harlow, M. C., and Brighton, C. T. (1971). *J. Trauma* **11**, 883.

Friedenberg, Z. B., Harlow, M. C., Heppenstall, R. B., and Brighton, C. T. (1973). *Calcif. Tissue Res.* **13,** 53.

Friedenberg, Z. B., Zemsky, I. M., Pollis, R. P., and Brighton, C. T. (1974). *J. Bone Joint Surg.* **56a,** 1023–1030.

Fukada, E., and Yasuda, I. (1957). *J. Phys. Soc. Jpn.* **12,** 1158–1162.

Gordon, M. S., Schmidt-Nielsen, K., and Kelly, H. M. (1961). *J. Exp. Biol.* **38,** 659–678.

Goss, R. J. (1969). *In* "Principles of Regeneration," pp. 223–255. Academic Press, New York.

Goss, R. J., and Grimes, L. N. (1972). *Am. Zool.* **12,** 151–157.

Goss, R. J., and Grimes, L. N. (1975). *J. Morphol.* **146,** 533–542.

Grafstein, B., and McQuarrie, I. G. (1978). *In* "Neuronal Plasticity" (C. W. Cotman, ed.), pp. 155–195. Raven, New York.

Graper, L. (1922). *Arch. Entwicklungsmech. Org.* **51,** 587–609.

Herlitzka, A. (1910). *Wilhelm Roux Arch.* **10,** 126–158.

Hinkle, L., McCaig, C. D., and Robinson, K. R. (1981). *J. Physiol. (London)* **314,** 121–136.

Hinsenkamp, M., Bourgois, R., Bassett, C. A. L., Chiabrera, A., Burny, F., and Ryaby, J. (1978). *Acta Orthopaed. Belg. (Brusells)* **44,** 671–698.

Huang, C. B.-Y., Vanable, J. W., Jr., and Jaffe, L. F. (1979). *Am Zool.* **19,** 923.

Illingworth, C. M. (1974). *J. Pediat. Surg.* **9,** 853–858.

Illingworth, C. M., and Barker, A. T. (1980). *Clin. Phys. Physiol. Meas.* **1,** 87–89.

Jaffe, L. A., Weisenseel, M. H., and Jaffe, L. F. (1975). *J. Cell Biol.* **67,** 488–492.

Jaffe, L. F. (1977). *Nature (London)* **265,** 600–602.

Jaffe, L. F. (1979). *In* "Membrane Transduction Mechanisms" (R. A. Cone and J. E. Dowling, eds.), pp. 199–231. Raven, New York.

Jaffe, L. F. (1980). *Ann. N.Y. Acad. Sci.* **339,** 86–101.

Jaffe, L. F., and Poo, M.-M. (1979). *J. Exp. Zool.* **209,** 115–127.

Jaffe, L. F., and Nuccitelli, R. (1974). *J. Cell Biol.* **63,** 614–628.

Jaffe, L. F., and Nuccitelli, R. (1977). *Annu. Rev. Biophys. Bioeng.* **6,** 445–476.

Jørgensen, C. B. (1949). *Acta. Physiol. Scand.* **18,** 171–180.

Kasanzeff, W. (1930). *Wilhelm Roux Arch. Entwicklungsmech. Org.* **121,** 658–707.

Kenner, G. H., Gabrielson, E. W., Lovell, J. E., Marshall, A. E., and Williams, W. S. (1975). *Calcif. Tissue Res.* **18,** 111–117.

Kirschner, L. B. (1970). *Am. Zool.* **10,** 365–376.

Kirschner, L. B. (1973). *In* "Transport Mechanisms in Epithelia" (H. H. Ussing and N. A. Thorn, eds.), pp. 447–460. Munksgaard, Copenhagen.

Korostoff, E. (1977). *J. Biomech.* **10,** 41–44.

Larsen, L. O. (1976). *In* "Physiology of the Amphibia" (B. Lofts, ed.), Vol. III, pp. 53–100. Academic Press, New York.

Lassalle, B. (1974a). *C. R. Acad. Sci. Paris, Ser. D* **278,** 483–486.

Lassalle, B. (1974b). *C. R. Acad. Sci. Paris, Ser. D* **278,** 1055–1058.

Lassalle, B. (1979). *J. Embryol. Exp. Morphol.* **53,** 213–223.

Lassalle, B. (1980). *Dev. Biol.* **75,** 460–466.

Libbin, R. M., Person, P., Papierman, S., Shah, D., Nevid, D., and Grob, H. (1979). *J. Morphol.* **159,** 439–452.

Lorente de Nó, R. (1947a). "Institute for Medical Research," Vol. 131. Rockefeller Inst. for Medical Research, New York.

Lorente de Nó, R. (1947b). "Institute for Medical Research," Vol. 132. Rockefeller Inst. for Medical Research, New York.

Lubińska, L. (1956). *Exp. Cell Res.* **10,** 40–47.

McAfee, R. D. (1972). *Science* **178,** 183–185.

McCaig, C. D., and Robinson, K. R. (1980). *J. Gen. Physiol.* **76,** 14a.

Marsh, G., and Beams, H. W. (1946). *J. Cell. Comp. Physiol.* **27,** 139–157.

Meiri, H., Spira, M. E., and Parnus, I. (1981). *Science* **211,** 709–712.

Mescher, A. L. (1976). *J. Exp. Zool.* **195,** 117–128.

Mizelle, M. (1968). *Science* **161,** 283–285.

Monroy, A. (1941). *Publ. Stn. Zool. Napoli* **18,** 265–281.

Myers, R. M., Bishop, W. R., and Scheer, B. T. (1961). *Am. J. Physiol.* **202,** 444–450.

Neufeld, D. A. (1980). *J. Exp. Zool.* **212,** 31–36.

Nielsen, R., and Tomilson, R. W. S. (1970). *Acta Physiol. Scand.* **79,** 238–243.

Nuccitelli, R. (1978). *Dev. Biol.* **62,** 13–33.

Nuccitelli, R., Poo, M.-M., and Jaffe, L. F. (1977). *J. Gen. Physiol.* **69,** 743–763.

Orida, N., and Poo, M.-M. (1978). *Nature (London)* **275,** 31–35.

Patterson, D. C., Lewis, G. N., and Cass, C. A. (1980). *Clin. Orthop. Relat. Res.* **148,** 129–135.

Polezhaev, L. V. (1972). "Loss and Restoration of Regenerative Capacity in Tissues and Organs of Animals." Harvard Univ. Press, Cambridge, Massachusetts.

Polezhaev, L. V., and Favorina, W. N. (1935). *Wilhelm Roux Arch. Entwicklungsmech. Org.* **133,** 701–727.

Poo, M.-M., and Robinson, K. R. (1977). *Nature (London)* **265,** 602–605.

Poo, M.-M., Poo, W.-J., and Lam, J. W. (1978). *J. Cell Biol.* **76,** 483–501.

Robinson, K. R., and McCaig, C. D. (1981). *Ann. N.Y. Acad. Sci.* **339,** 132–138.

Rose, S. M. (1942). *Soc. Exp. Biol. Med.* **49,** 408–410.

Rose, S. M. (1944). *J. Exp. Zool.* **95,** 149–170.

Rose, S. M. (1945). *J. Morphol.* **77,** 119–135.

Rose, S. M. (1970). *In* "Regeneration: Key to Understanding Normal and Abnormal Growth and Development." Appleton, New York.

Rose, S. M., and Rose, F. C. (1974). *Growth* **38,** 363–380.

Rovainen, C. M. (1967). *J. Neurophysiol.* **30,** 1000–1023.

Rovainen, C. M. (1974). *J. Comp. Neurol.* **168,** 545–554.

Rzehak, K., and Singer, M. (1966). *Anat. Rec.* **155,** 537–540.

Schackmann, R. W., Eddy, E. M., and Shapiro, B. M. (1978). *Dev. Biol.* **65,** 483–495.

Scharf, A. (1961). *Growth* **25,** 7–23.

Scharf, A. (1963). *Growth* **27,** 255–269.

Schlaepfer, W. W. (1974). *Brain Res.* **69,** 203–215.

Schlaepfer, W. W. (1977). *Brain Res.* **136,** 1–9.

Schlaepfer, W. W., and Bunge, R. P. (1973). *J. Cell Biol.* **59,** 456–470.

Schlaepfer, W. W., and Hasler, M. B. (1979). *Brain Res.* **168,** 299–309.

Schotté, O. E., and Harland, M. (1943). *J. Morphol.* **73,** 329–362.

Schotté, O. E., and Wilber, J. F. (1958). *J. Embryol. Exp. Morphol.* **6,** 247–261.

Schotté, O. E., and Smith, C. B. (1959). *Biol. Bull.* **117,** 546–561.

Schotté, O. E., and Smith, C. B. (1961). *J. Exp. Zool.* **146,** 209–229.

Schwan, H. P. (1963). *Biophyik* **1,** 198–208.

Shamos, M. H., Lavine, L. S., and Shamos, M. I. (1963). *Nature (London)* **197,** 81.

Singer, M. (1952). *Q. Rev. Biol.* **27,** 169–200.

Singer, M. (1954). *J. Exp. Zool.* **126,** 419–471.

Singer, M. (1965). *In* "Regeneration in Animals and Related Problems" (V. Kiortsis and H. A. L. Trampusch, eds.), pp. 20–32. North-Holland Publ., Amsterdam.

Singer, M., and Salpeter, M. M. (1961). "Growth in Living Systems." Basic Books, New York.

Singer, M., Rzehak, K., and Maire, C. S. (1967). *J. Exp. Zool.* **166,** 89–97.

Smith, S. D. (1967). *Anat. Rec.* **158,** 89–98.

Smith, S. D. (1974). *Ann. N.Y. Acad. Sci.* **238,** 500–507.

Spadaro, J. A. (1977). *Clin. Orthop. Related Res.* **122,** 325–332.

Spitzer, N. C. (1979). *Annu. Rev. Neurosci.* **2,** 363–397.

Steinberg, M. E., Labosky, D. A., Jimenez, S., Lane, J. M., Korostoff, E., and Pollack, S. R. (1977). *Trans. Orthop. Res. Soc.* **2,** 285.

Stombaugh, D. P., and Adams, T. (1971). *Am. J. Physiol.* **221,** 1014–1018.

Taylor, R. E., and Barker, S. B. (1965). *Science* **148,** 1612–1613.

Thornton, C. S. (1962). *J. Exp. Zool.* **150,** 5–15.

Thornton, C. S. (1968). *Adv. Morphogen.* **7,** 205–209.

Tilney, L. G. (1979). *In* "Membrane Transduction Mechanisms" (R. A. Cone and J. E. Dowling, eds.), pp. 163–186. Raven, New York.

Ussing, H. H. (1964). *Harvey Lect.* **59,** 1–30.

Ussing, H. H. (1965). *Acta Physiol. Scand.* **63,** 141–155.

Vanable, J. W., Jr., Hearson, L. L., and Jaffe, L. F. (1980). *Am. Zool.* **20,** 739.

Veraa, R. P., and Grafstein, B. (1981). *Exp. Neurol.* **71,** 6–75.

Viale, G. (1916). *Arch. Fisiol.* **14,** 113–146.

Watson, W. E. (1976). "Cell Biology of Brain," pp. 201–275. Chapman & Hall, London.

Weiss, A. B., Parsons, J. R., and Alexander, H. (1980). *J. Med. Soc. N.J.* **77,** 523–526.

Weiss, P. (1934). *J. Exp. Zool.* **168,** 393–448.

Wheeler, P. C., Wolcott, L. E., Morris, J. L., and Spangler, R. M. (1971). "Neuroelectric Research" (D. V. Reynolds and A. E. Sjoberg, eds.), pp. 83–99. Thomas, Springfield, Illinois.

Williams, W. S., and Breger, L. (1975). *J. Biomech.* **8,** 407–413.

Williams, W. S., Johnson, M., and Gross, D. (1979). "Electrical Properties of Bone Cartilage" (C. T. Brighton, J. Black, and S. R. Pollack, eds.), pp. 83–93. Grune & Stratton, New York.

Wood, M. R., and Cohen, M. J. (1979). *Science* **206,** 344–347.

Wood, M. R., and Cohen, M. J. (1981). *J. Neurocytol.* **10,** 57–79.

Woodruff, R. I., and Telfer, W. H. (1973). *J. Cell Biol.* **58,** 172–188.

Woodruff, R. I., and Telfer, W. H. (1974). *Ann. N.Y. Acad. Sci.* **238,** 408–419.

Woodruff, R. I., and Telfer, W. H. (1980). *Nature (London)* **286,** 84–86.

Yasuda, I. (1974). *Ann. N.Y. Acad. Sci.* **238,** 457–465.

Yasuda, I., Noguchi, W., and Sata, T. (1957). *J. Bone Jt. Surg.* **37A,** 1292–1293.

Zelena, J. (1969). *In* "Symposia of the International Society for Cell Biology: Cellular Dynamics of the Neuron" (S. Barondes, ed.), Vol. 8, pp. 73–94. Academic Press, New York.

Zelena, J., Lubinska, L., and Gutman, E. (1968). *Z. Zellforsch.* **91,** 200–219.

Zika, J., and Singer, M. (1965). *Anat. Rec.* **152,** 137–140.

Metabolism of Ethylene by Plants

JOHN H. DODDS* AND MICHAEL A. HALL†

*Department of Plant Biology, University of Birmingham, Birmingham, Enland, and
†Department of Botany and Microbiology, University College of Wales, Aberystwyth, Dyfed, Wales

I. Introduction

The discovery of the growth-regulating properties of ethylene can be attributed to Girardin (1864) who noted that leaks from illuminating gas mains were causing defoliation of shade trees in several German cities. Subsequently, Neljubov (1901, 1913) demonstrated that ethylene was the active compound in illuminating gas and he noted that as little as $0.06 \, \mu$l liter^{-1} (equivalent to $2.86 \times 10^{-10} \, M$ in the aqueous phase) was sufficient to cause horizontal growth of etiolated pea seedlings. This observation was followed by numerous reports about the gas involving it in a wide range of developmental phenomena, for example, fruit ripening (Cousins, 1910), seed germination (Nord and Weicherz, 1929), flowering (Rodriguez, 1932), and sex expression in cucurbits (Minina, 1938).

Gane (1934) first demonstrated that plants themselves produce ethylene and this discovery together with the observed effects on development led to the suggestion that ethylene be considered as a plant hormone (Crocker *et al.*, 1935). Partly because of the discovery of auxins and other plant hormones and partly because it is a gas, many workers were reluctant to accept that ethylene is a natural growth regulator.

However, the enormous body of evidence which has accumulated in the last 20 years has established beyond reasonable doubt that ethylene plays a major natural role in plant growth and development.

Until recently it was not thought that higher plants had the capability of metabolizing ethylene. This did not appear incongruous since ethylene diffuses rapidly into and out of plant tissue (see, for example, Hall, 1977) and thus there appeared to be an alternative to metabolism in the control of endogenous ethylene concentration. However, it has become clear in the last 5 or 6 years that these suppositions are incorrect and this article seeks to describe the metabolizing systems which have been discovered and to assess their functionality.

II. Metabolism of Ethylene by Microorganisms

In the soil environment ethylene is produced by many fungi; out of more than 200 species tested by Ilag and Curtis (1968) over 50 had this ability. The fungus *Mucor hiemalis* was found to be a major producer of ethylene (Lynch and Harper, 1974; Lynch, 1975) and a *Penicillium* sp. from soil was also found to produce ethylene by degradation of phenolic acids (Constradine and Patching, 1975).

It has been shown by many workers and is generally agreed that anaerobiosis favors the microbial formation of ethylene in the soil environment (Smith and Restall, 1971; Smith and Cook, 1974). Indeed, plants growing in anaerobic waterlogged soils show many symptoms characteristic of ethylene effects, such as leaf abscission, adventitious root formation, and leaf epinasty.

The metabolism of ethylene in the soil has received far less study than ethylene production. From the early work of Abeles *et al.* (1971) it was clear that in the presence of oxygen, levels of soil ethylene fall, and further, that sterilization of the soil abolished this effect, a result later confirmed by Smith *et al.* (1973).

Subsequently de Bont (1975) was able to isolate from soil a strain of *Mycobacterium* (E20) which was able to utilize ethylene. de Bont investigated the rate of loss of ethylene from the gas space above cultures of this microorganism and his results are shown in Fig. 1.

Transformation of this data yields the Lineweaver–Burk plot shown in Fig. 2. The apparent K_m for the metabolizing system was about 40 ppm (μl liter^{-1}) ethylene in the gaseous phase which represents a value of about 2×10^{-7} M in the liquid phase. The ability of the organisms to utilize ethylene at such low concentration is impressive; however, it must be remembered that, in the soil, even under waterlogged conditions, ethylene is only rarely present at concentrations above 10 μl liter^{-1} (Scott-Russell, 1969).

In further studies aimed at elucidating the pathway of ethylene metabolism, de Bont (1975, 1979) cultured *Mycobacterium* E20 on a range of substrates; growth occurred on both ethylene oxide and ethanolamine but not on ethylene glycol. Simultaneous adaptation studies also indicated that ethylene glycol is not a metabolite. This finding was somewhat unexpected since ethylene oxide is converted nonenzymatically to ethylene glycol quite readily. de Bont proposed that

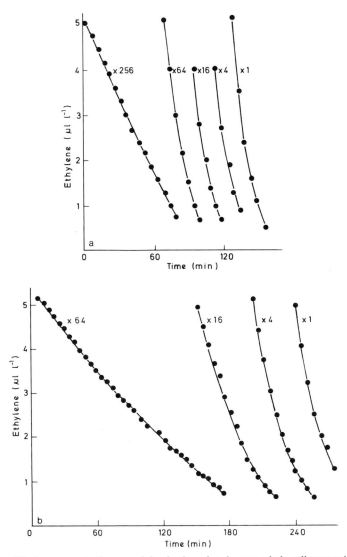

FIG. 1.   Ethylene concentrations remaining in air enclosed over a whole-cell suspension of strain E20. A culture of strain E20 pregrown in MS medium on ethylene was centrifuged and resuspended in 0.03 M sodium phosphate buffer, pH 6.8. An aliquot of this suspension (a) or one-fifth of such an aliquot (b) was put into a 100-ml Erlenmeyer flask. The final volume was brought to 20 ml with buffer solution. Flasks were sealed with suba-seals and ethylene was injected. The flasks were vigorously shaken and ethylene concentrations in the gas phase, starting at 1280 ppm (a) and 330 ppm (b), followed to complete oxidation of ethylene by periodically taking 0.05-ml samples for gas-chromatographic analysis.

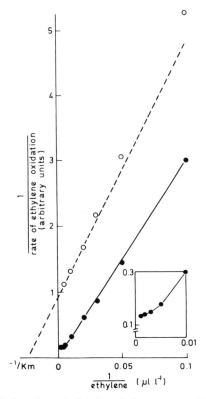

FIG. 2. Lineweaver–Burk reciprocal plot for ethylene concentration-dependent oxidation of ethylene by strain E20. From each of the two progress curves of Fig. 1a and b tangents at different ethylene concentrations were used to obtain reciprocal plots relating rate of ethylene oxidation to its concentration in the gas phase. (●) Curve obtained from Fig. 1a and (○) curve obtained from Fig. 1b.

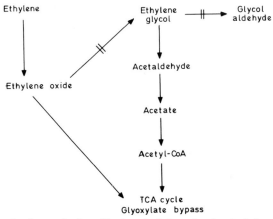

FIG. 3. Proposed pathways for the utilization of ethylene glycol and ethylene in *Mycobacterium* E44.

the glyoxylate cycle is involved in the metabolism of ethylene in *Mycobacterium* E20 yielding acetyl CoA.

More recently de Bont (1981) has isolated a new strain of *Mycobacterium* (E44) which is able to grow on ethylene glycol. A summary of the pathways of ethylene metabolism in *Mycobacterium* is illustrated in Fig. 3.

III. Ethylene Metabolism in Higher Plants

A. History

In the late 1950s and during the 1960s a number of reports appeared indicating that both fruit and vegetable tissue could metabolize radioactively labeled ethylene at low but significant rates. Buhler *et al.* (1957) looked at metabolism of ethylene by a variety of fruit tissues and Jansen (1963, 1964) studied metabolism of ethylene by avocado tissue. In 1968 a group of Japanese workers (Shimokana and Kasai, 1968, 1969) found incorporation of ethylene into RNA of Morning Glory (*Ipomea*) seedlings. However, not all the early studies on metabolism of ethylene by fruits were positive; Buhler *et al.* (1957) found that some fruit tissues gave negative results and Behmer (1958) could observe no incorporation in apple tissue.

Many of the early studies on ethylene metabolism did not prove very convincing because the rates of metabolism were very low, and no precautions had been taken to ensure the chemical purity of the radioactive ethylene used. Much of this early work used ethylene which had been regenerated from mercuric perchlorate traps and it was shown by Jansen (1969) that this procedure caused the formation of labeled impurities, such as toluene, which were readily incorporated into plant tissues. The discovery of these impurities led Abeles (1973) and other workers (McGlasson, 1970; Varner and Ho, 1976) to conclude that ethylene is probably not metabolized by plant tissues and that the ^{14}C incorporated represents impurities in the ethylene and is in effect an experimental artifact.

This early concern of Abeles and others that the ^{14}C incorporated was due to impurities led Beyer (1973) to begin a detailed study on the possible metabolism of ethylene using highly purified radiolabeled ethylene. This work, which is outlined below, has led to the general acceptance that ethylene metabolism occurs naturally in some higher plants at least.

B. Recent Studies

The early concern about the radiochemical purity of labeled ethylene appears to have had good cause since all stocks of the [^{14}C]ethylene analyzed by Beyer have proved to be contaminated when checked by FID gas chromatography (Beyer, 1980) and we have made similar observations. All subsequent studies carried out by Beyer used radioactive ethylene which had been rigorously purified. The purification procedure used is a two-step method (Beyer, 1975a). First, the ethylene is separated from impurities on a silver nitrate–ethylene glycol-coated Gas Chrom R column. The [^{14}C]ethylene is collected and is then rechromatographed on a Porapak T column. The [^{14}C]ethylene produced by this method has been shown by GC techniques to be of high purity. Although long-term storage leads to the formation of new impurities, freshly purified [^{14}C]ethylene can be stored for 2–3 months provided it is kept in an oxygen-free atmosphere over dilute NaOH.

1. *Studies on Pea Seedlings*

In some of Beyer's earliest work purified [^{14}C]ethylene was incubated with 0- to 5-day-old pea seedlings for 1 week at 30°C in modified Erlenmeyer flasks as shown in Fig. 4. Any CO_2 produced was collected by including NaOH in the flask. After incubation the amount of ^{14}C incorporated into the tissue and the amount of $^{14}CO_2$ were determined.

A significant and readily detectable amount of $^{14}C_2H_4$ was incorporated into the tissue and conversion to $^{14}CO_2$ was observed with pea seedlings treated with 5 μl liter^{-1} of $^{14}C_2H_4$ (Beyer, 1975c). For the first 20 hours of incubation the amount of $^{14}C_2H_4$ incorporated or converted to $^{14}CO_2$ was small, 55 and 30 dpm

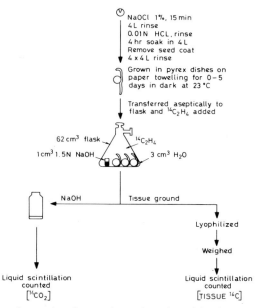

FIG. 4. Experimental procedure for growing and treating etiolated pea seedlings with purified $^{14}C_2H_4$ and for analyzing the tissue and NaOH for radioactivity.

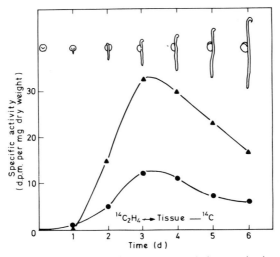

FIG. 5.   Changes in specific activity of $^{14}C_2H_4$ with time during germination of pea seedlings ▲, $^{14}CO_2$; ●, tissue incorporation.

gm^{-1} dwt hr^{-1}, respectively. Tissue incorporation and $^{14}CO_2$ production increased with time reaching a maximum on day 3 then gradually declining until day 5 (Fig. 5). Beyer notes that the level of incorporation observed in these experiments would only have been detected by the use of highly purified ethylene

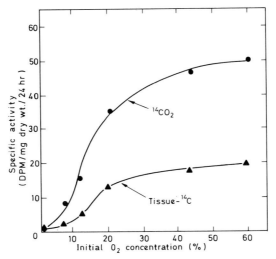

FIG. 6.   Effect of O_2 concentration on specific radioactivity of $^{14}C_2H_4$ to $^{14}CO_2$ conversion and its incorporation into 2-day-old pea seedlings during 24-hour exposure to 14 μl/liter of $^{14}C_2H_4$.

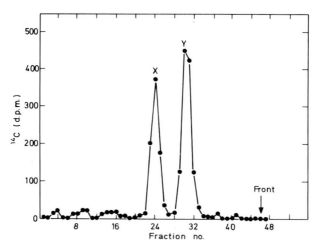

FIG. 7.   Separation of neutral metabolites of ethylene by paper chromatography with butanol:acetic acid:water, 3:3:2.

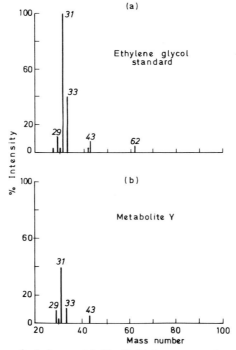

FIG. 8.   Identification of ethylene metabolite Y (b) by mass spectroscopy. Mass spectra of metabolite isolated from ethylene-exposed pea tissue and of authentic ethylene glycol (a) were obtained by scanning effluent peaks from a Chromosorb 101 GC column, heated from 75 to 190°C at 15°C min^{-1}, with a Vacuum Generators 16F mass spectrometer. Because of the low concentration of metabolite in the extract its three weakest mass peaks were below the detection threshold and hence were absent from its spectrum.

of high specific activity. Beyer also indicates that metabolism by these seedlings is almost certainly of a true biological nature as no radioactivity was recovered in the tissue or as $^{14}CO_2$ when seedlings were heated to 80°C for 1 minute prior to incubation.

It is interesting to note that it appears that some factor(s) essential for the maintenance of ethylene metabolism must be provided by the root–shoot axis since metabolism is much reduced if the root–shoot axis and the cotyledons are separated just prior to incubation in ethylene (Beyer, 1975a).

Beyer (1975b) also looked at the effect of oxygen concentration on $^{14}C_2H_4$ incorporation and its conversion to $^{14}CO_2$ in intact 2-day-old pea seedlings (Fig. 6). Under virtually anaerobic conditions (0.08% O_2) essentially no metabolism of $^{14}C_2H_4$ occurred. Increases in the O_2 concentration caused a rapid increase in $^{14}C_2H_4$ conversion to $^{14}CO_2$, reaching 35 dpm/mg.

The products of metabolism are of interest as they could give an indication of the significance of the mechanism to the whole plant. Early studies of the metabolic products were carried out by Giaquinta and Beyer (1977) but it was not until 1980 that Blomstrom and Beyer identified the metabolites in pea seedlings.

Two neutral metabolites were found on paper chromatography as shown in

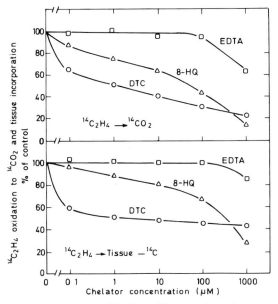

FIG. 9. Effect of metal chelators on $^{14}C_2H_4$ oxidation and on tissue incorporation. EDTA, ethylenediaminetetraacetic acid; 8-HQ, 8-hydroxyquinoline; DTC, diethyldithiocarbamic acid. Exposed pea hooks from 4-day-old seedlings exposed to 5 µl/liter of purified $^{14}C_2H_4$ (120 mCi/mmol) for 20 hours at 23°C.

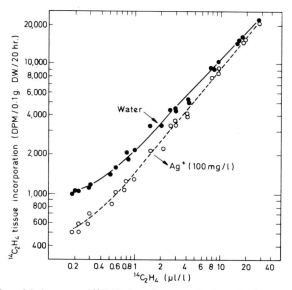

FIG. 10.   Effect of Ag⁺ on rate of $^{14}C_2H_4$ tissue incorporation in excised pea tips. All data points from three separate experiments have been plotted.

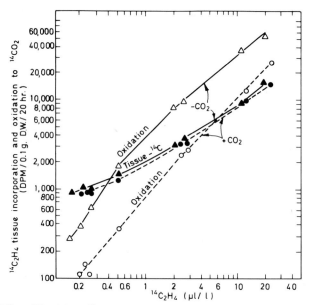

FIG. 11.   Effect of 7% CO_2 on $^{14}C_2H_4$ metabolism in excised pea tips. Data are from one experiment. Similar results were obtained in two other experiments.

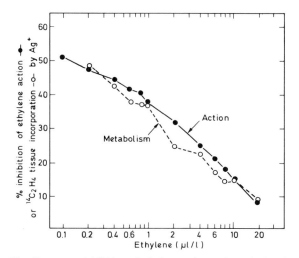

FIG. 12. Percentage inhibition of ethylene action and metabolism by Ag^+.

Fig. 7. Mass spectroscopic analysis of Peak Y showed this to be ethylene glycol (Fig. 8) while metabolite X was found to be a glucose conjugate of metabolite Y, namely, β-(20-hydroxyethyl)-D-glucoside.

The sensitivity of this metabolizing system to metal chelators (diethyl-dithiocarbamate, 8-hydroxyquinoline), CS_2, and COs (Beyer, 1977) (Fig. 9), gave support to the notion that a metal-containing receptor site (probably Cu) may be involved. It is perhaps significant that in nonbiological aqueous systems

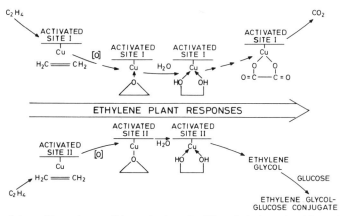

FIG. 13. Scheme illustrating possible mechanisms for CO_2 and ethylene glycol formation in plant tissues.

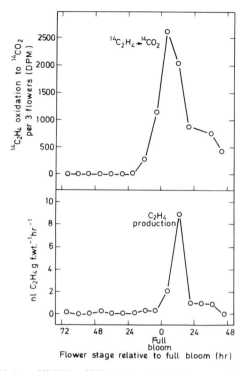

FIG. 15. Top: oxidation of $^{14}C_2H_4$ to $^{14}CO_2$ during bud development, flowering, and senescence in Morning Glory expressed as radioactivity recovered in the NaOH trap during 15 separate 8-hour enclosure periods of three flowers with 6 μl/liter of $^{14}C_2H_4$. Bottom: Ethylene production by buds and flowers.

C. THE *Vicia faba* SYSTEM

Apparently anomalous observations on the transport of ethylene in plants by Zeroni *et al.* (1977) led Jerie *et al.* (1978) to screen a number of species for their ability to incorporate ethylene. It was not at first apparent whether the observed incorporation was due to metabolism or compartmentation (binding) of the ethylene supplied (Jerie *et al.*, 1979), but further work demonstrated that in developing cotyledons of *Vicia faba* cv. Aquadulce at least, the former was the case (Jerie and Hall, 1978) (see Fig. 16).

    a. *Identification of Metabolic Products in V. faba.*   Cotyledons of *V. faba* cv. Aquadulce were incubated with $^{14}C_2H_4$. To prevent the samples from becoming anaerobic the incubation vessel was periodically tested, flushed with air, and reinjected with fresh ethylene. After 48 hours sufficient metabolite had accumulated in the tissue to become detectable. The metabolite was released by heating the tissue to 70°C for 15 minutes. The metabolite was identified as ethylene oxide by GCMS, thus confirming the tentative identification made on the basis of radio-gas chromatography (Fig. 17).

It was shown that the metabolism was a function of the tissue and was not due to the presence of microorganisms or to wounding. Moreover, the system was heat labile.

The most striking difference between this system and those reported by Beyer and his co-workers lies in the activity of the two systems which appears to be between three and five orders of magnitude greater in the case of *V. faba*. Other parts of the plant including stems, roots, leaves, and abscission zones also have the ability to metabolize ethylene but at much lower rates on a fresh weight basis.

b. *Further Metabolism of Ethylene Oxide.* In contrast to the systems described by Beyer referred to above, carbon dioxide does not appear to be a major product of ethylene metabolism in *Vicia faba* cotyledons. Thus, after a 6–7 hour incubation of developing cotyledons in [^{14}C]ethylene less than 1% of the total metabolized appeared as $^{14}CO_2$. Of the remainder more than 90% was extractable on refluxing the tissue with 95% ethanol. Of the radioactivity so extracted, ion-exchange chromatography revealed that 65–75% was present in the basic fraction, 15–25% in the neutral fraction, and 5–10% in the acidic fraction. The principal components of the neutral fraction were ethylene glycol with smaller amounts of a compound tentatively identified as glycerol. In the basic fraction

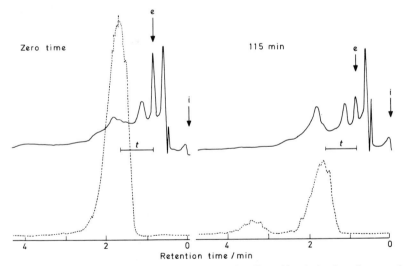

FIG. 16. Radio-gas chromatography traces showing the effect of incubating broadbean cotyledon segments in $^{14}C_2H_4$ at an initial concentration of 3.0 nl/ml. Solid line, H_2 flame ionization detector (H_2 f.i.d.); dotted line, isotope detector. The isotope trace is delayed relative to the H_2 f.i.d. trace by the length of the bar (t). Arrow (i) shows the point of injection and arrow (e) the retention of ethylene on the H_2 f.i.d. trace.

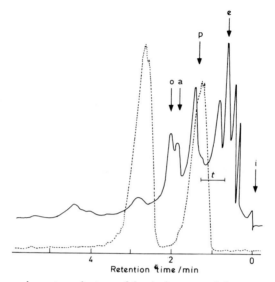

Fig. 17.   Radio-gas chromatography trace of the air above a cotyledon segment after venting and reinjection of ethylene for the fifth time. Sample, 0.9 ml from the vial plus 0.1 ml of a suitable dilution of acetaldehyde in air. Solid line, H₂ flame ionization detector; dotted line, isotope detector. The isotope trace is delayed relative to the H₂ f.i.d. trace by the length of the bar (t). Arrows show the following points: (i) injection; the retention on the H₂ f.i.d. trace of (e) ethylene, (p) propane, (a) acetaldehyde. (o) ethylene oxide.

ethanolamine represented more than 60% of the total radioactivity. The acidic fraction was very heterogeneous but contained significant amounts of oxalate and glycollate (Musa, 1981). While ethylene glycol can be formed nonenzymatically from ethylene oxide by hydoxylation, the observation in *V. faba* that [^{14}C]ethylene oxide is only slowly converted to other molecular species in steam-killed tissue while being relatively rapidly transferred in living cotyledons makes it likely that enzymatic processes are involved, directly or indirectly. Certainly it has been shown in microorganisms that epoxidases exist capable of converting epoxides to glycols (Jakoby and Fjellstedt, 1972).

The fact that much $^{14}C_2H_4$ is incorporated but that little $^{14}CO_2$ is evolved suggests that the products must be compartmented to some extent and do not rapidly enter respiratory pathways.

These observations suggest that in *Vicia faba* the CO_2-forming pathway proposed by Beyer (Fig. 13, site I) is either absent and/or that the site II activity predominates. Alternatively, unlike the situation in pea the system in *V. faba* which yields ethylene oxide may have a much higher affinity for ethylene than the CO_2-forming pathway if indeed the latter is present.

The principal metabolic transformations of ethylene in *V. faba* cotyledons are summarized in Fig. 18. The pathway for ethanolamine is uncertain since although formation from glycollate seems likely, [^{14}C]ethanolamine is readily detectable only 10 minutes after treating tissue with $^{14}C_2H_4$.

c. *Purification and Properties of the System.* It has proved relatively straightforward to prepare cell-free preparations from *V. faba* cotyledons capable of oxidizing ethylene. The preparations are particulate, more than half of the activity sediments after 45 minutes between 3000 and 10,000 *g*. The precise location of the activity in the cell is unclear as yet since efforts to purify cell-free extracts on density gradients are hampered by the fact that all the materials commonly used to prepare such gradients, e.g., sucrose, Ficoll, Percoll, etc., inhibit enzyme activity markedly. A Lineweaver–Burk plot of results from experiments with cell-free extracts is shown in Fig. 19 (Dodds *et al.*, 1979). The enzyme has a K_m of 4.17×10^{-10} *M* for ethylene, this corresponds to a concentration of 0.096 μl liter^{-1} in the gaseous phase and compares well with the values of ethylene concentration found in the tissue air spaces of the plant (Zeroni *et al.*, 1977). Allowing for the proportion of the total activity present in the particulate preparation the V_{max} was about 20 pmol hr^{-1} gm^{-1} fresh weight. The figure for intact tissue is much higher indicating fairly substantial loss of activity during extraction.

It had been shown that anaerobiosis inhibited ethylene metabolism *in vivo* and this also proved to be the case for cell-free preparations. Rates of oxidation increase with increased oxygen tension and analysis of such data yielded a K_m for

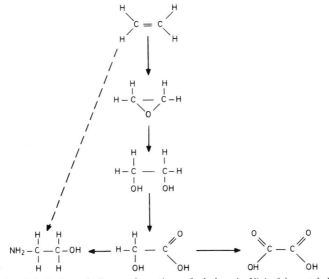

FIG. 18. Principal metabolic transformations of ethylene in *Vicia faba* cotyledons.

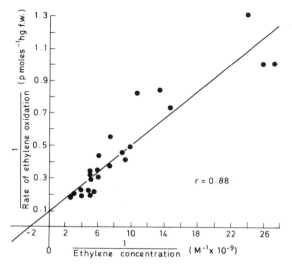

FIG. 19.   Lineweaver–Burk plot of rate of ethylene oxidation versus ehtylene concentration in the liquid phase.

oxygen of 3.63×10^{-4} M which is within the general range observed for oxygenases. Further work is in progress to determine whether molecular oxygen is incorporated directly in the formation of ethylene oxide, thus establishing whether or not the enzyme is indeed an oxygenase.

Addition of various possible cofactors (NADH, NADPH, Mg, Mn, Cu, Co, Fe^{2+}, Fe^{3+}, Ca) had no effect on the rate of oxidation and the system showed a sharp pH optimum at 8.5.

Various protein-reactive compounds were tested in the system (Table I). Dithiothreitol, dithioerythritol, and p-chloromercuribenzoate, even at very low concentrations (0.001 mM), significantly reduce oxidizing activity. Pretreatment of the preparations with diamide at 10 mM partially reversed the inhibitory effect of dithioerythritol. These results may imply the involvement of a cysteine residue and a disulfide bridge at the active site. Rather similar observations have been made with an ethylene-binding system from *Phaseolus vulgaris* (Bengochea *et al.*, 1980b).

There exist a large number of structural analogs of ethylene—far more than for any other naturally occurring plant growth regulator—and this has permitted a comprehensive investigation of the specificity of the enzyme system (Dodds *et al.*, 1980). Additionally, as the effects of most of these analogs on a number of developmental systems is known (Burg and Burg, 1967; Abeles and Gahagan, 1968) meaningful comparisons can be made between the affinity of the enzyme for the analogs and their relative effectiveness in the control of ethylene-dependent processes.

Of the analogs tested none of the alkanes (methane, ethane, propane, and cyclo-propane) had any effect on ethylene metabolism even at concentrations up to 10^5 μl liter^{-1} (4.5 mM) in the gas phase. All the alkenes alkynes, vinyl ethers, carbon monoxide, and carbon dioxide inhibited metabolism to varying extents at appropriate concentrations.

Lineweaver–Burk analyses of the data revealed that the inhibition was purely competitive in all cases. The results with two of the analogs, namely, propylene and propyne, are shown in Fig. 20.

Table II shows a comparison of the effects of the various analogs in terms of their K_i values. The data are expressed as a molar concentration in the gaseous and liquid phases and as μl liter^{-1} in the gaseous phase. The effectiveness of the analogs relative to ethylene is obtained by taking the ratio of the K_m for ethylene and the K_i for each analog and this is compared with the figures of Burg and Burg (1967) for the relative effectiveness of the analogs in the pea growth test. Also included are comparable figures for the ethylene binding site from *Phaseolus vulgaris* cotyledons (see below).

The similarity between the relative effectiveness of the analogs in the metabolizing system and that in the pea growth test is striking. Thus, the alkanes and *cis* and *trans* 2-butene do not affect developmental processes and do not inhibit metabolism. Similarly, with the exception of carbon monoxide, the order

TABLE I

EFFECTS OF SULFHYDRYL REACTIVE COMPOUNDS ON ETHYLENE OXIDATION BY CELL-FREE PREPARATIONS FROM COTYLEDONS OF *Vicia faba*[a]

Compound	Concentration (mM)	Percentage of control enzyme activity
p-Chloromercuribenzoate	1.0	0
	0.1	8
	0.01	30
	0.001	49
Dithiothreitol	1.0	0
	0.1	0
	0.01	25
	0.001	55
Dithioerythritol	1.0	0
	0.1	10
	0.01	35
	0.001	42
Dithioerythritol + 10 mm diamide	0.1	45

[a] Compounds were added at appropriate concentrations prior to incubations with [^{14}C]ethylene (1 μl liter^{-1}). Diamide was added during isolation of cell-free extracts and maintained throughout. Samples were assayed after 1 hour at 25°C.

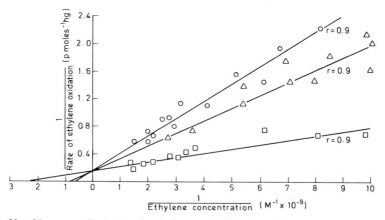

FIG. 20. Lineweaver–Burk plots for ethylene metabolism by cell-free preparations from *Vicia faba*. Samples were treated with $^{14}C_2H_4$ at a range of concentrations in the presence or absence of propylene or propyne for 1 hour at 25°C (○) + 500 μl/liter propylene; (△) + 230 μl/liter propyne; (□) control.

of effectiveness both for the enzyme and for physiological response is almost identical. Carbon monoxide is about five times less effective in inhibiting metabolism than it is in the pea growth test. There could be a number of reasons for this observation but the most likely appears to be that *in vivo*, other processes upon which the overall ethylene effect may be dependent—for example respiration—are also inhibited by carbon monoxide. On the other hand in the *Phaseolus* system carbon monoxide has about the same relative effectiveness in inhibiting binding as in affecting development (Bengochea *et al.*, 1980b).

In the experiments just described it was observed that the analogs causing inhibition were also metabolized to some extent. This is in agreement with the work of Beyer (1978) who observed metabolism of propylene in the same pea seedling system described earlier in relation to ethylene. Beyer (1980) has identified two neutral products as 1,2-propanediol and its glucose conjugate. In the *Vicia faba* system we have tentatively identified propylene oxide as a product of propylene metabolism. From this data it seems likely that the systems in pea and bean which convert ethylene to ethylene oxide and ethylene glycol are also capable of metabolizing propylene in the same way.

Two curious features emerge form the work on pea. First, that where propylene is the substrate tissue incorporation proceeds more rapidly than CO_2 production and second such tissue incorporation is much more rapid than that observed for ethylene (Fig. 21). Since ethylene is produced naturally by higher plants whereas propylene is not, these results are unexpected.

It is possible to make some interesting comparisons between the pea and bean systems by using the results of Beyer (1978) and those of Jerie and Hall (1978)

TABLE II

COMPARISON OF INHIBITOR CONSTANTS FOR STRUCTURAL ANALOGS OF ETHYLENE IN THE *Vicia faba* OXIDIZING SYSTEM AND THE *Phaseolus vulgaris* BINDING SYSTEM[a]

		K_i (gas) (M) (*Vicia*)	K_i (liquid) (M) (*Vicia*)	K_i (gas) (μl liter^{-1}) (*Vicia*)	K_i (gas) relative (*Vicia*)	K_i (gas) relative (*Phaseolus*)	Relative μl liter^{-1} for half maximal activity on pea growth[b]
Ethylene	$CH_2{=}CH_2$	5.21×10^{-9}	5.73×10^{-10}	0.12	1	1	1
Propylene	$CH_2{=}CH \cdot CH_3$	3.30×10^{-6}	1.80×10^{-6}	74	615	128	128
Vinyl chloride	$CH_2{=}CHCl$	3.19×10^{-6}	1.50×10^{-6}	72	595	466	1,400
Acetylene	$CH{\equiv}CH$	7.97×10^{-6}	9.97×10^{-6}	178	1,493	1,013	2,800
Vinyl fluoride	$CH_2{=}CHF$	1.34×10^{-5}	6.29×10^{-6}	300	2,500	1,139	4,300
Propyne	$CH{\equiv}C \cdot CH_3$	6.84×10^{-5}	2.12×10^{-5}	1,533	12,775	2,651	8,000
Carbon monoxide	CO	7.82×10^{-5}	1.80×10^{-6}	1,753	14,610	1,068	2,700
Vinyl methyl ether	$CH_2{=}CH \cdot O \cdot CH_3$	4.50×10^{-4}	4.46×10^{-5}	10,000	83,330	136,196	100,000
Vinyl ethyl ether	$CH_2{=}CH \cdot O \cdot C_2H_5$	—	6.92×10^{-3}	155,000	1,291,670	601,227	300,000
Carbon dioxide	CO_2	2.08×10^{-3}	—	46,760	389,700	141,104	300,000

[a] Oxidation was measured in the presence and absence of the analog and the K_i for the analog calculated from Lineweaver–Burk plots of the data. The K_i (liquid) was calculated from the K_i (gas) from the partition coefficient of the analog in the assay medium.

[b] From Burg and Burg (1967), reproduced with permission. Inactive at concentrations (μl liter^{-1}) in parentheses: Methane, ethane, propane, cyclopropane (10^5); cis-2-butene, trans-2-butene (10^4).

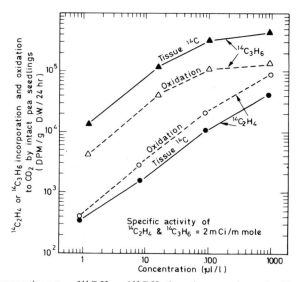

FIG. 21.   Comparative rates of $^{14}C_3H_6$ and $^{14}C_2H_4$ tissue incorporation and oxidation to $^{14}CO_2$ by intact aseptically grown 3-day-old pea seedlings exposed to various concentrations of either gas. Specific radioactivity of $^{14}C_2H_4$ was reduced from 119 mCi/mmol by adding enough nonlabeled C_2H_4 to match exactly the initial 2 mCi/mmol specific radioactivity of $^{14}C_3H_6$. Specific radioactivities were experimentally determined with the use of a Cary ionization chamber-vibrating reed electrometer system.

and Dodds *et al.* (1979, 1980). Thus, if it is assumed that the tissue incorporation measured by Beyer corresponds to the tissue incorporation measured by Jerie and Hall (i.e., that the products are ethylene oxide and propylene oxide and their subsequent metabolites) then it is observable that the K_m for propylene in the pea system is around 50 μl liter^{-1} in the gaseous phase (1.22 × 10^{-6} M in the liquid phase) compared with a figure for the K_i of propylene in bean of 74 μl liter^{-1} (1.8 × 10^{-6} M in the liquid phase). Further calculations indicate a V_{max} for propylene in the pea system of about 67 nl gm^{-1} f.m. hr^{-1} compared with about 50 nl gm^{-1} f.m. hr^{-1} in the bean system. Although there are a number of assumptions and approximations in these calculations the figures are so close that the suggestions that the enzymes in both systems are the same and are, however, present in approximately similar amounts are not unreasonable.

This assumption is further supported by experiments that show that the V_{max} and K_m for ethylene and propylene in peas are indeed very similar to those for the same gases in broad beans. Moreover, propylene does competitively inhibit ethylene metabolism in pea with K_i of 7 × 10^{-7} M (Evans and Hall, unpublished).

This makes the situation in pea vis-à-vis ethylene all the more curious since if similar calculations are made, the K_m for ethylene in pea would be around 450 μl liter^{-1} in the gaseous phase (1.94×10^{-6} M in the liquid phase). The question therefore arises as to why, if the enzyme systems in pea and broad bean are similar if not identical as regards their activity in relation to propylene, the two systems should not also behave in the same way in relation to ethylene.

The only obvious difference lies in the fact that whereas pea can oxidize both propylene and ethylene to CO_2, this is not the case in broad bean. Calculations such as those above indicate that as with tissue incorporation the K_ms for CO_2 productions for ethylene and propylene are similar, though somewhat higher than those for the same gases in relation to tissue incorporation. This is an agreement with the observations of Beyer (1981) that tissue incorporation occurs at a greater rate than CO_2 production at low ethylene concentrations with a reversal of this situation at high ethylene concentrations.

It should be noted in relation to the calculations above that Beyer (1980b) has shown that some of the radioactivity in NaOH traps is due to ethylene oxide and not to CO_2; this would also presumably be the case with propylene oxide. This will have the effect of altering the relative values of the V_{max} for CO_2 production and that for tissue incorporation in both cases but is unlikely to significantly affect the calculations on the relative affinities of the system for ethylene and propylene.

These contradictions remain unresolved, but it seems unlikely that competition between the CO_2 producing system and tissue incorporation is the cause of the disparity in the results for ethylene and propylene in pea or between the two systems in pea and bean.

IV. The Role of Ethylene Metabolism

We have already referred to the fact that in the past the apparent lack of any system for ethylene metabolism in higher plants did not appear incongruous because it was felt that the balance between rates of biosynthesis and outward diffusion would constitute a sufficient mechanism for the control of endogenous concentration of the growth regulator. In the main, there is still no reason to doubt this contention, since the system in *Vicia faba* does appear to be exceptional, indeed, other varieties of *V. faba* do not appear to metabolize ethylene at anything like the rates observed in cv. Aquadulce (Dodds, Smith, and Hall, unpublished), and no other species of higher plant has been shown to metabolize ethylene as rapidly.

In the pea system Beyer (1981) has calculated that the rate of total metabolism would only account for about 0.2% of the rate of ethylene biosynthesis and this led him to propose another role for metabolism, namely, that it is related to mode of action. Thus, it is proposed that ethylene binds to the active site(s) of the enzyme(s) which initiates a change in conformation of the complex in the same way as envisaged for other hormone–receptor complexes in animals and plants. This is then followed by metabolism of the bound ligand. The general contention is supported by an impressive body of evidence correlating rates of metabolism in tissues with their sensitivity to ethylene in terms of developmental response (see above). The effects of the antiethylene reagents Ag and CO_2 also seem to indicate a relationship between metabolism and mode of action.

While the authors would not in general dissent from this general hypothesis there are a number of observations both in our own work and that of Beyer which appear to cast doubt on the proposals.

The most troublesome of these relate to the concentration dependence and specificity of the systems which Beyer has described. We have analyzed some of these above. Whether or not it is suggested that metabolism is involved in the mode of action of a growth regulator, all other hormone-binding site systems in plants or animals appear to involve receptors (which might or might not be enzymes) with a high affinity and specificity for the hormone (see Dodds and Hall, 1980, and references therein for a discussion of this point). Thus, the dissociation contant of the hormone–receptor complex should correspond to the concentration of growth regulator which induces a half-maximal response in developmental systems and the effectiveness of analogs in competing with the growth regulator for the active site should correspond to their relative effectiveness in minimizing the effects of the growth regulator in developmental systems. This is even true for ethylene since we have isolated a binding site for ethylene from *Phaseolus vulgaris* which possesses the appropriate properties and which does not metabolize the growth regulator (see Table II). Similar results have been obtained by Sisler (1979, 1980).

In the case of peas (and apparently in other cases) the affinity of the metabolizing system for ethylene appears to be lower by about three orders of magnitude than the physiologically effective concentration of the growth regulator. Moreover, propylene has a higher affinity for the metabolizing system than ethylene whereas it is about three orders of magnitude less effective than ethylene in bringing about physiological responses (see Table II). Nevertheless, in *Vicia faba* the affinities of the enzyme for ethylene and propylene are proportional to biological effectiveness and this is also true in absolute terms for propylene in pea. It can, and has (Beyer, 1981), been argued that there is no prima facie reason why the whole potential of an ethylene metabolizing system linked to a developmental response should be used although this is not the situation in any animal hormone–receptor system and does not appear to be the case for other plant hormones, or indeed for ethylene in *Vicia faba* or *Phaseolus vulgaris*. Moreover, if the system is operating in the same way as normal hormone–receptor interactions then using the figures derived above it is possible to calculate that over the concentration range 0.01–0.1–1.0 μl liter^{-1} ethylene (4.5×10^{-11}–4.5×10^{-10}–4.5×10^{-9} M liquid phase), which represents the range from threshold through half maximal to maximal effect in most developmental systems, the percentage occupancy of the site by ethylene would only range from about 0.0025–0.025–0.25%.

In other words if the magnitude of the response controlled by the system is proportional to the number of sites occupied then one would expect an approximately 10-fold increase in rate between 0.1 and 1 μl liter^{-1} instead of the approximate doubling observed. It is, moreover, difficult to comprehend why a system exists, 99% of the capacity of which is likely to remain unused.

The situation is further complicated by the fact that the metabolizing system in *V. faba* and the ethylene binding systems in tobacco and *Phaseolus vulgaris* do fulfill the expected kinetic criteria for receptors.

Despite these reservations the correlative evidence presented by Beyer does necessitate that serious consideration be given to the hypothesis he has outlined. It may well be that we are wrong in extrapolating from the well-characterized hormone–receptor systems in animals to the situation in plants—at least in all cases. Beyer (1979a) has recently demonstrated that ethylene oxide acts synergistically with ethylene in the "triple response" of peas and this has led him to suggest that the alkylation of key cellular components by ethylene oxide is a feature of ethylene action. If this were indeed the case, i.e., that the product of the hormone–receptor interaction also interacted with the same or different receptor complex, then calculations such as these above, which assume a simple hormone–receptor interaction would be invalidated and would not lead to correct interpretation.

The situation remains very confused and much further work is needed both on the characterization of the systems involved and on their functionality.

REFERENCES

Abeles, F. B. (1971). *Planta* **97**, 89–91.
Abeles, F. B. (1972). *Annu. Rev. Plant Physiol.* **23**, 259–292.
Abeles, F. B. (1973). "Ethylene in Plant Biology." Academic Press, New York.
Abeles, F. B., and Gahagan, H. E. (1968). *Plant Physiol.* **43**, 1255–1258.
Behmer, M. (1958). *Klosterneuberg, Austria, Hoehere Bundeslehr Versuchanstalt Wein, Obst Gartenbau, Ser. B* **8**, 257–274.
Bengochea, T., Dodds, J. H., Evans, D. E., Jerie, P. H., Niepel, B., Shaari, A. R., and Hall, M. A. (1980a). *Planta* **148**, 397–406.
Bengochea, T., Acaster, M. A., Evans, D. E., Jerie, P. H., and Hall, M. A. (1980b). *Planta* **148**, 407–411.
Beyer, E. M. (1973). *Plant Physiol.* **49** 672–677.
Beyer, E. M. (1975a). *Plant Physiol.* **55**, 845–848.
Beyer, E. M. (1975b). *Plant Physiol.* **56**, 273–278.
Beyer, E. M. (1975c). *Nature (London)* **255**, 144–147.
Beyer, E. M. (1975d). *Proc. Beltwide Cotton Prod. Conf., 29th* pp. 51–52.

Beyer, E. M. (1977). *Plant Physiol.* **60**, 203–206.

Beyer, E. M. (1978). *Plant Physiol.* **61**, 893–895.

Beyer, E. M. (1979a). *Plant Physiol.* **63**, 169–173.

Beyer, E. M. (1979b). *Plant Physiol.* **64**, 971–974.

Beyer, E. M. (1980). *DPGRG/BPGRG. Monogr* **6**, 35–69.

Beyer, E. M. (1981). *Proc. Phytochem. Soc. 1980. Norwich*

Beyer, E. M., and Blomstrom, D. C. (1980). *Proc. Int. Conf. Plant Growth Subs., 10th.*

Beyer, E. M., and Sudin, O. (1978). *Plant Physiol.* **61**, 896–899.

Blomstrom, D. C., and Beyer, E. M., (1980). *Nature (London)* **283**, 66–67.

Buhler, D. R., Hansen, E., and Wang, C. H. (1957). *Nature (London)* **174**, 48–49.

Burg, S. P., and Burg, E. A. (1967). *Plant Physiol.* **42**, 144–152.

Buxton, G. V., Green, J. C., and Sellers, R. M. J. (1976). *J. Chem. Soc. Dalton Trans.* 2160.

Constradine, P. J., and Patching, J. W., (1975). *Ann. Appl. Biol.* **81**, 115–119.

Cousins, H. H. (1910). *Ann. Rep. Dept. Agric. Jamacia.*

Crocker, W., Hitchcock, A. E., and Zimmerman, A. (1935). *Contri Boyce Thompson Inst.* **7**, 231–248.

de Bont, J. A. M. (1975). *Ann. Appl. Biol.* **81**, 119–121.

de Bont, J. A. M., Attwood, M. M., Primrose, S. B., and Harder, W. (1979). *FEMS Lett.* **6**, 183–188.

Dodds, J. H., and Hall, M. A. (1980). *Sci. Prog.* **66**, 513–535.

Dodds, J. H., Musa, S. K., Jerie, P. H., and Hall, M. A. (1979). *Plant Sci. Lett.* **17**, 109–114.

Dodds, J. H., Heslop-Harrison, J. S., and Hall, M. A. (1980). *Plant Sci. Lett.* **19**, 175–180.

Gane, R. (1934). *Nature (London)* **134**, 1008.

Giaquinta, R., and Beyer, E. M. (1977). *Plant Cell Physiol.* **18**, 141–148.

Girardin, J. P. L. (1864). *Jahresber. Agrik. Chem. Versuchanst.* **7**, 199.

Hall, M. A. (1977). *Ann. Appl. Biol.* **85**, 424.

Ilag, L., and Curtis, R. W., (1968). *Science.* **159**, 1357–1358.

Jakoby, W. B., and Fjellstedt, J. A. (1972). *In* ''The Enzymes'' (P. D. Boyer ed.), p. 199–212. Academic Press, New York.

Jansen, E. F. (1963). *J. Biol. Chem.* **238**, 1552–1555.

Jansen, E. F. (1964). *J. Biol. Chem.* **239**, 1664–1667.

Jansen, E. F. (1969). *Food Sci. Technol.* **1**, 475–481.

Jerie, P. H., and Hall, M. A. (1978). *Proc. R. Soc. Ser. B* **200**, 87–91.

Jerie, P. H., Shaari, A. R., Zeroni, M., and Hall, M. A. (1978). *New Phytol.* **81**, 499–506.

Jerie, P. H., Shaari, A. R., and Hall, M. A. (1979). *Planta* **144**, 503–507.

Kende, H., and Hanson, A. D. (1976). *Plant Physiol.* **57**, 523–527.

Lynch, J. M. (1975). *Nature (London)* **256**, 576–577.

Lynch, J. M., and Harper, S. H. T. (1974). *J. Gen. Microbiol.* **80**, 187–195.

McGlasson, W. B. (1970). ''Food Science and Technology Monographs.'' Academic Press, New York.

Minina, E. G. (1938). *Dokl. Akad Nank. SSSR* **21**, 298.

Musa, S. K. (1981). Ph.D., Thesis. Univ. Coll. Wales, Aberystwyth, U.K.

Neljubov, D. (1901). *Beih. Bot. Zentralbl.* **10**, 128.

Neljubov, D. (1913). *Imp. Acad. Sci. (St. Petersburg)* **31**, 1.

Nord, F. F., and Weicherz, J. (1929). *Hoppe-Seylers Z. Physiol. Chem.* **183**, 218.

Rodriquez, A. B. (1932). *J. Dept. Agric. P.R.* **26**, 5.

Scott-Russell, R. (1969). *Nature (London)* **222**, 769–771.

Shimokana, K., and Kasai, Z. (1968). *Agric. Biol. Chem.* **32**, 680.

Shimokana, K., and Kasai, Z. (1969). *Mem. Res. Inst. Food. Sci. Kyoto* **30**, 1.

Sisler, E. C. (1979). *Plant Physiol.* **64**, 538–542.

Sisler, E. C. (1980). *Plant Physiol.* **66,** 404–406.

Smith, A. M., and Cook, R. J. (1974). *Nature (London)* **252,** 703–705.

Smith, K. A., and Restall, S. W. F. (1971). *J. Soil Sci.* **22.** 430–443.

Smith, K. A., Bremner, J. M., and Tabatabai, M. A. (1973). *Soil Sci.* **116,** 313–319.

Varner, J. E., and Ho, D. T. (1976). "Plant Biochemistry." Academic Press, New York.

Zeroni, M., Jerie, P. H., and Hall, J. A. (1977). *Planta* **134,** 119–125.

Index

Contents of Recent Volumes and Supplements